COMPLEX NUMBERS IN N DIMENSIONS

NORTH-HOLLAND MATHEMATICS STUDIES 190

(Continuation of the Notas de Matemática)

Editor: Saul LUBKIN
University of Rochester
New York, U.S.A.

N·H

2002
ELSEVIER
Amsterdam – Boston – London – New York – Oxford –
Paris – San Diego – San Francisco – Singapore – Sydney – Tokyo

COMPLEX NUMBERS
IN N DIMENSIONS

Silviu OLARIU
Institute of Physics and Nuclear Engineering
Magurele, Bucharest, Romania

N·H

2002
ELSEVIER
Amsterdam – Boston – London – New York – Oxford –
Paris – San Diego – San Francisco – Singapore – Sydney – Tokyo

ELSEVIER SCIENCE B.V.
Sara Burgerhartstraat 25
P.O. Box 211, 1000 AE Amsterdam, The Netherlands

First edition 2002

Library of Congress Cataloging in Publication Data
A catalog record from the Library of Congress has been applied for.

British Library Cataloguing in Publication Data
A catalogue record from the British Library has been applied for.

ISBN: 0-444-51123-7
ISSN: 0304-0208

♾ The paper used in this publication meets the requirements of ANSI/NISO Z39.48-1992 (Permanence of Paper).

Transferred to digital printing 2006

To Agata

Preface

A regular, two-dimensional complex number $x + iy$ can be represented geometrically by the modulus $\rho = (x^2 + y^2)^{1/2}$ and by the polar angle $\theta = \arctan(y/x)$. The modulus ρ is multiplicative and the polar angle θ is additive upon the multiplication of ordinary complex numbers.

The quaternions of Hamilton are a system of hypercomplex numbers defined in four dimensions, the multiplication being a noncommutative operation, [1] and many other hypercomplex systems are possible, [2]-[4] but these interesting hypercomplex systems do not have all the required properties of regular, two-dimensional complex numbers which rendered possible the development of the theory of functions of a complex variable.

Two distinct systems of hypercomplex numbers in n dimensions will be described in this work, for which the multiplication is associative and commutative, and which are rich enough in properties such that exponential and trigonometric forms exist and the concepts of analytic n-complex function, contour integration and residue can be defined. [5] The n-complex numbers described in this work have the form $u = x_0 + h_1 x_1 + \cdots + h_{n-1} x_{n-1}$, where $h_1, ..., h_{n-1}$ are the hypercomplex bases and the variables $x_0, ..., x_{n-1}$ are real numbers, unless otherwise stated. If the n-complex number u is represented by the point A of coordinates $x_0, x_1, ..., x_{n-1}$, the position of the point A can be described with the aid of the modulus $d = (x_0^2 + x_1^2 + \cdots + x_{n-1}^2)^{1/2}$ and of $n - 1$ angular variables.

The first type of hypercomplex numbers described in this work is characterized by the presence, in an even number of dimensions $n \geq 4$, of two polar axes, and by the presence, in an odd number of dimensions, of one polar axis. Therefore, these numbers will be called polar hypercomplex numbers in n dimensions. One polar axis is the normal through the origin O to the hyperplane $v_+ = 0$, where $v_+ = x_0 + x_1 + \cdots + x_{n-1}$. In an even number n of dimensions, the second polar axis is the normal through the origin O to the hyperplane $v_- = 0$, where $v_- = x_0 - x_1 + \cdots + x_{n-2} - x_{n-1}$. Thus, in addition to the distance d, the position of the point A can be specified, in an even number of dimensions, by 2 polar angles θ_+, θ_-, by $n/2 - 2$

vii

planar angles ψ_k, and by $n/2 - 1$ azimuthal angles ϕ_k. In an odd number of dimensions, the position of the point A is specified by d, by 1 polar angle θ_+, by $(n-3)/2$ planar angles ψ_{k-1}, and by $(n-1)/2$ azimuthal angles ϕ_k. The multiplication rules for the polar hypercomplex bases $h_1, ..., h_{n-1}$ are $h_j h_k = h_{j+k}$ if $0 \leq j + k \leq n - 1$, and $h_j h_k = h_{j+k-n}$ if $n \leq j + k \leq 2n - 2$, where $h_0 = 1$.

The other type of hypercomplex numbers described in this work exists as a distinct entity only when the number of dimensions n of the space is even. The position of the point A is specified, in addition to the distance d, by $n/2 - 1$ planar angles ψ_k and by $n/2$ azimuthal angles ϕ_k. These numbers will be called planar hypercomplex numbers. The multiplication rules for the planar hypercomplex bases $h_1, ..., h_{n-1}$ are $h_j h_k = h_{j+k}$ if $0 \leq j + k \leq n - 1$, and $h_j h_k = -h_{j+k-n}$ if $n \leq j + k \leq 2n - 2$, where $h_0 = 1$. For $n = 2$, the planar hypercomplex numbers become the usual 2-dimensional complex numbers $x + iy$.

The development of analytic functions of hypercomplex variables was rendered possible by the existence of an exponential form of the n-complex numbers. The azimuthal angles ϕ_k, which are cyclic variables, appear in these forms at the exponent, and lead to the concept of n-dimensional hypercomplex residue. Expressions are given for the elementary functions of n-complex variable. In particular, the exponential function of an n-complex number is expanded in terms of functions called in this work n-dimensional cosexponential functions of the polar and respectively planar type. The polar cosexponential functions are a generalization to n dimensions of the hyperbolic functions $\cosh y, \sinh y$, and the planar cosexponential functions are a generalization to n dimensions of the trigonometric functions $\cos y, \sin y$. Addition theorems and other relations are obtained for the n-dimensional cosexponential functions.

Many of the properties of 2-dimensional complex functions can be extended to hypercomplex numbers in n dimensions. Thus, the functions $f(u)$ of an n-complex variable which are defined by power series have derivatives independent of the direction of approach to the point under consideration. If the n-complex function $f(u)$ of the n-complex variable u is written in terms of the real functions $P_k(x_0, ..., x_{n-1}), k = 0, ..., n - 1$, then relations of equality exist between the partial derivatives of the functions P_k. The integral $\int_A^B f(u)du$ of an n-complex function between two points A, B is independent of the path connecting A, B, in regions where f is regular. If $f(u)$ is an analytic n-complex function, then $\oint_\Gamma f(u)du/(u - u_0)$ is expressed in this work in terms of the n-dimensional hypercomplex residue $f(u_0)$.

In the case of polar complex numbers, a polynomial can be written as a product of linear or quadratic factors, although several factorizations are in general possible. In the case of planar hypercomplex numbers, a polynomial can always be written as a product of linear factors, although, again, several factorizations are in general possible.

The work presents a detailed analysis of the hypercomplex numbers in 2, 3 and 4 dimensions, then presents the properties of hypercomplex numbers in 5 and 6 dimensions, and it continues with a detailed analysis of polar and planar hypercomplex numbers in n dimensions. The essence of this work is the interplay between the algebraic, the geometric and the analytic facets of the relations.

Contents

Chapter 1

Hyperbolic Complex Numbers in Two Dimensions

A system of hypercomplex numbers in 2 dimensions is described in this chapter, for which the multiplication is associative and commutative, and for which an exponential form and the concepts of analytic twocomplex function and contour integration can be defined. The twocomplex numbers introduced in this chapter have the form $u = x + \delta y$, the variables x, y being real numbers. The multiplication rules for the complex units $1, \delta$ are $1 \cdot \delta = \delta, \delta^2 = 1$. In a geometric representation, the twocomplex number u is represented by the point A of coordinates (x, y). The product of two twocomplex numbers is equal to zero if both numbers are equal to zero, or if one of the twocomplex numbers lies on the line $x = y$ and the other on the line $x = -y$.

The exponential form of a twocomplex number, defined for $x+y > 0, x-y > 0$, is $u = \rho \exp(\delta \lambda / 2)$, where the amplitude is $\rho = (x^2 - y^2)^{1/2}$ and the argument is $\lambda = \ln \tan \theta$, $\tan \theta = (x+y)/(x-y), 0 < \theta < \pi/2$. The trigonometric form of a twocomplex number is $u = d\sqrt{\sin 2\theta} \exp\{(1/2)\delta \ln \tan \theta\}$, where $d^2 = x^2 + y^2$. The amplitude ρ is equal to zero on the lines $x = \pm y$. The division $1/(x+\delta y)$ is possible provided that $\rho \neq 0$. If $u_1 = x_1 + \delta y_1, u_2 = x_2 + \delta y_2$ are twocomplex numbers of amplitudes and arguments ρ_1, λ_1 and respectively ρ_2, λ_2, then the amplitude and the argument ρ, λ of the product twocomplex number $u_1 u_2 = x_1 x_2 + y_1 y_2 + \delta(x_1 y_2 + y_1 x_2)$ are $\rho = \rho_1 \rho_2, \lambda = \lambda_1 + \lambda_2$. Thus, the amplitude ρ is a multiplicative quantity and the argument λ is an additive quantity upon the multiplication of twocomplex numbers, which reminds the properties of ordinary, two-

1

dimensional complex numbers.

Expressions are given for the elementary functions of twocomplex variable. Moreover, it is shown that the region of convergence of series of powers of twocomplex variables is a rectangle having the sides parallel to the bisectors $x = \pm y$.

A function $f(u)$ of the twocomplex variable $u = x + \delta y$ can be defined by a corresponding power series. It will be shown that the function f has a derivative $\lim_{u \to u_0} [f(u) - f(u_0)]/(u - u_0)$ independent of the direction of approach of u to u_0. If the twocomplex function $f(u)$ of the twocomplex variable u is written in terms of the real functions $P(x,y), Q(x,y)$ of real variables x, y as $f(u) = P(x,y) + \delta Q(x,y)$, then relations of equality exist between partial derivatives of the functions P, Q, and the functions P, Q are solutions of the two-dimensional wave equation.

It will also be shown that the integral $\int_A^B f(u) du$ of a twocomplex function between two points A, B is independent of the path connecting the points A, B.

A polynomial $u^n + a_1 u^{n-1} + \cdots + a_{n-1} u + a_n$ can be written as a product of linear or quadratic factors, although the factorization may not be unique.

The twocomplex numbers described in this chapter are a particular case for $n = 2$ of the polar complex numbers in n dimensions discussed in Sec. 6.1.

1.1   Operations with hyperbolic twocomplex numbers

A hyperbolic complex number in two dimensions is determined by its two components (x, y). The sum of the hyperbolic twocomplex numbers (x, y) and (x', y') is the hyperbolic twocomplex number $(x + x', y + y')$. The product of the hyperbolic twocomplex numbers (x, y) and (x', y') is defined in this chapter to be the hyperbolic twocomplex number $(xx' + yy', xy' + yx')$.

Twocomplex numbers and their operations can be represented by writing the twocomplex number (x, y) as $u = x + \delta y$, where δ is a basis for which the multiplication rules are

$$1 \cdot \delta = \delta, \ \delta^2 = 1. \tag{1.1}$$

Two twocomplex numbers $u = x + \delta y, u' = x' + \delta y'$ are equal, $u = u'$, if and only if $x = x', y = y'$. If $u = x + \delta y, u' = x' + \delta y'$ are twocomplex numbers, the sum $u + u'$ and the product uu' defined above can be obtained by applying the usual algebraic rules to the sum $(x + \delta y) + (x' + \delta y')$ and to the product $(x + \delta y)(x' + \delta y')$, and grouping of the resulting terms,

$$u + u' = x + x' + \delta(y + y'), \tag{1.2}$$

$$uu' = xx' + yy' + \delta(xy' + yx'). \tag{1.3}$$

If u, u', u'' are twocomplex numbers, the multiplication is associative

$$(uu')u'' = u(u'u'') \tag{1.4}$$

and commutative

$$uu' = u'u, \tag{1.5}$$

as can be checked through direct calculation. The twocomplex zero is $0 + \delta \cdot 0$, denoted simply 0, and the twocomplex unity is $1 + \delta \cdot 0$, denoted simply 1.

The inverse of the twocomplex number $u = x + \delta y$ is a twocomplex number $u' = x' + \delta y'$ having the property that

$$uu' = 1. \tag{1.6}$$

Written on components, the condition, Eq. (1.6), is

$$\begin{aligned} xx' + yy' &= 1, \\ yx' + xy' &= 0. \end{aligned} \tag{1.7}$$

The system (1.7) has the solution

$$x' = \frac{x}{\nu}, \tag{1.8}$$

$$y' = -\frac{y}{\nu}, \tag{1.9}$$

provided that $\nu \neq 0$, where

$$\nu = x^2 - y^2. \tag{1.10}$$

The quantity ν can be written as

$$\nu = v_+ v_-, \tag{1.11}$$

where

$$v_+ = x + y, \quad v_- = x - y. \tag{1.12}$$

The variables v_+, v_- will be called canonical hyperbolic twocomplex variables. Then a twocomplex number $u = x + \delta y$ has an inverse, unless

$$v_+ = 0, \text{ or } v_- = 0. \tag{1.13}$$

For arbitrary values of the variables x, y, the quantity ν can be positive or negative. If $\nu \geq 0$, the quantity

$$\rho = \nu^{1/2}, \ \nu > 0, \tag{1.14}$$

will be called amplitude of the twocomplex number $x + \delta y$. The normals of the lines in Eq. (1.13) are orthogonal to each other. Because of conditions (1.13) these lines will be also called the nodal lines. It can be shown that if $uu' = 0$ then either $u = 0$, or $u' = 0$, or one of the twocomplex numbers u, u' is of the form $x + \delta x$ and the other is of the form $x - \delta x$.

1.2   Geometric representation of hyperbolic twocomplex numbers

The twocomplex number $x + \delta y$ can be represented by the point A of coordinates (x, y). If O is the origin of the two-dimensional space x, y, the distance from A to the origin O can be taken as

$$d^2 = x^2 + y^2. \tag{1.15}$$

The distance d will be called modulus of the twocomplex number $x + \delta y$.
  Since

$$(x + y)^2 + (x - y)^2 = 2d^2, \tag{1.16}$$

$x + y$ and $x - y$ can be written as

$$x + y = \sqrt{2}d \sin\theta, \ x - y = \sqrt{2}d \cos\theta, \tag{1.17}$$

so that

$$x = d \sin(\theta + \pi/4), \ y = -d \cos(\theta + \pi/4). \tag{1.18}$$

If $u = x + \delta y, u_1 = x_1 + \delta y_1, u_2 = x_2 + \delta y_2$, and $u = u_1 u_2$, and if

$$v_{j+} = x_j + y_j, \ v_{j-} = x_j - y_j, \ 2d_j^2 = v_{j+}^2 + v_{j-}^2,$$
$$x_j + y_j = \sqrt{2}d_j \sin\theta_j, \ x_j - y_j = d_j \sqrt{2} \cos\theta_j, \tag{1.19}$$

for $j = 1, 2$, it can be shown that

$$v_+ = v_{1+}v_{2+}, \ v_- = v_{1-}v_{2-}, \ \tan\theta = \tan\theta_1 \tan\theta_2. \tag{1.20}$$

The relations (1.20) are a consequence of the identities

$$(x_1 x_2 + y_1 y_2) + (x_1 y_2 + y_1 x_2) = (x_1 + y_1)(x_2 + y_2), \tag{1.21}$$

$$(x_1 x_2 + y_1 y_2) - (x_1 y_2 + y_1 x_2) = (x_1 - y_1)(x_2 - y_2). \tag{1.22}$$

A consequence of Eqs. (1.20) is that if $u = u_1 u_2$, then

$$\nu = \nu_1 \nu_2, \tag{1.23}$$

where

$$\nu_j = v_{j+} v_{j-}, \tag{1.24}$$

for $j = 1, 2$. If $\nu > 0, \nu_1 > 0, \nu_2 > 0$, then

$$\rho = \rho_1 \rho_2, \tag{1.25}$$

where

$$\rho_j = \nu_j^{1/2}, \tag{1.26}$$

for $j = 1, 2$.

The twocomplex numbers

$$e_+ = \frac{1 + \delta}{2}, \ e_- = \frac{1 - \delta}{2}, \tag{1.27}$$

are orthogonal,

$$e_+ e_- = 0, \tag{1.28}$$

and have also the property that

$$e_+^2 = e_+, \ e_-^2 = e_-. \tag{1.29}$$

The ensemble e_+, e_- will be called the canonical hyperbolic twocomplex base. The twocomplex number $u = x + \delta y$ can be written as

$$x + \delta y = (x + y)e_+ + (x - y)e_-, \tag{1.30}$$

or, by using Eq. (1.12),

$$u = v_+ e_+ + v_- e_-, \tag{1.31}$$

which will be called the canonical form of the hyperbolic twocomplex number. Thus, if $u_j = v_{j+} e_+ + v_{j-} e_-, \ j = 1, 2$, and $u = u_1 u_2$, then the multiplication of the hyperbolic twocomplex numbers is expressed by the relations (1.20).

The relation (1.23) for the product of twocomplex numbers can be demonstrated also by using a representation of the multiplication of the twocomplex numbers by matrices, in which the twocomplex number $u = x + \delta y$ is represented by the matrix

$$\begin{pmatrix} x & y \\ y & x \end{pmatrix}.$$ (1.32)

The product $u = x + \delta y$ of the twocomplex numbers $u_1 = x_1 + \delta y_1, u_2 = x_2 + \delta y_2$, can be represented by the matrix multiplication

$$\begin{pmatrix} x & y \\ y & x \end{pmatrix} = \begin{pmatrix} x_1 & y_1 \\ y_1 & x_1 \end{pmatrix}\begin{pmatrix} x_2 & y_2 \\ y_2 & x_2 \end{pmatrix}.$$ (1.33)

It can be checked that

$$\det\begin{pmatrix} x & y \\ y & x \end{pmatrix} = \nu.$$ (1.34)

The identity (1.23) is then a consequence of the fact the determinant of the product of matrices is equal to the product of the determinants of the factor matrices.

1.3 Exponential and trigonometric forms of a twocomplex number

The exponential function of the hypercomplex variable u can be defined by the series

$$\exp u = 1 + u + u^2/2! + u^3/3! + \cdots.$$ (1.35)

It can be checked by direct multiplication of the series that

$$\exp(u + u') = \exp u \cdot \exp u'.$$ (1.36)

The series for the exponential function and the addition theorem have the same form for all systems of commutative hypercomplex numbers discussed in this work. If $u = x + \delta y$, then $\exp u$ can be calculated as $\exp u = \exp x \cdot \exp(\delta y)$. According to Eq. (1.1),

$$\delta^{2m} = 1, \delta^{2m+1} = \delta,$$ (1.37)

where m is a natural number, so that $\exp(\delta y)$ can be written as

$$\exp(\delta y) = \cosh y + \delta \sinh y.$$ (1.38)

From Eq. (1.38) it can be inferred that

$$(\cosh t + \delta \sinh t)^m = \cosh mt + \delta \sinh mt. \tag{1.39}$$

The twocomplex numbers $u = x + \delta y$ for which $v_+ = x + y > 0$, $v_- = x - y > 0$ can be written in the form

$$x + \delta y = e^{x_1 + \delta y_1}. \tag{1.40}$$

The expressions of x_1, y_1 as functions of x, y can be obtained by developing $e^{\delta y_1}$ with the aid of Eq. (1.38) and separating the hypercomplex components,

$$x = e^{x_1} \cosh y_1, \tag{1.41}$$

$$y = e^{x_1} \sinh y_1. \tag{1.42}$$

It can be shown from Eqs. (1.41)-(1.42) that

$$x_1 = \frac{1}{2} \ln(v_+ v_-), \; y_1 = \frac{1}{2} \ln \frac{v_+}{v_-}. \tag{1.43}$$

The twocomplex number u can thus be written as

$$u = \rho \exp(\delta \lambda), \tag{1.44}$$

where the amplitude is $\rho = (x^2 - y^2)^{1/2}$ and the argument is $\lambda = (1/2) \ln\{(x + y)/(x - y)\}$, for $x + y > 0, x - y > 0$. The expression (1.44) can be written with the aid of the variables d, θ, Eq. (1.17), as

$$u = \rho \exp\left(\frac{1}{2} \delta \ln \tan \theta\right), \tag{1.45}$$

which is the exponential form of the twocomplex number u, where $0 < \theta < \pi/2$.

The relation between the amplitude ρ and the distance d is

$$\rho = d \sin^{1/2} 2\theta. \tag{1.46}$$

Substituting this form of ρ in Eq. (1.45) yields

$$u = d \sin^{1/2} 2\theta \exp\left(\frac{1}{2} \delta \ln \tan \theta\right), \tag{1.47}$$

which is the trigonometric form of the twocomplex number u.

1.4   Elementary functions of a twocomplex variable

The logarithm u_1 of the twocomplex number u, $u_1 = \ln u$, can be defined for $v_+ > 0, v_- > 0$ as the solution of the equation

$$u = e^{u_1}, \tag{1.48}$$

for u_1 as a function of u. From Eq. (1.45) it results that

$$\ln u = \ln \rho + \frac{1}{2} \delta \ln \tan \theta. \tag{1.49}$$

It can be inferred from Eqs. (1.49) and (1.20) that

$$\ln(u_1 u_2) = \ln u_1 + \ln u_2. \tag{1.50}$$

The explicit form of Eq. (1.49) is

$$\ln(x + \delta y) = \frac{1}{2}(1 + \delta) \ln(x + y) + \frac{1}{2}(1 - \delta) \ln(x - y), \tag{1.51}$$

so that the relation (1.49) can be written with the aid of Eq. (1.27) as

$$\ln u = e_+ \ln v_+ + e_- \ln v_-. \tag{1.52}$$

The power function u^n can be defined for $v_+ > 0, v_- > 0$ and real values of n as

$$u^n = e^{n \ln u}. \tag{1.53}$$

It can be inferred from Eqs. (1.53) and (1.50) that

$$(u_1 u_2)^n = u_1^n \, u_2^n. \tag{1.54}$$

Using the expression (1.52) for $\ln u$ and the relations (1.28) and (1.29) it can be shown that

$$(x + \delta y)^n = \frac{1}{2}(1 + \delta)(x + y)^n + \frac{1}{2}(1 - \delta)(x - y)^n. \tag{1.55}$$

For integer n, the relation (1.55) is valid for any x, y. The relation (1.55) for $n = -1$ is

$$\frac{1}{x + \delta y} = \frac{1}{2}\left(\frac{1 + \delta}{x + y} + \frac{1 - \delta}{x - y}\right). \tag{1.56}$$

The trigonometric functions $\cos u$ and $\sin u$ of the hypercomplex variable u are defined by the series

$$\cos u = 1 - u^2/2! + u^4/4! + \cdots, \tag{1.57}$$

$$\sin u = u - u^3/3! + u^5/5! + \cdots. \tag{1.58}$$

It can be checked by series multiplication that the usual addition theorems hold for the hypercomplex numbers u_1, u_2,

$$\cos(u_1 + u_2) = \cos u_1 \cos u_2 - \sin u_1 \sin u_2, \tag{1.59}$$

$$\sin(u_1 + u_2) = \sin u_1 \cos u_2 + \cos u_1 \sin u_2. \tag{1.60}$$

The series for the trigonometric functions and the addition theorems have the same form for all systems of commutative hypercomplex numbers discussed in this work. The cosine and sine functions of the hypercomplex variables δy can be expressed as

$$\cos \delta y = \cos y, \ \sin \delta y = \delta \sin y. \tag{1.61}$$

The cosine and sine functions of a twocomplex number $x + \delta y$ can then be expressed in terms of elementary functions with the aid of the addition theorems Eqs. (1.59), (1.60) and of the expressions in Eq. (1.61).

The hyperbolic functions $\cosh u$ and $\sinh u$ of the hypercomplex variable u are defined by the series

$$\cosh u = 1 + u^2/2! + u^4/4! + \cdots, \tag{1.62}$$

$$\sinh u = u + u^3/3! + u^5/5! + \cdots. \tag{1.63}$$

It can be checked by series multiplication that the usual addition theorems hold for the hypercomplex numbers u_1, u_2,

$$\cosh(u_1 + u_2) = \cosh u_1 \cosh u_2 + \sinh u_1 \sinh u_2, \tag{1.64}$$

$$\sinh(u_1 + u_2) = \sinh u_1 \cosh u_2 + \cosh u_1 \sinh u_2. \tag{1.65}$$

The series for the hyperbolic functions and the addition theorems have the same form for all systems of hypercomplex numbers discussed in this work. The cosh and sinh functions of the hypercomplex variable δy can be expressed as

$$\cosh \delta y = \cosh y, \ \sinh \delta y = \delta \sinh y. \tag{1.66}$$

The hyperbolic cosine and sine functions of a twocomplex number $x + \delta y$ can then be expressed in terms of elementary functions with the aid of the addition theorems Eqs. (1.64), (1.65) and of the expressions in Eq. (1.66).

1.5 Twocomplex power series

A twocomplex series is an infinite sum of the form

$$a_0 + a_1 + a_2 + \cdots + a_n + \cdots, \tag{1.67}$$

where the coefficients a_n are twocomplex numbers. The convergence of the series (1.67) can be defined in terms of the convergence of its 2 real components. The convergence of a twocomplex series can however be studied using twocomplex variables. The main criterion for absolute convergence remains the comparison theorem, but this requires a number of inequalities which will be discussed further.

The modulus of a twocomplex number $u = x + \delta y$ can be defined as

$$|u| = (x^2 + y^2)^{1/2}, \tag{1.68}$$

so that according to Eq. (1.15) $d = |u|$. Since $|x| \leq |u|, |y| \leq |u|$, a property of absolute convergence established via a comparison theorem based on the modulus of the series (1.67) will ensure the absolute convergence of each real component of that series.

The modulus of the sum $u_1 + u_2$ of the twocomplex numbers u_1, u_2 fulfils the inequality

$$||u_1| - |u_2|| \leq |u_1 + u_2| \leq |u_1| + |u_2|. \tag{1.69}$$

For the product the relation is

$$|u_1 u_2| \leq \sqrt{2}|u_1||u_2|, \tag{1.70}$$

which replaces the relation of equality extant for regular complex numbers. The equality in Eq. (1.70) takes place for $x_1 = y_1, x_2 = y_2$ or $x_1 = -y_1, x_2 = -y_2$. In particular

$$|u^2| \leq \sqrt{2}|u|^2. \tag{1.71}$$

The inequality in Eq. (1.70) implies that

$$|u^m| \leq 2^{(m-1)/2}|u|^m. \tag{1.72}$$

From Eqs. (1.70) and (1.72) it results that

$$|au^m| \leq 2^{m/2}|a||u|^m. \tag{1.73}$$

A power series of the twocomplex variable u is a series of the form

$$a_0 + a_1 u + a_2 u^2 + \cdots + a_l u^l + \cdots. \tag{1.74}$$

Since

$$\left| \sum_{l=0}^{\infty} a_l u^l \right| \leq \sum_{l=0}^{\infty} 2^{l/2} |a_l| |u|^l, \tag{1.75}$$

a sufficient condition for the absolute convergence of this series is that

$$\lim_{l \to \infty} \frac{\sqrt{2}|a_{l+1}||u|}{|a_l|} < 1. \tag{1.76}$$

Thus the series is absolutely convergent for

$$|u| < c_0, \tag{1.77}$$

where

$$c_0 = \lim_{l \to \infty} \frac{|a_l|}{\sqrt{2}|a_{l+1}|}. \tag{1.78}$$

The convergence of the series (1.74) can be also studied with the aid of the formula (1.55) which, for integer values of l, is valid for any x, y, z, t. If $a_l = a_{lx} + \delta a_{ly}$, and

$$A_{l+} = a_{lx} + a_{ly}, \quad A_{l-} = a_{lx} - a_{ly}, \tag{1.79}$$

it can be shown with the aid of relations (1.28) and (1.29) that

$$a_l e_+ = A_{l+} e_+, \quad a_l e_- = A_{l-} e_-, \tag{1.80}$$

so that the expression of the series (1.74) becomes

$$\sum_{l=0}^{\infty} \left(A_{l+} v_+^l e_+ + A_{l-} v_-^l e_- \right), \tag{1.81}$$

where the quantities v_+, v_- have been defined in Eq. (1.12). The sufficient conditions for the absolute convergence of the series in Eq. (1.81) are that

$$\lim_{l \to \infty} \frac{|A_{l+1,+}||v_+|}{|A_{l+}|} < 1, \quad \lim_{l \to \infty} \frac{|A_{l+1,-}||v_-|}{|A_{l-}|} < 1. \tag{1.82}$$

Thus the series in Eq. (1.81) is absolutely convergent for

$$|x + y| < c_+, \quad |x - y| < c_-, \tag{1.83}$$

where

$$c_+ = \lim_{l \to \infty} \frac{|A_{l+}|}{|A_{l+1,+}|}, \quad c_- = \lim_{l \to \infty} \frac{|A_{l-}|}{|A_{l+1,-}|}. \tag{1.84}$$

The relations (1.83) show that the region of convergence of the series (1.81) is a rectangle having the sides parallel to the bisectors $x = \pm y$. It can be shown that $c_0 = (1/\sqrt{2}) \min(c, c')$, where $\min(c, c')$ designates the smallest of the numbers c, c'. Since $|u|^2 = (v_+^2 + v_-^2)/2$, it can be seen that the circular region of convergence defined in Eqs. (1.77), (1.78) is included in the parallelogram defined in Eqs. (1.83) and (1.84).

1.6   Analytic functions of twocomplex variables

The derivative of a function $f(u)$ of the hypercomplex variables u is defined as a function $f'(u)$ having the property that

$$|f(u) - f(u_0) - f'(u_0)(u - u_0)| \to 0 \text{ as } |u - u_0| \to 0. \tag{1.85}$$

If the difference $u - u_0$ is not parallel to one of the nodal hypersurfaces, the definition in Eq. (1.85) can also be written as

$$f'(u_0) = \lim_{u \to u_0} \frac{f(u) - f(u_0)}{u - u_0}. \tag{1.86}$$

The derivative of the function $f(u) = u^m$, with m an integer, is $f'(u) = mu^{m-1}$, as can be seen by developing $u^m = [u_0 + (u - u_0)]^m$ as

$$u^m = \sum_{p=0}^{m} \frac{m!}{p!(m-p)!} u_0^{m-p} (u - u_0)^p, \tag{1.87}$$

and using the definition (1.85).

If the function $f'(u)$ defined in Eq. (1.85) is independent of the direction in space along which u is approaching u_0, the function $f(u)$ is said to be analytic, analogously to the case of functions of regular complex variables. [6] The function u^m, with m an integer, of the hypercomplex variable u is analytic, because the difference $u^m - u_0^m$ is always proportional to $u - u_0$, as can be seen from Eq. (1.87). Then series of integer powers of u will also be analytic functions of the hypercomplex variable u, and this result holds in fact for any commutative algebra.

If an analytic function is defined by a series around a certain point, for example $u = 0$, as

$$f(u) = \sum_{k=0}^{\infty} a_k u^k, \tag{1.88}$$

an expansion of $f(u)$ around a different point u_0,

$$f(u) = \sum_{k=0}^{\infty} c_k (u - u_0)^k, \tag{1.89}$$

can be obtained by substituting in Eq. (1.88) the expression of u^k according to Eq. (1.87). Assuming that the series are absolutely convergent so that the order of the terms can be modified and ordering the terms in the resulting expression according to the increasing powers of $u - u_0$ yields

$$f(u) = \sum_{k,l=0}^{\infty} \frac{(k+l)!}{k!l!} a_{k+l} u_0^l (u - u_0)^k. \tag{1.90}$$

Since the derivative of order k at $u = u_0$ of the function $f(u)$, Eq. (1.88), is

$$f^{(k)}(u_0) = \sum_{l=0}^{\infty} \frac{(k+l)!}{l!} a_{k+l} u_0^l, \tag{1.91}$$

the expansion of $f(u)$ around $u = u_0$, Eq. (1.90), becomes

$$f(u) = \sum_{k=0}^{\infty} \frac{1}{k!} f^{(k)}(u_0)(u - u_0)^k, \tag{1.92}$$

which has the same form as the series expansion of 2-dimensional complex functions. The relation (1.92) shows that the coefficients in the series expansion, Eq. (1.89), are

$$c_k = \frac{1}{k!} f^{(k)}(u_0). \tag{1.93}$$

The rules for obtaining the derivatives and the integrals of the basic functions can be obtained from the series of definitions and, as long as these series expansions have the same form as the corresponding series for the 2-dimensional complex functions, the rules of derivation and integration remain unchanged. The relations (1.85)-(1.93) have the same form for all systems of commutative hypercomplex numbers discussed in this work.

If the twocomplex function $f(u)$ of the twocomplex variable u is written in terms of the real functions $P(x,y), Q(x,y)$ of real variables x, y as

$$f(u) = P(x,y) + \delta Q(x,y), \tag{1.94}$$

then relations of equality exist between partial derivatives of the functions P, Q. These relations can be obtained by writing the derivative of the function f as

$$\lim_{\Delta x, \Delta y \to 0} \frac{1}{\Delta x + \delta \Delta y} \left[\frac{\partial P}{\partial x} \Delta x + \frac{\partial P}{\partial y} \Delta y + \delta \left(\frac{\partial Q}{\partial x} \Delta x + \frac{\partial Q}{\partial y} \Delta y \right) \right], \tag{1.95}$$

where the difference $u - u_0$ in Eq. (1.86) is $u - u_0 = \Delta x + \delta \Delta y$. The relations between the partials derivatives of the functions P, Q are obtained by setting successively in Eq. (1.95) $\Delta x \to 0, \Delta y = 0$; then $\Delta x = 0, \Delta y \to 0$. The relations are

$$\frac{\partial P}{\partial x} = \frac{\partial Q}{\partial y}, \tag{1.96}$$

$$\frac{\partial Q}{\partial x} = \frac{\partial P}{\partial y}. \tag{1.97}$$

The relations (1.96)-(1.97) are analogous to the Riemann relations for the real and imaginary components of a complex function. It can be shown from Eqs. (1.96)-(1.97) that the components P, Q are solutions of the equations

$$\frac{\partial^2 P}{\partial x^2} - \frac{\partial^2 P}{\partial y^2} = 0, \tag{1.98}$$

$$\frac{\partial^2 Q}{\partial x^2} - \frac{\partial^2 Q}{\partial y^2} = 0. \tag{1.99}$$

As can be seen from Eqs. (1.98)-(1.99), the components P, Q of an analytic function of twocomplex variable are solutions of the wave equation with respect to the variables x, y.

1.7   Integrals of twocomplex functions

The singularities of twocomplex functions arise from terms of the form $1/(u - u_0)^m$, with $m > 0$. Functions containing such terms are singular not only at $u = u_0$, but also at all points of the lines passing through u_0 and which are parallel to the nodal lines.

The integral of a twocomplex function between two points A, B along a path situated in a region free of singularities is independent of path, which means that the integral of an analytic function along a loop situated in a region free from singularities is zero,

$$\oint_\Gamma f(u)du = 0. \tag{1.100}$$

Using the expression, Eq. (1.94) for $f(u)$ and the fact that $du = dx + \delta dy$, the explicit form of the integral in Eq. (1.100) is

$$\oint_\Gamma f(u)du = \oint_\Gamma [(Pdx + Qdy) + \delta(Qdx + Pdy)]. \tag{1.101}$$

If the functions P, Q are regular on the surface Σ enclosed by the loop Γ, the integral along the loop Γ can be transformed with the aid of the theorem of Stokes in an integral over the surface Σ of terms of the form $\partial P/\partial y - \partial Q/\partial x$ and $\partial P/\partial x - \partial Q/\partial y$ which are equal to zero by Eqs. (1.96)-(1.97), and this proves Eq. (1.100).

The exponential form of the twocomplex numbers, Eq. (1.44), contains no cyclic variable, and therefore the concept of residue is not applicable to the twocomplex numbers defined in Eqs. (1.1).

1.8   Factorization of twocomplex polynomials

A polynomial of degree m of the twocomplex variable $u = x + \delta y$ has the form

$$P_m(u) = u^m + a_1 u^{m-1} + \cdots + a_{m-1} u + a_m, \tag{1.102}$$

where the constants are in general twocomplex numbers. If $a_m = a_{mx} + \delta a_{my}$, and with the notations of Eqs. (1.12) and (1.79) applied for $0, 1, \cdots, m$, the polynomial $P_m(u)$ can be written as

$$\begin{aligned} P_m = & \left[v_+^m + A_{1+} v_+^{m-1} + \cdots + A_{m-1,+} v_+ + A_{m+} \right] e_+ \\ & + \left[v_-^m + A_{1-} v_-^{m-1} + \cdots + A_{m-1,-} v_- + A_{m-} \right] e_-. \end{aligned} \tag{1.103}$$

Each of the polynomials of degree m with real coefficients in Eq. (1.103) can be written as a product of linear or quadratic factors with real coefficients, or as a product of linear factors which, if imaginary, appear always in complex conjugate pairs. Using the latter form for the simplicity of notations, the polynomial P_m can be written as

$$P_m = \prod_{l=1}^{m} (v_+ - v_{l+}) e_+ + \prod_{l=1}^{m} (v_- - v_{l-}) e_-, \tag{1.104}$$

where the quantities v_{l+} appear always in complex conjugate pairs, and the same is true for the quantities v_{l-}. Due to the properties in Eqs. (1.28) and (1.29), the polynomial $P_m(u)$ can be written as a product of factors of the form

$$P_m(u) = \prod_{l=1}^{m} \left[(v_+ - v_{l+}) e_+ + (v_- - v_{l-}) e_- \right]. \tag{1.105}$$

This relation can be written with the aid of Eqs. (1.31) as

$$P_m(u) = \prod_{l=1}^{m} (u - u_l), \tag{1.106}$$

where

$$u_l = e_+ v_{l+} + e_- v_{l-}, \tag{1.107}$$

for $l = 1, ..., m$. The roots v_{l+} and the roots v_{l-} defined in Eq. (1.104) may be ordered arbitrarily, which means that Eq. (1.107) gives sets of m roots $u_1, ..., u_m$ of the polynomial $P_m(u)$, corresponding to the various ways in which the roots v_{l+}, v_{l-} are ordered according to l in each group.

Thus, while the hypercomplex components in Eq. (1.104) taken separately have unique factorizations, the polynomial $P_m(u)$ can be written in many different ways as a product of linear factors.

If $P(u) = u^2 - 1$, the degree is $m = 2$, the coefficients of the polynomial are $a_1 = 0, a_2 = -1$, the twocomplex components of a_2 are $a_{2x} = -1, a_{2y} = 0$, the components A_{2+}, A_{2-} are $A_{2+} = -1, A_{2-} = -1$. The expression, Eq. (1.103), of $P(u)$ is $P(u) = e_+(v_+^2 - 1) + e_-(v_-^2 - 1)$, and the factorization in Eq. (1.106) is $u^2 - 1 = (u - u_1)(u - u_2)$, where $u_1 = \pm e_+ \pm e_-, u_2 = -u_1$. The factorizations are thus $u^2 - 1 = (u+1)(u-1)$ and $u^2 - 1 = (u+\delta)(u-\delta)$. It can be checked that $(\pm e_+ \pm e_-)^2 = e_+ + e_- = 1$.

1.9   Representation of hyperbolic twocomplex numbers by irreducible matrices

If the matrix in Eq. (1.32) representing the twocomplex number u is called U, and

$$T = \begin{pmatrix} \frac{1}{\sqrt{2}} & \frac{1}{\sqrt{2}} \\ -\frac{1}{\sqrt{2}} & \frac{1}{\sqrt{2}} \end{pmatrix}, \tag{1.108}$$

it can be checked that

$$TUT^{-1} = \begin{pmatrix} x+y & 0 \\ 0 & x-y \end{pmatrix}. \tag{1.109}$$

The relations for the variables $v_+ = x+y, v_- = x-y$ for the multiplication of twocomplex numbers have been written in Eq. (1.20). The matrix TUT^{-1} provides an irreducible representation [7] of the twocomplex numbers $u = x + \delta y$, in terms of matrices with real coefficients.

Chapter 2

Complex Numbers in Three Dimensions

A system of hypercomplex numbers in three dimensions is described in this chapter, for which the multiplication is associative and commutative, which have exponential and trigonometric forms, and for which the concepts of analytic tricomplex function, contour integration and residue can be defined. The tricomplex numbers introduced in this chapter have the form $u = x + hy + kz$, the variables x, y and z being real numbers. The multiplication rules for the complex units h, k are $h^2 = k$, $k^2 = h$, $hk = 1$. In a geometric representation, the tricomplex number u is represented by the point P of coordinates (x, y, z). If O is the origin of the x, y, z axes, (t) the trisector line $x = y = z$ of the positive octant and Π the plane $x + y + z = 0$ passing through the origin (O) and perpendicular to (t), then the tricomplex number u can be described by the projection s of the segment OP along the line (t), by the distance D from P to the line (t), and by the azimuthal angle ϕ of the projection of P on the plane Π, measured from an angular origin defined by the intersection of the plane determined by the line (t) and the x axis, with the plane Π. The amplitude ρ of a tricomplex number is defined as $\rho = (x^3 + y^3 + z^3 - 3xyz)^{1/3}$, the polar angle θ of OP with respect to the trisector line (t) is given by $\tan \theta = D/s$, and $d^2 = x^2 + y^2 + z^2$. The amplitude ρ is equal to zero on the trisector line (t) and on the plane Π. The division $1/(x + hy + kz)$ is possible provided that $\rho \neq 0$. The product of two tricomplex numbers is equal to zero if both numbers are equal to zero, or if one of the tricomplex numbers lies in the Π plane and the other on the (t) line.

If $u_1 = x_1 + hy_1 + kz_1, u_2 = x_2 + hy_2 + kz_2$ are tricomplex numbers of amplitudes and angles ρ_1, θ_1, ϕ_1 and respectively ρ_2, θ_2, ϕ_2, then

the amplitude and the angles ρ, θ, ϕ for the product tricomplex number
$$u_1 u_2 = x_1 x_2 + y_1 z_2 + y_2 z_1 + h(z_1 z_2 + x_1 y_2 + y_1 x_2) + k(y_1 y_2 + x_1 z_2 + z_1 x_2)$$
are $\rho = \rho_1 \rho_2, \tan\theta = \tan\theta_1 \tan\theta_2 / \sqrt{2}, \phi = \phi_1 + \phi_2$. Thus, the amplitude ρ
and $(\tan\theta)/\sqrt{2}$ are multiplicative quantities and the angle ϕ is an additive
quantity upon the multiplication of tricomplex numbers, which reminds the
properties of ordinary, two-dimensional complex numbers.

For the description of the exponential function of a tricomplex variable, it is useful to define the cosexponential functions $cx(\xi) = 1 + \xi^3/3! +$
$\xi^6/6! \cdots, mx(\xi) = \xi + \xi^4/4! + \xi^7/7! \cdots, px(\xi) = \xi^2/2 + \xi^5/5! + \xi^8/8! \cdots,$
where p and m stand for plus and respectively minus, as a reference to the
sign of a phase shift in the expressions of these functions. These functions
fulfil the relation $cx^3\xi + px^3\xi + mx^3\xi - 3cx\xi\,px\xi\,mx\xi = 1$.

The exponential form of a tricomplex number is $u = \rho \exp[(1/3)(h + k)$
$\ln(\sqrt{2}/\tan\theta) + (1/3)(h - k)\phi]$, and the trigonometric form of a tricomplex
number is $u = d\sqrt{3/2}\left\{(1/3)(2 - h - k)\sin\theta + (1/3)(1 + h + k)\sqrt{2}\cos\theta\right\}$
$\exp\left\{(h - k)\phi/\sqrt{3}\right\}$.

Expressions are given for the elementary functions of tricomplex variable. Moreover, it is shown that the region of convergence of series of powers
of tricomplex variables are cylinders with the axis parallel to the trisector
line. A function $f(u)$ of the tricomplex variable $u = x + hy + kz$ can be
defined by a corresponding power series. It will be shown that the function
$f(u)$ has a derivative at u_0 independent of the direction of approach of u to
u_0. If the tricomplex function $f(u)$ of the tricomplex variable u is written
in terms of the real functions $F(x, y, z), G(x, y, z), H(x, y, z)$ of real variables x, y, z as $f(u) = F(x, y, z) + hG(x, y, z) + kH(x, y, z)$, then relations
of equality exist between partial derivatives of the functions F, G, H, and
the differences $F - G, F - H, G - H$ are solutions of the equation of Laplace.

It will be shown that the integral $\int_A^B f(u)du$ of a regular tricomplex
function between two points A, B is independent of the three-dimensional
path connecting the points A, B. If $f(u)$ is an analytic tricomplex function,
then $\oint_\Gamma f(u)du/(u - u_0) = 2\pi(h - k)f(u_0)$ if the integration loop is threaded
by the parallel through u_0 to the line (t).

A tricomplex polynomial $u^m + a_1 u^{m-1} + \cdots + a_{m-1}u + a_m$ can be written
as a product of linear or quadratic factors, although the factorization may
not be unique.

The tricomplex numbers described in this chapter are a particular case
for $n = 3$ of the polar complex numbers in n dimensions discussed in Sec.
6.1.

2.1 Operations with tricomplex numbers

A tricomplex number is determined by its three components (x, y, z). The sum of the tricomplex numbers (x, y, z) and (x', y', z') is the tricomplex number $(x + x', y + y', z + z')$. The product of the tricomplex numbers (x, y, z) and (x', y', z') is defined in this chapter to be the tricomplex number $(xx' + yz' + zy', zz' + xy' + yx', yy' + xz' + zx')$.

Tricomplex numbers and their operations can be represented by writing the tricomplex number (x, y, z) as $u = x + hy + kz$, where h and k are bases for which the multiplication rules are

$$h^2 = k, \; k^2 = h, \; 1 \cdot h = h, \; 1 \cdot k = k, \; hk = 1. \tag{2.1}$$

Two tricomplex numbers $u = x + hy + kz, u' = x' + hy' + kz'$ are equal, $u = u'$, if and only if $x = x', y = y', z = z'$. If $u = x + hy + kz, u' = x' + hy' + kz'$ are tricomplex numbers, the sum $u + u'$ and the product uu' defined above can be obtained by applying the usual algebraic rules to the sum $(x + hy + kz) + (x' + hy' + kz')$ and to the product $(x + hy + kz)(x' + hy' + kz')$, and grouping of the resulting terms,

$$u + u' = x + x' + h(y + y') + k(z + z'), \tag{2.2}$$

$$uu' = xx' + yz' + zy' + h(zz' + xy' + yx') + k(yy' + xz' + zx'). \tag{2.3}$$

If u, u', u'' are tricomplex numbers, the multiplication is associative

$$(uu')u'' = u(u'u'') \tag{2.4}$$

and commutative

$$uu' = u'u, \tag{2.5}$$

as can be checked through direct calculation. The tricomplex zero is $0 + h \cdot 0 + k \cdot 0$, denoted simply 0, and the tricomplex unity is $1 + h \cdot 0 + k \cdot 0$, denoted simply 1.

The inverse of the tricomplex number $u = x + hy + kz$ is a tricomplex number $u' = x' + y' + z'$ having the property that

$$uu' = 1. \tag{2.6}$$

Written on components, the condition, Eq. (2.6), is

$$\begin{aligned}
xx' + zy' + yz' &= 1, \\
yx' + xy' + zz' &= 0, \\
zx' + yy' + xz' &= 0.
\end{aligned} \tag{2.7}$$

The system (2.7) has the solution

$$x' = \frac{x^2 - yz}{x^3 + y^3 + z^3 - 3xyz},$$ (2.8)

$$y' = \frac{z^2 - xy}{x^3 + y^3 + z^3 - 3xyz},$$ (2.9)

$$z' = \frac{y^2 - xz}{x^3 + y^3 + z^3 - 3xyz},$$ (2.10)

provided that $x^3 + y^3 + z^3 - 3xyz \neq 0$. Since

$$x^3 + y^3 + z^3 - 3xyz = (x + y + z)(x^2 + y^2 + z^2 - xy - xz - yz), \text{(2.11)}$$

a tricomplex number $x + hy + kz$ has an inverse, unless

$$x + y + z = 0$$ (2.12)

or

$$x^2 + y^2 + z^2 - xy - xz - yz = 0.$$ (2.13)

The relation in Eq. (2.12) represents the plane Π perpendicular to the trisector line (t) of the x, y, z axes, and passing through the origin O of the axes. The plane Π, shown in Fig. 2.1. intersects the xOy plane along the line $z = 0, x + y = 0$, it intersect the yOz plane along the line $x = 0, y + z = 0$, and it intersects the xOz plane along the line $y = 0, x + z = 0$. The condition (2.13) is equivalent to $(x - y)^2 + (x - z)^2 + (y - z)^2 = 0$, which for real x, y, z means that $x = y = z$, which represents the trisector line (t) of the axes x, y, z. The trisector line (t) is perpendicular to the plane Π. Because of conditions (2.12) and (2.13). the trisector line (t) and the plane Π will be also called nodal line and respectively nodal plane. It can be shown that if $uu' = 0$ then either $u = 0$, or $u' = 0$, or one of the tricomplex numbers u, u' belongs to the trisector line (t) and the other belongs to the nodal plane Π.

2.2 Geometric representation of tricomplex numbers

The tricomplex number $x + hy + kz$ can be represented by the point P of coordinates (x, y, z). If O is the origin of the axes, then the projection

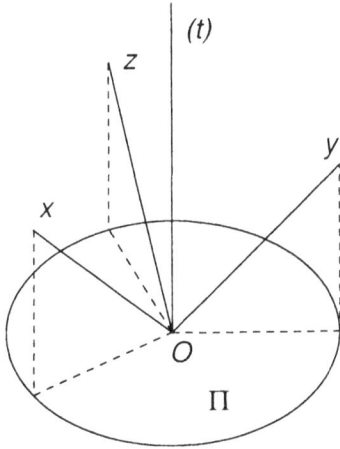

Figure 2.1: Nodal plane Π, of equation $x + y + z = 0$, and trisector line (t), of equation $x = y = z$, both passing through the origin O of the rectangular axes x, y, z.

$s = OQ$ of the line OP on the trisector line $x = y = z$, which has the unit tangent $(1/\sqrt{3}, 1/\sqrt{3}, 1/\sqrt{3})$, is

$$s = \frac{1}{\sqrt{3}}(x + y + z). \tag{2.14}$$

The distance $D = PQ$ from P to the trisector line $x = y = z$, calculated as the distance from the point (x, y, z) to the point Q of coordinates $[(x + y + z)/3, (x + y + z)/3, (x + y + z)/3]$, is

$$D^2 = \frac{2}{3}(x^2 + y^2 + z^2 - xy - xz - yz). \tag{2.15}$$

The quantities s and D are shown in Fig. 2.2, where the plane through the point P and perpendicular to the trisector line (t) intersects the x axis at point A of coordinates $(x + y + z, 0, 0)$, the y axis at point B of coordinates $(0, x + y + z, 0)$, and the z axis at point C of coordinates $(0, 0, x + y + z)$. The azimuthal angle ϕ of the tricomplex number $x + hy + kz$ is defined as the angle in the plane Π of the projection of P on this plane, measured from the line of intersection of the plane determined by the line (t) and the x axis with the plane Π, $0 \leq \phi < 2\pi$. The expression of ϕ in terms of x, y, z can be obtained in a system of coordinates defined by the unit vectors

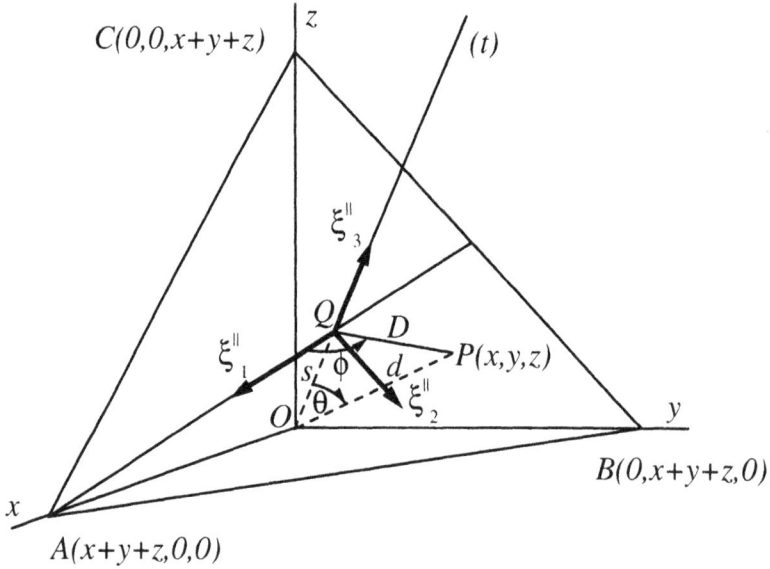

Figure 2.2: Tricomplex variables s, d, θ, ϕ for the tricomplex number $x + hy + kz$, represented by the point $P(x, y, z)$. The azimuthal angle ϕ is shown in in the plane parallel to Π, passing through P, which intersects the trisector line (t) at Q and the axes of coordinates x, y, z at the points A, B, C. The orthogonal axes $\xi_1^{\parallel}, \xi_2^{\parallel}, \xi_3^{\parallel}$ have the origin at Q.

$$\xi_1 : \frac{1}{\sqrt{6}}(2, -1, -1); \ \xi_2 : \frac{1}{\sqrt{2}}(0, 1, -1); \ \xi_3 : \frac{1}{\sqrt{3}}(1, 1, 1), \qquad (2.16)$$

and having the point O as origin. The relation between the coordinates of P in the systems ξ_1, ξ_2, ξ_3 and x, y, z can be written in the form

$$\begin{pmatrix} \xi_1 \\ \xi_2 \\ \xi_3 \end{pmatrix} = \begin{pmatrix} \frac{2}{\sqrt{6}} & -\frac{1}{\sqrt{6}} & -\frac{1}{\sqrt{6}} \\ 0 & \frac{1}{\sqrt{2}} & -\frac{1}{\sqrt{2}} \\ \frac{1}{\sqrt{3}} & \frac{1}{\sqrt{3}} & \frac{1}{\sqrt{3}} \end{pmatrix} \begin{pmatrix} x \\ y \\ z \end{pmatrix}. \qquad (2.17)$$

The components of the vector OP in the system ξ_1, ξ_2, ξ_3 can be obtained with the aid of Eq. (2.17) as

$$(\xi_1, \xi_2, \xi_3) = \left(\frac{1}{\sqrt{6}}(2x - y - z), \frac{1}{\sqrt{2}}(y - z), \frac{1}{\sqrt{3}}(x + y + z) \right). \qquad (2.18)$$

The expression of the angle ϕ as a function of x, y, z is then

$$\cos\phi = \frac{2x - y - z}{2(x^2 + y^2 + z^2 - xy - xz - yz)^{1/2}}, \tag{2.19}$$

$$\sin\phi = \frac{\sqrt{3}(y - z)}{2(x^2 + y^2 + z^2 - xy - xz - yz)^{1/2}}. \tag{2.20}$$

It can be seen from Eqs. (2.19),(2.20) that the angle of points on the x axis is $\phi = 0$, the angle of points on the y axis is $\phi = 2\pi/3$, and the angle of points on the z axis is $\phi = 4\pi/3$. The angle ϕ is shown in Fig. 2.2 in the plane parallel to Π, passing through P. The axis $Q\xi_1^{\parallel}$ is parallel to the axis $O\xi_1$, the axis $Q\xi_2^{\parallel}$ is parallel to the axis $O\xi_2$, and the axis $Q\xi_3^{\parallel}$ is parallel to the axis $O\xi_3$, so that, in the plane ABC, the angle ϕ is measured from the line QA. The angle θ between the line OP and the trisector line (t) is given by

$$\tan\theta = \frac{D}{s}, \tag{2.21}$$

where $0 \leq \theta \leq \pi$. It can be checked that

$$d^2 = D^2 + s^2, \tag{2.22}$$

where

$$d^2 = x^2 + y^2 + z^2, \tag{2.23}$$

so that

$$D = d\sin\theta, \quad s = d\cos\theta. \tag{2.24}$$

The relations (2.14), (2.15), (2.19)-(2.21) can be used to determine the associated projection s, the distance D, the polar angle θ with the trisector line (t) and the angle ϕ in the Π plane for the tricomplex number $x + hy + kz$. It can be shown that if $u_1 = x_1 + hy_1 + kz_1, u_2 = x_2 + hy_2 + kz_2$ are tricomplex numbers of projections, distances and angles $s_1, D_1, \theta_1, \phi_1$ and respectively $s_2, D_2, \theta_2, \phi_2$, then the projection s, distance D and the angle θ, ϕ for the product tricomplex number $u_1 u_2 = x_1 x_2 + y_1 z_2 + y_2 z_1 + h(z_1 z_2 + x_1 y_2 + y_1 x_2) + k(y_1 y_2 + x_1 z_2 + z_1 x_2)$ are

$$s = \sqrt{3}s_1 s_2, \quad D = \sqrt{\frac{3}{2}}D_1 D_2, \quad \tan\theta = \frac{1}{\sqrt{2}}\tan\theta_1\tan\theta_2, \quad \phi = \phi_1 + \phi_2. \tag{2.25}$$

The relations (2.25) are consequences of the identities

$$(x_1x_2 + y_1z_2 + y_2z_1) + (z_1z_2 + x_1y_2 + y_1x_2) + (y_1y_2 + x_1z_2 + z_1x_2)$$
$$= (x_1 + y_1 + z_1)(x_2 + y_2 + z_2), \tag{2.26}$$

$$(x_1x_2 + y_1z_2 + y_2z_1)^2 + (z_1z_2 + x_1y_2 + y_1x_2)^2 + (y_1y_2 + x_1z_2 + z_1x_2)^2$$
$$-(x_1x_2 + y_1z_2 + y_2z_1)(z_1z_2 + x_1y_2 + y_1x_2)$$
$$-(x_1x_2 + y_1z_2 + y_2z_1)(y_1y_2 + x_1z_2 + z_1x_2)$$
$$-(z_1z_2 + x_1y_2 + y_1x_2) + (y_1y_2 + x_1z_2 + z_1x_2)$$
$$= (x_1^2 + y_1^2 + z_1^2 - x_1y_1 - x_1z_1 - y_1z_1)$$
$$(x_2^2 + y_2^2 + z_2^2 - x_2y_2 - x_2z_2 - y_2z_2), \tag{2.27}$$

$$\frac{2x_1 - y_1 - z_1}{2}\frac{2x_2 - y_2 - z_2}{2} - \frac{\sqrt{3}}{2}(y_1 - z_1)\frac{\sqrt{3}}{2}(y_2 - z_2)$$
$$= \frac{1}{2}[2(x_1x_2 + y_1z_2 + z_1y_2) - (z_1z_2 + x_1y_2 + y_1x_2)$$
$$-(y_1y_2 + x_1z_2 + z_1x_2)], \tag{2.28}$$

$$\frac{\sqrt{3}}{2}(y_1 - z_1)\frac{2x_2 - y_2 - z_2}{2} + \frac{\sqrt{3}}{2}(y_2 - z_2)\frac{2x_1 - y_1 - z_1}{2}$$
$$= \frac{\sqrt{3}}{2}[(z_1z_2 + x_1y_2 + y_1x_2) - (y_1y_2 + x_1z_2 + z_1x_2)]. \tag{2.29}$$

The relation (2.26) shows that if u is in the plane Π, such that $x+y+z = 0$, then the product uu' is also in the plane Π for any u'. The relation (2.27) shows that if u is on the trisector line (t), such that $x^2 + y^2 + z^2 - xy - xz - yz = 0$, then uu' is also on the trisector line (t) for any u'. If u, u' are points in the plane $x + y + z = 1$, then the product uu' is also in that plane, and if u, u' are points of the cylindrical surface $x^2 + y^2 + z^2 - xy - xz - yz = 1$, then uu' is also in that cylindrical surface. This means that if u, u' are points on the circle $x + y + z = 1, x^2 + y^2 + z^2 - xy - xz - yz = 1$, which is perpendicular to the trisector line, is situated at a distance $1/\sqrt{3}$ from the origin and has the radius $\sqrt{2/3}$, then the tricomplex product uu' is also on the same circle. This invariant circle for the multiplication of tricomplex numbers is described by the equations

$$x = \frac{1}{3} + \frac{2}{3}\cos\phi, \ y = \frac{1}{3} - \frac{1}{3}\cos\phi + \frac{1}{\sqrt{3}}\sin\phi,$$
$$z = \frac{1}{3} - \frac{1}{3}\cos\phi - \frac{1}{\sqrt{3}}\sin\phi. \tag{2.30}$$

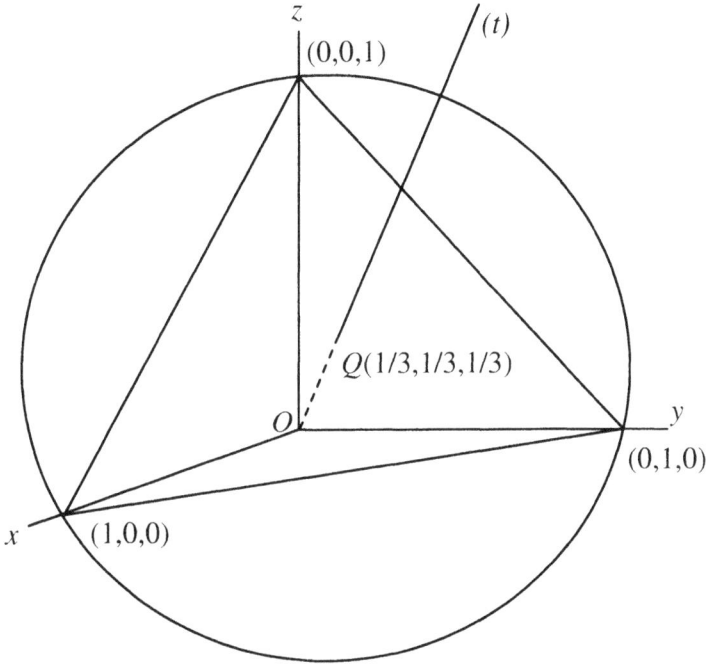

Figure 2.3: Invariant circle for the multiplication of tricomplex numbers, lying in a plane perpendicular to the trisector line and passing through the points (1,0,0), (0,1,0) and (0,0,1). The center of the circle is at the point $(1/3,1/3,1/3)$, and its radius is $(2/3)^{1/2}$.

It has the center at the point $(1/3,1/3,1/3)$ and passes through the points (1,0,0), (0,1,0) and (0,0,1), as shown in Fig. 2.3.

An important quantity is the amplitude ρ defined as $\rho = \nu^{1/3}$, so that

$$\rho^3 = x^3 + y^3 + z^3 - 3xyz. \tag{2.31}$$

The amplitude ρ of the product $u_1 u_2$ of the tricomplex numbers u_1, u_2 of amplitudes ρ_1, ρ_2 is

$$\rho = \rho_1 \rho_2, \tag{2.32}$$

as can be seen from the identity

$$(x_1x_2 + y_1z_2 + y_2z_1)^3 + (z_1z_2 + x_1y_2 + y_1x_2)^3 + (y_1y_2 + x_1z_2 + z_1x_2)^3$$
$$-3(x_1x_2 + y_1z_2 + y_2z_1)(z_1z_2 + x_1y_2 + y_1x_2)(y_1y_2 + x_1z_2 + z_1x_2)$$
$$= (x_1^3 + y_1^3 + z_1^3 - 3x_1y_1z_1)(x_2^3 + y_2^3 + z_2^3 - 3x_2y_2z_2). \qquad (2.33)$$

The identity in Eq. (2.33) can be demonstrated with the aid of Eqs. (2.11), (2.26) and (2.27). Another method would be to use the representation of the multiplication of the tricomplex numbers by matrices, in which the tricomplex number $u = x + hy + kz$ is represented by the matrix

$$\begin{pmatrix} x & y & z \\ z & x & y \\ y & z & x \end{pmatrix}. \qquad (2.34)$$

The product $u = x + hy + kz$ of the tricomplex numbers $u_1 = x_1 + hy_1 + kz_1, u_2 = x_2 + hy_2 + kz_2$, is represented by the matrix multiplication

$$\begin{pmatrix} x & y & z \\ z & x & y \\ y & z & x \end{pmatrix} = \begin{pmatrix} x_1 & y_1 & z_1 \\ z_1 & x_1 & y_1 \\ y_1 & z_1 & x_1 \end{pmatrix} \begin{pmatrix} x_2 & y_2 & z_2 \\ z_2 & x_2 & y_2 \\ y_2 & z_2 & x_2 \end{pmatrix}. \qquad (2.35)$$

If

$$\nu = \det \begin{pmatrix} x & y & z \\ z & x & y \\ y & z & x \end{pmatrix}, \qquad (2.36)$$

it can be checked that

$$\nu = x^3 + y^3 + z^3 - 3xyz. \qquad (2.37)$$

The identity (2.33) is then a consequence of the fact the determinant of the product of matrices is equal to the product of the determinants of the factor matrices.

It can be seen from Eqs. (2.14) and (2.15) that

$$x^3 + y^3 + z^3 - 3xyz = \frac{3\sqrt{3}}{2} s D^2, \qquad (2.38)$$

which can be written with the aid of relations (2.24) and (2.31) as

$$\rho = \frac{3^{1/2}}{2^{1/3}} d \sin^{2/3} \theta \cos^{1/3} \theta. \qquad (2.39)$$

This means that the surfaces of constant ρ are surfaces of rotation having the trisector line (t) as axis, as shown in Fig. 2.4.

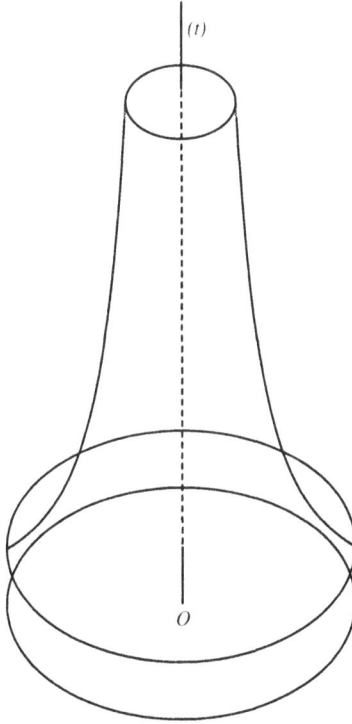

Figure 2.4: Surfaces of constant ρ, which are surfaces of rotation having the trisector line (t) as axis.

2.3   The tricomplex cosexponential functions

The exponential function of a hypercomplex variable u and the addition theorem for the exponential function have been written in Eqs. (1.35)-(1.36). If $u = x + hy + kz$, then $\exp u$ can be calculated as $\exp u = \exp x \cdot \exp(hy) \cdot \exp(kz)$. According to Eq. (2.1), $h^2 = k, h^3 = 1, k^2 = h, k^3 = 1$, and in general

$$h^{3m} = 1, h^{3m+1} = h, h^{3m+2} = k, k^{3m} = 1, k^{3m+1} = k, k^{3m+2} = h,$$

$$(2.40)$$

where n is a natural number, so that $\exp(hy)$ and $\exp(kz)$ can be written as

$$\exp(hy) = \operatorname{cx} y + h \operatorname{mx} y + k \operatorname{px} y, \tag{2.41}$$

$$\exp(kz) = \operatorname{cx} z + h \operatorname{px} z + k \operatorname{mx} z, \tag{2.42}$$

where the functions cx, mx, px, which will be called in this chapter polar cosexponential functions, are defined by the series

$$\operatorname{cx} y = 1 + y^3/3! + y^6/6! + \cdots, \tag{2.43}$$

$$\operatorname{mx} y = y + y^4/4! + y^7/7! + \cdots, \tag{2.44}$$

$$\operatorname{px} y = y^2/2! + y^5/5! + y^8/8! + \cdots. \tag{2.45}$$

From the series definitions it can be seen that $\operatorname{cx} 0 = 1, \operatorname{mx} 0 = 0, \operatorname{px} 0 = 0$. The tridimensional polar cosexponential functions belong to the class of the polar n-dimensional cosexponential functions g_{nk}, and $\operatorname{cx} = g_{30}, \operatorname{mx} = g_{31}, \operatorname{px} = g_{32}$. It can be checked that

$$\operatorname{cx} y + \operatorname{px} y + \operatorname{mx} y = \exp y. \tag{2.46}$$

By expressing the fact that $\exp(hy+hz) = \exp(hy) \cdot \exp(hz)$ with the aid of the cosexponential functions (2.43)-(2.45) the following addition theorems can be obtained

$$\operatorname{cx}(y+z) = \operatorname{cx} y \operatorname{cx} z + \operatorname{mx} y \operatorname{px} z + \operatorname{px} y \operatorname{mx} z, \tag{2.47}$$

$$\operatorname{mx}(y+z) = \operatorname{px} y \operatorname{px} z + \operatorname{cx} y \operatorname{mx} z + \operatorname{mx} y \operatorname{cx} z, \tag{2.48}$$

$$\operatorname{px}(y+z) = \operatorname{mx} y \operatorname{mx} z + \operatorname{cx} y \operatorname{px} z + \operatorname{px} y \operatorname{cx} z. \tag{2.49}$$

For $y = z$, Eqs. (2.47)-(2.49) yield

$$\operatorname{cx} 2y = \operatorname{cx}^2 y + 2 \operatorname{mx} y \operatorname{px} z, \tag{2.50}$$

$$\operatorname{mx} 2y = \operatorname{px}^2 y + 2 \operatorname{cx} y \operatorname{mx} z, \tag{2.51}$$

$$\operatorname{px} 2y = \operatorname{mx}^2 y + 2 \operatorname{cx} y \operatorname{px} z. \tag{2.52}$$

The cosexponential functions are neither even nor odd functions. For $z = -y$, Eqs. (2.47)-(2.49) yield

$$\operatorname{cx} y \operatorname{cx}(-y) + \operatorname{mx} y \operatorname{px}(-y) + \operatorname{px} y \operatorname{mx}(-y) = 1, \tag{2.53}$$

$$\mathrm{px}\, y\, \mathrm{px}\, (-y) + \mathrm{cx}\, y\, \mathrm{mx}\, (-y) + \mathrm{mx}\, y\, \mathrm{cx}\, (-y) = 0, \tag{2.54}$$

$$\mathrm{mx}\, y\, \mathrm{mx}\, (-y) + \mathrm{cx}\, y\, \mathrm{px}\, (-y) + \mathrm{px}\, y\, \mathrm{cx}\, (-y) = 0. \tag{2.55}$$

Expressions of the cosexponential functions in terms of regular exponential and cosine functions can be obtained by considering the series expansions for $e^{(h+k)y}$ and $e^{(h-k)y}$. These expressions can be obtained by calculating first $(h+k)^m$ and $(h-k)^m$. It can be shown that

$$(h+k)^m = \frac{1}{3}\left[(-1)^{m-1} + 2^m\right](h+k) + \frac{2}{3}\left[(-1)^m + 2^{m-1}\right], \tag{2.56}$$

$$\begin{aligned} (h-k)^{2m} &= (-1)^{m-1}3^{m-1}(k+k-2), \\ (h-k)^{2m+1} &= (-1)^m 3^m (h-k), \end{aligned} \tag{2.57}$$

where m is a natural number. Then

$$e^{(h+k)y} = (h+k)\left(-\frac{1}{3}e^{-y} + \frac{1}{3}e^{2y}\right) + \frac{2}{3}e^{-y} + \frac{1}{3}e^{2y}. \tag{2.58}$$

As a corollary, the following identities can be obtained from Eq. (2.58) by writing $e^{(h+k)y} = e^{hy}e^{ky}$ and expressing e^{hy} and e^{ky} in terms of cosexponential functions via Eqs. (2.41) and (2.42),

$$\mathrm{cx}^2\, y + \mathrm{mx}^2\, y + \mathrm{px}^2\, y = \frac{2}{3}e^{-y} + \frac{1}{3}e^{2y}, \tag{2.59}$$

$$\mathrm{cx}\, y\, \mathrm{mx}\, y + \mathrm{cx}\, y\, \mathrm{px}\, y + \mathrm{mx}\, y\, \mathrm{px}\, y = -\frac{1}{3}e^{-y} + \frac{1}{3}e^{2y}. \tag{2.60}$$

From Eqs. (2.59) and (2.60) it results that

$$\begin{aligned} \mathrm{cx}^2\, y + \mathrm{mx}^2\, y + \mathrm{px}^2\, y \\ -\mathrm{cx}\, y\, \mathrm{mx}\, y - \mathrm{cx}\, y\, \mathrm{px}\, y - \mathrm{mx}\, y\, \mathrm{px}\, y = \exp(-y). \end{aligned} \tag{2.61}$$

Then from Eqs. (2.11), (2.46) and (2.61) it follows that

$$\mathrm{cx}^3\, y + \mathrm{mx}^3\, y + \mathrm{px}^3\, y - 3\mathrm{cx}\, y\, \mathrm{mx}\, y\, \mathrm{px}\, y = 1. \tag{2.62}$$

Similarly,

$$\begin{aligned} e^{(h-k)y} &= \frac{1}{3}(1 + h + k) + \frac{1}{3}(2 - h - k)\cos(\sqrt{3}y) \\ &\quad + \frac{1}{\sqrt{3}}(h-k)\sin(\sqrt{3}y). \end{aligned} \tag{2.63}$$

The last relation can also be written as

$$e^{(h-k)y} = \frac{1}{3} + \frac{2}{3}\cos(\sqrt{3}y) + h\left[\frac{1}{3} + \frac{2}{3}\cos\left(\sqrt{3}y - \frac{2\pi}{3}\right)\right]$$
$$+ k\left[\frac{1}{3} + \frac{2}{3}\cos\left(\sqrt{3}y + \frac{2\pi}{3}\right)\right]. \tag{2.64}$$

As a corollary, the following identities can be obtained from Eq. (2.63) by writing $e^{(h-k)y} = e^{hy}e^{-ky}$ and expressing e^{hy} and e^{-ky} in terms of cosexponential functions via Eqs. (2.41) and (2.42),

$$\operatorname{cx} y \operatorname{cx}(-y) + \operatorname{mx} y \operatorname{mx}(-y) + \operatorname{px} y \operatorname{px}(-y) = \frac{1}{3} + \frac{2}{3}\cos(\sqrt{3}y), \tag{2.65}$$

$$\operatorname{cx} y \operatorname{px}(-y) + \operatorname{mx} y \operatorname{cx}(-y) + \operatorname{px} y \operatorname{mx}(-y)$$
$$= \frac{1}{3} + \frac{2}{3}\cos\left(\sqrt{3}y - \frac{2\pi}{3}\right), \tag{2.66}$$

$$\operatorname{cx} y \operatorname{mx}(-y) + \operatorname{mx} y \operatorname{px}(-y) + \operatorname{px} y \operatorname{cx}(-y)$$
$$= \frac{1}{3} + \frac{2}{3}\cos\left(\sqrt{3}y + \frac{2\pi}{3}\right). \tag{2.67}$$

Expressions of e^{2hy} in terms of the regular exponential and cosine functions can be obtained by the multiplication of the expressions of $e^{(h+k)y}$ and $e^{(h-k)y}$ from Eqs. (2.58) and (2.63). At the same time, Eq. (2.41) gives an expression of e^{2hy} in terms of cosexponential functions. By equating the real and hypercomplex parts of these two forms of e^{2y} and then replacing $2y$ by y gives the expressions of the cosexponential functions as

$$\operatorname{cx} y = \frac{1}{3}e^y + \frac{2}{3}\cos\left(\frac{\sqrt{3}}{2}y\right)e^{-y/2}, \tag{2.68}$$

$$\operatorname{mx} y = \frac{1}{3}e^y + \frac{2}{3}\cos\left(\frac{\sqrt{3}}{2}y - \frac{2\pi}{3}\right)e^{-y/2}, \tag{2.69}$$

$$\operatorname{px} y = \frac{1}{3}e^y + \frac{2}{3}\cos\left(\frac{\sqrt{3}}{2}y + \frac{2\pi}{3}\right)e^{-y/2}. \tag{2.70}$$

It is remarkable that the series in Eqs. (2.43)-(2.45), in which the terms are either of the form y^{3m}, or y^{3m+1}, or y^{3m+2}, can be expressed in terms of elementary functions whose power series are not subject to such restrictions. The cosexponential functions differ by the phase of the cosine function in their expression, and the designation of the functions in Eqs. (2.69) and

(2.70) as mx and px refers respectively to the minus or plus sign of the phase term $2\pi/3$. The graphs of the cosexponential functions are shown in Fig. 2.5.

It can be checked that the cosexponential functions are solutions of the third-order differential equation

$$\frac{d^3\zeta}{du^3} = \zeta, \tag{2.71}$$

whose solutions are of the form $\zeta(u) = A\operatorname{cx} u + B\operatorname{mx} u + C\operatorname{px} u$. It can also be checked that the derivatives of the cosexponential functions are related by

$$\frac{d\operatorname{px}}{du} = \operatorname{mx}, \quad \frac{d\operatorname{mx}}{du} = \operatorname{cx}, \quad \frac{d\operatorname{cx}}{du} = \operatorname{px}. \tag{2.72}$$

2.4 Exponential and trigonometric forms of tricomplex numbers

If for a tricomplex number $u = x + ky + kz$ another tricomplex number $u_1 = x_1 + hy_1 + kz_1$ exists such that

$$x + hy + kz = e^{x_1 + hy_1 + kz_1}, \tag{2.73}$$

then u_1 is said to be the logarithm of u,

$$u_1 = \ln u. \tag{2.74}$$

The expressions of x_1, y_1, z_1 as functions of x, y, z can be obtained by developing e^{hy_1} and e^{kz_1} with the aid of Eqs. (2.41) and (2.42), by multiplying these expressions and separating the hypercomplex components,

$$x = e^{x_1}[\operatorname{cx} y_1 \ \operatorname{cx} z_1 + \operatorname{mx} y_1 \ \operatorname{mx} z_1 + \operatorname{px} y_1 \ \operatorname{px} z_1], \tag{2.75}$$

$$y = e^{x_1}[\operatorname{cx} y_1 \ \operatorname{px} z_1 + \operatorname{mx} y_1 \ \operatorname{cx} z_1 + \operatorname{px} y_1 \ \operatorname{mx} z_1], \tag{2.76}$$

$$z = e^{x_1}[\operatorname{cx} y_1 \ \operatorname{mx} z_1 + \operatorname{px} y_1 \ \operatorname{cx} z_1 + \operatorname{mx} y_1 \ \operatorname{px} z_1], \tag{2.77}$$

Using Eq. (2.33) with the substitutions $x_1 \to \operatorname{cx} y_1, y_1 \to \operatorname{mx} y_1, z_1 \to \operatorname{px} y_1, x_2 \to \operatorname{cx} z_1, y_2 \to \operatorname{px} z_1, z_2 \to \operatorname{mx} z_1$ and then the identity (2.62) yields

$$x^3 + y^3 + z^3 - 3xyz = e^{3x_1}, \tag{2.78}$$

whence

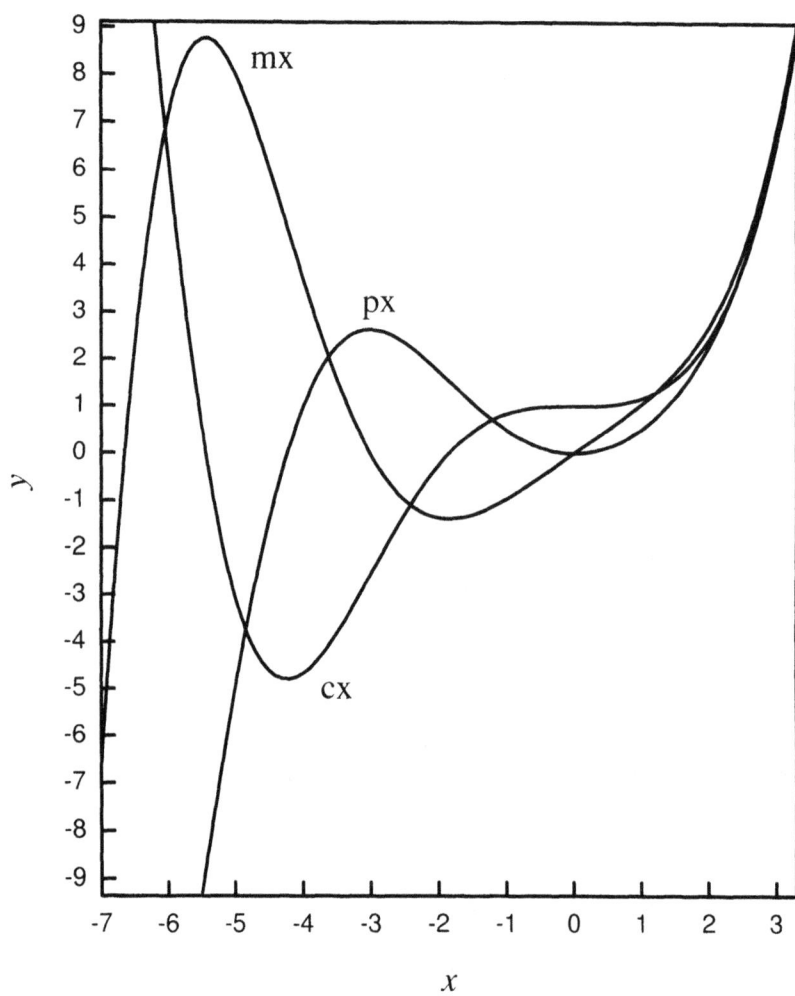

Figure 2.5: Graphs of the cosexponential functions cx, mx, px.

$$x_1 = \frac{1}{3}\ln(x^3 + y^3 + z^3 - 3xyz). \tag{2.79}$$

The logarithm in Eq. (2.79) exists as a real function for $x + y + z > 0$. A further relation can be obtained by summing Eqs. (2.75)-(2.77) and then using the addition theorems (2.47)-(2.49)

$$\frac{x + y + z}{(x^3 + y^3 + z^3 - 3xyz)^{1/3}} = \mathrm{cx}(y_1+z_1)+\mathrm{mx}(y_1+z_1)+\mathrm{px}(y_1+z_1). \tag{2.80}$$

The sum in Eq. (2.80) is according to Eq. (2.46) $e^{y_1+z_1}$, so that

$$y_1 + z_1 = \ln \frac{x + y + z}{(x^3 + y^3 + z^3 - 3xyz)^{1/3}}. \tag{2.81}$$

The logarithm in Eq. (2.81) is defined for points which are not on the trisector line (t), so that $x^2 + y^2 + z^2 - xy - xz - yz \neq 0$. Substituting in Eq. (2.73) the expression of x_1, Eq. (2.79), and of z_1 as a function of x, y, z, y_1, Eq. (2.81), yields

$$\frac{u}{\rho} \exp\left[-k\ln\left(\frac{\sqrt{2}s}{D}\right)^{2/3}\right] = e^{(h-k)y_1}, \tag{2.82}$$

where the quantities ρ, s and D have been defined in Eqs. (2.31),(2.14) and (2.15). Developing the exponential functions in the left-hand side of Eq. (2.82) with the aid of Eq. (2.42) and using the expressions of the cosexponential functions, Eqs. (2.68)-(2.70), and using the relation (2.63) for the right-hand side of Eq. (2.82) yields for the real part

$$\frac{\left(x - \frac{y+z}{2}\right)\cos\left[\frac{1}{\sqrt{3}}\ln\left(\frac{\sqrt{2}s}{D}\right)\right] - \frac{\sqrt{3}}{2}(y - z)\sin\left[\frac{1}{\sqrt{3}}\ln\left(\frac{\sqrt{2}s}{D}\right)\right]}{(x^2 + y^2 + z^2 - xy - xz - yz)^{1/2}}$$
$$= \cos(\sqrt{3}y_1), \tag{2.83}$$

which can also be written as

$$\cos\left[\frac{1}{\sqrt{3}}\ln\left(\frac{\sqrt{2}s}{D}\right) + \phi\right] = \cos(\sqrt{3}y_1). \tag{2.84}$$

where ϕ is the angle defined in Eqs. (2.19) and (2.20). Thus

$$y_1 = \frac{1}{3}\ln\left(\frac{\sqrt{2}s}{D}\right) + \frac{1}{\sqrt{3}}\phi. \tag{2.85}$$

The exponential form of the tricomplex number u is then

$$u = \rho \exp\left[\frac{1}{3}(h + k)\ln\frac{\sqrt{2}}{\tan\theta} + \frac{1}{\sqrt{3}}(h - k)\phi\right], \tag{2.86}$$

where θ is the angle between the line OP connecting the origin to the point P of coordinates (x, y, z) and the trisector line (t), defined in Eq. (2.21) and shown in Fig. 2.2. The exponential in Eq. (2.86) can be expanded with the aid of Eq. (2.58) and (2.64) as

$$\exp\left[\frac{1}{3}(h+k)\ln\frac{\sqrt{2}}{\tan\theta}\right] = \frac{2-h-k}{3}\left(\frac{\tan\theta}{\sqrt{2}}\right)^{1/3}$$
$$+\frac{1+h+k}{3}\left(\frac{\sqrt{2}}{\tan\theta}\right)^{2/3}, \tag{2.87}$$

so that

$$x + hy + kz = \rho\left[\frac{2-h-k}{3}\left(\frac{\tan\theta}{\sqrt{2}}\right)^{1/3} + \frac{1+h+k}{3}\left(\frac{\sqrt{2}}{\tan\theta}\right)^{2/3}\right]$$
$$\exp\left(\frac{h-k}{\sqrt{3}}\phi\right). \tag{2.88}$$

Substituting in Eq. (2.88) the expression of the amplitude ρ, Eq. (2.39), yields

$$u = d\sqrt{\frac{3}{2}}\left(\frac{2-h-k}{3}\sin\theta + \frac{1+h+k}{3}\sqrt{2}\cos\theta\right)\exp\left(\frac{h-k}{\sqrt{3}}\phi\right), \tag{2.89}$$

which is the trigonometric form of the tricomplex number u. As can be seen from Eq. (2.89), the tricomplex number $x + hy + kz$ is written as the product of the modulus d, of a factor depending on the polar angle θ with respect to the trisector line, and of a factor depending of the azimuthal angle ϕ in the plane Π perpendicular to the trisector line. The exponential in Eq. (2.89) can be expanded further with the aid of Eq. (2.64) as

$$\exp\left(\frac{1}{\sqrt{3}}(h-k)\phi\right) = \frac{1+h+k}{3} + \frac{2-h-k}{3}\cos\phi + \frac{h-k}{\sqrt{3}}\sin\phi, \tag{2.90}$$

so that the tricomplex number $x + hy + kz$ can also be written, after multiplication of the factors, in the form

$$x + hy + kz = \frac{2-h-k}{3}(x^2 + y^2 + z^2 - xy - xz - yz)^{1/2}\cos\phi$$
$$+\frac{h-k}{\sqrt{3}}(x^2 + y^2 + z^2 - xy - xz - yz)^{1/2}\sin\phi$$
$$+\frac{1+h+k}{3}(x+y+z). \tag{2.91}$$

The validity of Eq. (2.91) can be checked by substituting the expressions of $\cos\phi$ and $\sin\phi$ from Eqs. (2.19) and (2.20).

2.5 Elementary functions of a tricomplex variable

It can be shown with the aid of Eq. (2.86) that

$$(x + hy + kz)^m = \rho^m \left[\frac{2 - h - k}{3} \left(\frac{\tan\theta}{\sqrt{2}} \right)^{m/3} \right.$$
$$\left. + \frac{1 + h + k}{3} \left(\frac{\sqrt{2}}{\tan\theta} \right)^{2m/3} \right] \exp\left(\frac{h - k}{\sqrt{3}} m\phi \right), \qquad (2.92)$$

or equivalently

$$(x + hy + kz)^m = \frac{2 - h - k}{3}(x^2 + y^2 + z^2 - xy - xz - yz)^{m/2} \cos(m\phi)$$
$$+ \frac{h - k}{\sqrt{3}}(x^2 + y^2 + z^2 - xy - xz - yz)^{m/2} \sin(m\phi)$$
$$+ \frac{1 + h + k}{3}(x + y + z)^m, \qquad (2.93)$$

which are valid for real values of m. Thus Eqs. (2.92) or (2.93) define the power function u^m of the tricomplex variable u.

The power function is multivalued unless m is an integer. It can be inferred from Eq. (2.86) that, for integer values of m,

$$(uu')^m = u^m \, u'^m. \qquad (2.94)$$

For natural m, Eq. (2.93) can be checked with the aid of relations (2.19) and (2.20). For example if $m = 2$, it can be checked that the right-hand side of Eq. (2.93) is equal to $(x + hy + kz)^2 = x^2 + 2yz + h(z^2 + 2xz) + k(y^2 + 2xz)$.

The logarithm u_1 of the tricomplex number u, $u_1 = \ln u$, can be defined as the solution of Eq. (2.73) for u_1 as a function of u. For $x + y + z > 0$, from Eq. (2.86) it results that

$$\ln u = \ln\rho + \frac{1}{3}(h + k)\ln\left(\frac{\tan\theta}{\sqrt{2}} \right) + \frac{1}{\sqrt{3}}(h - k)\phi. \qquad (2.95)$$

It can be checked with the aid of Eqs. (2.25) and (2.32) that

$$\ln(uu') = \ln u + \ln u', \qquad (2.96)$$

which is valid up to integer multiples of $2\pi(h - k)/\sqrt{3}$.

The trigonometric functions of the hypercomplex variable u and the addition theorems for these functions have been written in Eqs. (1.57)-(1.60). The trigonometric functions of the hypercomplex variables hy, ky

can be expressed in terms of the cosexponential functions as

$$\cos(hy) = \frac{1}{2}[\operatorname{cx}(iy) + \operatorname{cx}(-iy)] + \frac{1}{2}h[\operatorname{mx}(iy) + \operatorname{mx}(-iy)]$$
$$+ \frac{1}{2}k[\operatorname{px}(iy) + \operatorname{px}(-iy)], \tag{2.97}$$

$$\cos(ky) = \frac{1}{2}[\operatorname{cx}(iy) + \operatorname{cx}(-iy)] + \frac{1}{2}h[\operatorname{px}(iy) + \operatorname{px}(-iy)]$$
$$+ \frac{1}{2}k[\operatorname{mx}(iy) + \operatorname{mx}(-iy)], \tag{2.98}$$

$$\sin(hy) = \frac{1}{2i}[\operatorname{cx}(iy) - \operatorname{cx}(-iy)] + \frac{1}{2i}h[\operatorname{mx}(iy) - \operatorname{mx}(-iy)]$$
$$+ \frac{1}{2i}k[\operatorname{px}(iy) - \operatorname{px}(-iy)], \tag{2.99}$$

$$\sin(ky) = \frac{1}{2i}[\operatorname{cx}(iy) - \operatorname{cx}(-iy)] + \frac{1}{2i}h[\operatorname{px}(iy) - \operatorname{px}(-iy)]$$
$$+ \frac{1}{2i}k[\operatorname{mx}(iy) - \operatorname{mx}(-iy)], \tag{2.100}$$

where i is the imaginary unit. Using the expressions of the cosexponential functions in Eqs. (2.68)-(2.70) gives expressions of the trigonometric functions of hy, hz as

$$\cos(hy) = \frac{1}{3}\cos y + \frac{2}{3}\cosh\left(\frac{\sqrt{3}}{2}y\right)\cos\frac{y}{2} +$$
$$+ h\left[\frac{1}{3}\cos y - \frac{1}{3}\cosh\left(\frac{\sqrt{3}}{2}y\right)\cos\frac{y}{2} + \frac{1}{\sqrt{3}}\sinh\left(\frac{\sqrt{3}}{2}y\right)\sin\frac{y}{2}\right]$$
$$+ k\left[\frac{1}{3}\cos y - \frac{1}{3}\cosh\left(\frac{\sqrt{3}}{2}y\right)\cos\frac{y}{2} - \frac{1}{\sqrt{3}}\sinh\left(\frac{\sqrt{3}}{2}y\right)\sin\frac{y}{2}\right], \tag{2.101}$$

$$\cos(ky) = \frac{1}{3}\cos y + \frac{2}{3}\cosh\left(\frac{\sqrt{3}}{2}y\right)\cos\frac{y}{2} +$$
$$+ h\left[\frac{1}{3}\cos y - \frac{1}{3}\cosh\left(\frac{\sqrt{3}}{2}y\right)\cos\frac{y}{2} - \frac{1}{\sqrt{3}}\sinh\left(\frac{\sqrt{3}}{2}y\right)\sin\frac{y}{2}\right]$$
$$+ k\left[\frac{1}{3}\cos y - \frac{1}{3}\cosh\left(\frac{\sqrt{3}}{2}y\right)\cos\frac{y}{2} + \frac{1}{\sqrt{3}}\sinh\left(\frac{\sqrt{3}}{2}y\right)\sin\frac{y}{2}\right], \tag{2.102}$$

$$\sin(hy) = \frac{1}{3}\sin y - \frac{2}{3}\cosh\left(\frac{\sqrt{3}}{2}y\right)\sin\frac{y}{2}$$

$$+h\left[\frac{1}{3}\sin y + \frac{1}{3}\cosh\left(\frac{\sqrt{3}}{2}y\right)\sin\frac{y}{2} + \frac{1}{\sqrt{3}}\sinh\left(\frac{\sqrt{3}}{2}y\right)\cos\frac{y}{2}\right]$$

$$+k\left[\frac{1}{3}\sin y + \frac{1}{3}\cosh\left(\frac{\sqrt{3}}{2}y\right)\sin\frac{y}{2} - \frac{1}{\sqrt{3}}\sinh\left(\frac{\sqrt{3}}{2}y\right)\cos\frac{y}{2}\right],$$

$$(2.103)$$

$$\sin(ky) = \frac{1}{3}\sin y - \frac{2}{3}\cosh\left(\frac{\sqrt{3}}{2}y\right)\sin\frac{y}{2}$$

$$+h\left[\frac{1}{3}\sin y + \frac{1}{3}\cosh\left(\frac{\sqrt{3}}{2}y\right)\sin\frac{y}{2} - \frac{1}{\sqrt{3}}\sinh\left(\frac{\sqrt{3}}{2}y\right)\cos\frac{y}{2}\right]$$

$$+k\left[\frac{1}{3}\sin y + \frac{1}{3}\cosh\left(\frac{\sqrt{3}}{2}y\right)\sin\frac{y}{2} + \frac{1}{\sqrt{3}}\sinh\left(\frac{\sqrt{3}}{2}y\right)\cos\frac{y}{2}\right].$$

$$(2.104)$$

The trigonometric functions of a tricomplex number $x + hy + kz$ can then be expressed in terms of elementary functions with the aid of the addition theorems Eqs. (1.59), (1.60) and of the expressions in Eqs. (2.101)-(2.104).

The hyperbolic functions of the hypercomplex variable u and the addition theorems for these functions have been written in Eqs. (1.62)-(1.65). The hyperbolic functions of the hypercomplex variables hy, ky can be expressed in terms of the elementary functions as

$$\cosh(hy) = \frac{1}{3}\cosh y + \frac{2}{3}\cos\left(\frac{\sqrt{3}}{2}y\right)\cosh\frac{y}{2}+$$

$$+h\left[\frac{1}{3}\cosh y - \frac{1}{3}\cos\left(\frac{\sqrt{3}}{2}y\right)\cosh\frac{y}{2} - \frac{1}{\sqrt{3}}\sin\left(\frac{\sqrt{3}}{2}y\right)\sinh\frac{y}{2}\right]$$

$$+k\left[\frac{1}{3}\cosh y - \frac{1}{3}\cos\left(\frac{\sqrt{3}}{2}y\right)\cosh\frac{y}{2} + \frac{1}{\sqrt{3}}\sin\left(\frac{\sqrt{3}}{2}y\right)\sinh\frac{y}{2}\right],$$

$$(2.105)$$

$$\cosh(ky) = \frac{1}{3}\cosh y + \frac{2}{3}\cos\left(\frac{\sqrt{3}}{2}y\right)\cosh\frac{y}{2}+$$

$$+h\left[\frac{1}{3}\cosh y - \frac{1}{3}\cos\left(\frac{\sqrt{3}}{2}y\right)\cosh\frac{y}{2} + \frac{1}{\sqrt{3}}\sin\left(\frac{\sqrt{3}}{2}y\right)\sinh\frac{y}{2}\right]$$

$$+k\left[\frac{1}{3}\cosh y - \frac{1}{3}\cos\left(\frac{\sqrt{3}}{2}y\right)\cosh\frac{y}{2} - \frac{1}{\sqrt{3}}\sin\left(\frac{\sqrt{3}}{2}y\right)\sinh\frac{y}{2}\right],$$

$$(2.106)$$

$$\sinh(hy) = \frac{1}{3}\sinh y - \frac{2}{3}\cos\left(\frac{\sqrt{3}}{2}y\right)\sinh\frac{y}{2}$$

$$+h\left[\frac{1}{3}\sinh y + \frac{1}{3}\cos\left(\frac{\sqrt{3}}{2}y\right)\sinh\frac{y}{2} + \frac{1}{\sqrt{3}}\sin\left(\frac{\sqrt{3}}{2}y\right)\cosh\frac{y}{2}\right]$$

$$+k\left[\frac{1}{3}\sinh y + \frac{1}{3}\cos\left(\frac{\sqrt{3}}{2}y\right)\sinh\frac{y}{2} - \frac{1}{\sqrt{3}}\sin\left(\frac{\sqrt{3}}{2}y\right)\cosh\frac{y}{2}\right],$$

$$(2.107)$$

$$\sinh(ky) = \frac{1}{3}\sinh y - \frac{2}{3}\cos\left(\frac{\sqrt{3}}{2}y\right)\sinh\frac{y}{2}$$

$$+h\left[\frac{1}{3}\sinh y + \frac{1}{3}\cos\left(\frac{\sqrt{3}}{2}y\right)\sinh\frac{y}{2} - \frac{1}{\sqrt{3}}\sin\left(\frac{\sqrt{3}}{2}y\right)\cosh\frac{y}{2}\right]$$

$$+k\left[\frac{1}{3}\sinh y + \frac{1}{3}\cos\left(\frac{\sqrt{3}}{2}y\right)\sinh\frac{y}{2} + \frac{1}{\sqrt{3}}\sin\left(\frac{\sqrt{3}}{2}y\right)\cosh\frac{y}{2}\right].$$

$$(2.108)$$

The hyperbolic functions of a tricomplex number $x + hy + kz$ can then be expressed in terms of the elementary functions with the aid of the addition theorems Eqs. (1.64), (1.65) and of the expressions in Eqs. (2.105)-(2.108).

2.6   Tricomplex power series

A tricomplex series is an infinite sum of the form

$$a_0 + a_1 + a_2 + \cdots + a_l + \cdots, \tag{2.109}$$

where the coefficients a_n are tricomplex numbers. The convergence of the series (2.109) can be defined in terms of the convergence of its 3 real components. The convergence of a tricomplex series can however be studied using tricomplex variables. The main criterion for absolute convergence remains the comparison theorem, but this requires a number of inequalities which will be discussed further.

The modulus of a tricomplex number $u = x + hy + kz$ can be defined as

$$|u| = (x^2 + y^2 + z^2)^{1/2}. \tag{2.110}$$

Since $|x| \leq |u|, |y| \leq |u|, |z| \leq |u|$, a property of absolute convergence established via a comparison theorem based on the modulus of the series (2.109) will ensure the absolute convergence of each real component of that series.

The modulus of the sum $u_1 + u_2$ of the tricomplex numbers u_1, u_2 fulfils the inequality

$$||u_1| - |u_2|| \leq |u_1 + u_2| \leq |u_1| + |u_2|. \tag{2.111}$$

For the product the relation is

$$|u_1 u_2| \leq \sqrt{3}|u_1||u_2|, \tag{2.112}$$

which replaces the relation of equality extant for regular complex numbers. The equality in Eq. (2.112) takes place for $x_1 = y_1 = z_1$ and $x_2 = y_2 = z_2$, i.e when both tricomplex numbers lie on the trisector line (t). Using Eq. (2.91), the relation (2.112) can be written equivalently as

$$\frac{2}{3}\delta_1^2\delta_2^2 + \frac{1}{3}\sigma_1^2\sigma_2^2 \leq 3\left(\frac{2}{3}\delta_1^2 + \frac{1}{3}\sigma_1^2\right)\left(\frac{2}{3}\delta_2^2 + \frac{1}{3}\sigma_2^2\right), \tag{2.113}$$

where $\delta_j^2 = x_j^2 + y_j^2 + z_j^2 - x_j y_j - x_j z_j - y_j z_j, \sigma_j = x_j + y_j + z_j, j = 1, 2$, the equality taking place for $\delta_1 = 0, \delta_2 = 0$. A particular form of Eq. (2.112) is

$$|u^2| \leq \sqrt{3}|u|^2, \tag{2.114}$$

and it can be shown that

$$|u^l| \leq 3^{(l-1)/2}|u|^l, \tag{2.115}$$

the equality in Eqs. (2.114) and (2.115) taking place for $x = y = z$. It can be shown from Eq. (2.93) that

$$|u^l|^2 = \frac{2}{3}\delta^{2l} + \frac{1}{3}\sigma^{2l}, \tag{2.116}$$

where $\delta^2 = x^2 + y^2 + z^2 - xy - xz - yz, \sigma = x + y + z$. Then Eq. (2.115) can also be written as

$$\frac{2}{3}\delta^{2l} + \frac{1}{3}\sigma^{2l} \leq 3^{l-1}\left(\frac{2}{3}\delta^2 + \frac{1}{3}\sigma^2\right)^l, \tag{2.117}$$

the equality taking place for $\delta = 0$. From Eqs. (2.112) and (2.115) it results that

$$|au^l| \leq 3^{l/2}|a||u|^l. \tag{2.118}$$

It can also be shown that

$$\left|\frac{1}{u}\right| \geq \frac{1}{|u|},$$

(2.119)

the equality taking place for $\sigma^2 = \delta^2$, or $xy + xz + yz = 0$.

A power series of the tricomplex variable u is a series of the form

$$a_0 + a_1 u + a_2 u^2 + \cdots + a_l u^l + \cdots .$$

(2.120)

Since

$$\left|\sum_{l=0}^{\infty} a_l u^l\right| \leq \sum_{l=0}^{\infty} 3^{l/2} |a_l| |u|^l,$$

(2.121)

a sufficient condition for the absolute convergence of this series is that

$$\lim_{n\to\infty} \frac{\sqrt{3}|a_{l+1}||u|}{|a_l|} < 1.$$

(2.122)

Thus the series is absolutely convergent for

$$|u| < c_0,$$

(2.123)

where

$$c_0 = \lim_{l\to\infty} \frac{|a_l|}{\sqrt{3}|a_{l+1}|}.$$

(2.124)

The convergence of the series (2.120) can be also studied with the aid of a transformation which explicits the transverse and longitudinal parts of the variable u and of the constants a_l,

$$x + hy + kz = v_1 e_1 + \tilde{v}_1 \tilde{e}_1 + v_+ e_+,$$

(2.125)

where

$$v_1 = \frac{2x - y - z}{2}, \quad \tilde{v}_1 = \frac{\sqrt{3}}{2}(y - z), \quad v_+ = x + y + z,$$

(2.126)

and

$$e_1 = \frac{2 - h - k}{3}, \quad \tilde{e}_1 = \frac{h - k}{\sqrt{3}}, \quad e_+ = \frac{1 + h + k}{3}.$$

(2.127)

The variables $v_1, \tilde{v}_1, v_+$ will be called canonical tricomplex variables, $e_1, \tilde{e}_1, e_+$ will be called the canonical tricomplex base, and Eq. (2.125) gives the canonical form of a tricomplex number. In the geometric representation of

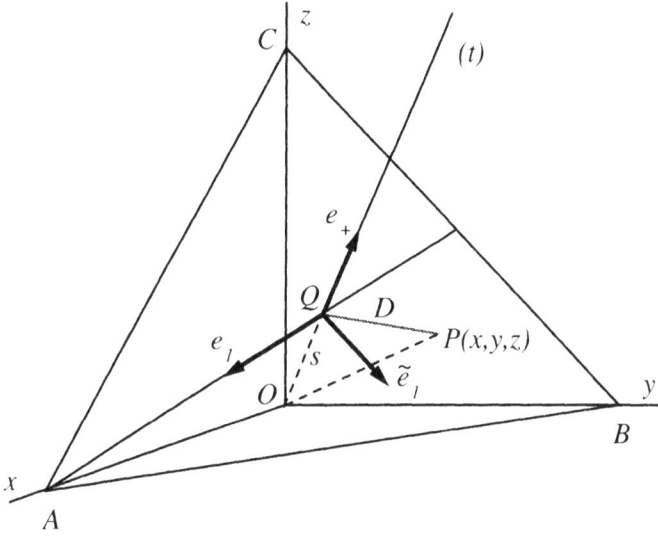

Figure 2.6: Unit vectors $e_1, \tilde{e}_1, e_+$ of the orthogonal system of coordinates with origin at Q. The plane parallel to Π passing through P intersects the trisector line (t) at Q and the axes of coordinates x, y, z at the points A, B, C.

Fig. 2.6, $e_1, \tilde{e}_1$ are situated in the plane Π, and e_+ is lying on the trisector line (t). It can be checked that

$$e_1^2 = e_1, \ \tilde{e}_1^2 = -e_1, \ e_1\tilde{e}_1 = \tilde{e}_1, \ e_1 e_+ = 0, \ \tilde{e}_1 e_+ = 0, \ e_+^2 = e_+. \quad (2.128)$$

The moduli of the bases in Eq. (2.128) are

$$|e_1| = \sqrt{\frac{2}{3}}, \ |\tilde{e}_1| = \sqrt{\frac{2}{3}}, \ |e_+| = \sqrt{\frac{1}{3}}, \quad (2.129)$$

and it can be checked that

$$|x + hy + kz|^2 = \frac{2}{3}(v_1^2 + \tilde{v}_1^2) + \frac{1}{3}v_+^2. \quad (2.130)$$

If $u = u'u''$, the transverse and longitudinal components are related by

$$v_1 = v_1'v_1'' - \tilde{v}_1'\tilde{v}_1'', \ \tilde{v}_1 = v_1'\tilde{v}_1'' + \tilde{v}_1'v_1'', \ v_+ = v_+'v_+'', \quad (2.131)$$

which show that, upon multiplication, the transverse components obey the same rules as the real and imaginary components of usual, two-dimensional complex numbers, and the rule for the longitudinal component is that of the regular multiplication of numbers.

If the constants in Eq. (2.120) are $a_l = p_l + hq_l + kr_l$, and

$$a_{l1} = \frac{2p_l - q_l - r_l}{2}, \quad \tilde{a}_{l1} = \frac{\sqrt{3}}{2}(q_l - r_l), \quad a_{l+} = p_l + q_l + r_l, \qquad (2.132)$$

where $p_0 = 1, q_0 = 0, r_0 = 0$, the series (2.120) can be written as

$$\sum_{l=0}^{\infty} \left[a_{l1}e_1 + \tilde{a}_{l1}\tilde{e}_1)(v_1 e_1 + \tilde{v}_1 \tilde{e}_1)^l + e_+ a_{l+} v_+^l \right]. \qquad (2.133)$$

The series in Eq. (2.133) is absolutely convergent for

$$|v_+| < c_+, \quad (v_1^2 + \tilde{v}_1^2)^{1/2} < c_1, \qquad (2.134)$$

where

$$c_+ = \lim_{l \to \infty} \frac{|a_{l+}|}{|a_{l+1,u}|}, \quad c_1 = \lim_{l \to \infty} \frac{\left(a_{l1}^2 + \tilde{a}_{l1}^2\right)^{1/2}}{\left(a_{l+1,1}^2 + a_{l+1,2}^2\right)^{1/2}}. \qquad (2.135)$$

The relations (2.134) and (2.130) show that the region of convergence of the series (2.120) is a cylinder of radius $c_1\sqrt{2/3}$ and height $2c_+/\sqrt{3}$, having the trisector line (t) as axis and the origin as center, which can be called cylinder of convergence, as shown in Fig. 2.7.

It can be shown that $c_1 = (1/\sqrt{3}) \min(c_+, c_1)$, where min designates the smallest of the numbers c_+, c_1. Using the expression of $|u|$ in Eq. (2.128), it can be seen that the spherical region of convergence defined in Eqs. (2.123), (2.124) is a subset of the cylindrical region of convergence defined in Eqs. (2.134) and (2.135).

2.7 Analytic functions of tricomplex variables

The analytic functions of the hypercomplex variable u and the series expansion of functions have been discussed in Eqs. (1.85)-(1.93). If the tricomplex function $f(u)$ of the tricomplex variable u is written in terms of the real functions $F(x, y, z), G(x, y, z), H(x, y, z)$ of real variables x, y, z as

$$f(u) = F(x, y, z) + hG(x, y, z) + kH(x, y, z), \qquad (2.136)$$

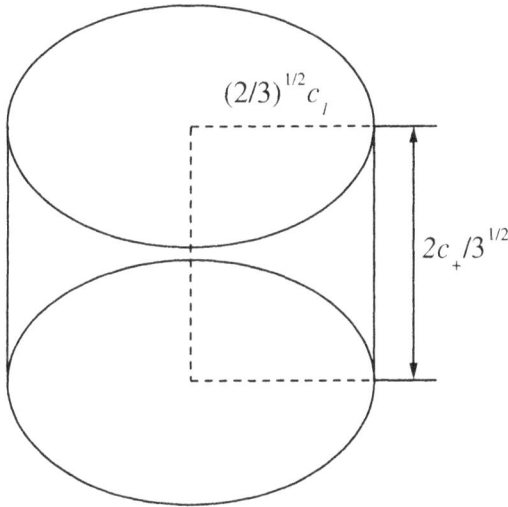

Figure 2.7: Cylinder of convergence of tricomplex series, of radius $(2/3)^{1/2}c_1$ and height $2c_+/3^{1/2}$, having the axis parallel to the trisector line, and the regions of space delimited by the plane, cylindrical and conical surfaces.

then relations of equality exist between partial derivatives of the functions F, G, H. These relations can be obtained by writing the derivative of the function f as

$$\frac{1}{\Delta x + h\Delta y + k\Delta z}\left[\frac{\partial F}{\partial x}\Delta x + \frac{\partial F}{\partial y}\Delta y + \frac{\partial F}{\partial z}\Delta z\right.$$
$$+ h\left(\frac{\partial G}{\partial x}\Delta x + \frac{\partial G}{\partial y}\Delta y + \frac{\partial G}{\partial z}\Delta z\right)$$
$$\left. + k\left(\frac{\partial H}{\partial x}\Delta x + \frac{\partial H}{\partial y}\Delta y + \frac{\partial H}{\partial z}\Delta z\right)\right], \tag{2.137}$$

where the difference appearing in Eq. (98) is $u - u_0 = \Delta x + h\Delta y + k\Delta z$. The relations between the partials derivatives of the functions F, G, H are obtained by setting successively in Eq. (2.137) $\Delta x \to 0, \Delta y = 0, \Delta z = 0$; then $\Delta x = 0, \Delta y \to 0, \Delta z = 0$; and $\Delta x = 0, \Delta y = 0, \Delta z \to 0$. The relations are

$$\frac{\partial F}{\partial x} = \frac{\partial G}{\partial y}, \frac{\partial G}{\partial x} = \frac{\partial H}{\partial y}, \frac{\partial H}{\partial x} = \frac{\partial F}{\partial y}, \tag{2.138}$$

$$\frac{\partial F}{\partial x} = \frac{\partial H}{\partial z}, \quad \frac{\partial G}{\partial x} = \frac{\partial F}{\partial z}, \quad \frac{\partial H}{\partial x} = \frac{\partial G}{\partial z}, \tag{2.139}$$

$$\frac{\partial G}{\partial y} = \frac{\partial H}{\partial z}, \quad \frac{\partial H}{\partial y} = \frac{\partial F}{\partial z}, \quad \frac{\partial F}{\partial y} = \frac{\partial G}{\partial z}. \tag{2.140}$$

The relations (2.138)-(2.140) are analogous to the Riemann relations for the real and imaginary components of a complex function. It can be shown from Eqs. (2.138)-(2.140) that the components F solutions of the equations

$$\frac{\partial^2 F}{\partial x^2} - \frac{\partial^2 F}{\partial y \partial z} = 0, \quad \frac{\partial^2 F}{\partial y^2} - \frac{\partial^2 F}{\partial x \partial z} = 0, \quad \frac{\partial^2 F}{\partial z^2} - \frac{\partial^2 F}{\partial x \partial y} = 0, \tag{2.141}$$

$$\frac{\partial^2 G}{\partial x^2} - \frac{\partial^2 G}{\partial y \partial z} = 0, \quad \frac{\partial^2 G}{\partial y^2} - \frac{\partial^2 G}{\partial x \partial z} = 0, \quad \frac{\partial^2 G}{\partial z^2} - \frac{\partial^2 G}{\partial x \partial y} = 0, \tag{2.142}$$

$$\frac{\partial^2 H}{\partial x^2} - \frac{\partial^2 H}{\partial y \partial z} = 0, \quad \frac{\partial^2 H}{\partial y^2} - \frac{\partial^2 H}{\partial x \partial z} = 0, \quad \frac{\partial^2 H}{\partial z^2} - \frac{\partial^2 H}{\partial x \partial y} = 0. \tag{2.143}$$

It can also be shown that the differences $F - G, F - H, G - H$ are solutions of the equation of Laplace,

$$\Delta(F - G) = 0, \quad \Delta(F - H) = 0, \quad \Delta(G - H) = 0,$$
$$\Delta = \frac{\partial^2}{\partial x^2} + \frac{\partial^2}{\partial y^2} + \frac{\partial^2}{\partial z^2}. \tag{2.144}$$

If a geometric transformation is considered in which to a point u is associated the point $f(u)$, it can be shown that the tricomplex function $f(u)$ transforms a straight line parallel to the trisector line (t) in a straight line parallel to (t), and transforms a plane parallel to the nodal plane Π in a plane parallel to Π. A transformation generated by a tricomplex function $f(u)$ does not conserve in general the angle of intersecting lines.

2.8   Integrals of tricomplex functions

The singularities of tricomplex functions arise from terms of the form $1/(u - a)^m$, with $m > 0$. Functions containing such terms are singular not only at $u = a$, but also at all points of a plane (Π_a) through the point a and parallel to the nodal plane Π and at all points of a straight line (t_a) passing through a and parallel to the trisector line (t).

The integral of a tricomplex function between two points A, B along a path situated in a region free of singularities is independent of path, which means that the integral of an analytic function along a loop situated in a region free from singularities is zero,

$$\oint_\Gamma f(u)du = 0, \tag{2.145}$$

where it is supposed that a surface S spanning the closed loop Γ is not intersected by any of the planes and is not threaded by any of the lines associated with the singularities of the function $f(u)$. Using the expression, Eq. (2.136) for $f(u)$ and the fact that $du = dx + hdy + kdz$, the explicit form of the integral in Eq. (2.145) is

$$\oint_\Gamma f(u)du = \oint_\Gamma [Fdx + Hdy + Gdz + h(Gdx + Fdy + Hdz) \\ +k(Hdx + Gdy + Fdz)]. \tag{2.146}$$

If the functions F, G, H are regular on a surface S spanning the loop Γ, the integral along the loop Γ can be transformed with the aid of the theorem of Stokes in an integral over the surface S of terms of the form $\partial H/\partial x - \partial F/\partial y$, $\partial G/\partial x - \partial F/\partial z$, $\partial G/\partial y - \partial H/\partial z, \ldots$ which are equal to zero by Eqs. (2.138)-(2.140), and this proves Eq. (2.145).

If there are singularities on the surface S, the integral $\oint f(u)du$ is not necessarily equal to zero. If $f(u) = 1/(u-a)$ and the loop Γ_a is situated in the half-space above the plane (Π_a) and encircles once the line (t_a), then

$$\oint_{\Gamma_a} \frac{du}{u-a} = \frac{2\pi}{\sqrt{3}}(h-k). \tag{2.147}$$

This is due to the fact that the integral of $1/(u-a)$ along the loop Γ_a is equal to the integral of $1/(u-a)$ along a circle (C_a) with the center on the line (t_a) and perpendicular to this line, as shown in Fig. 2.8.

$$\oint_{\Gamma_a} \frac{du}{u-a} = \oint_{C_a} \frac{du}{u-a}, \tag{2.148}$$

this being a corollary of Eq. (2.145). The integral on the right-hand side of Eq. (2.148) can be evaluated with the aid of the trigonometric form Eq. (2.88) of the tricomplex quantity $u - a$, so that

$$\frac{du}{u-a} = \frac{h-k}{\sqrt{3}}d\phi, \tag{2.149}$$

which by integration over $d\phi$ from 0 to 2π yields Eq. (2.147).

The integral $\oint_{\Gamma_a} du(u-a)^m$, with m an integer number not equal to -1, is equal to zero, because $\int du(u-a)^m = (u-a)^{m+1}/(m+1)$, and $(u-a)^{m+1}/(m+1)$ is singlevalued,

$$\oint_{\Gamma_a} du(u-a)^m = 0, \text{ for } m \text{ integer}, m \neq -1. \tag{2.150}$$

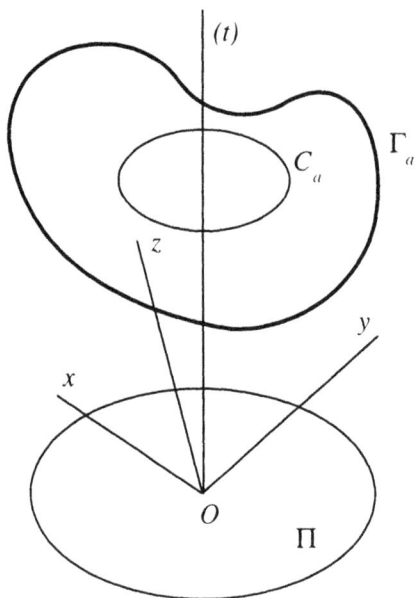

Figure 2.8: The integral of $1/(u - a)$ along the loop Γ_a is equal to the integral of $1/(u - a)$ along a circle (C_a) with the center on the line (t_a) and perpendicular to this line.

If $f(u)$ is an analytic tricomplex function which can be expanded in a series as written in Eq. (1.89), with $u_0 = a$, and the expansion holds on the curve Γ and on a surface spanning Γ, then from Eqs. (2.149) and (2.150) it follows that

$$\oint_\Gamma \frac{f(u)du}{u - a} = \frac{2\pi}{\sqrt{3}}(h - k)f(a). \tag{2.151}$$

Substituting in the right-hand side of Eq. (2.151) the expression of $f(u)$ in terms of the real components F, G, H, Eq. (2.136), at $u = a$, yields

$$\oint_\Gamma \frac{f(u)du}{u - a} = \frac{2\pi}{\sqrt{3}}[H - G + h(F - H) + k(G - F)]. \tag{2.152}$$

Since the sum of the real components in the paranthesis from the right-hand side of Eq. (2.152) is equal to zero, this equation determines only the differences between the components F, G, H. If $f(u)$ can be expanded as written in Eq. (1.89), with $u_0 = a$, on Γ and on a surface spanning Γ, then from Eqs. (2.147) and (2.150) it also results that

$$\oint_\Gamma \frac{f(u)du}{(u-a)^{m+1}} = \frac{2\pi}{\sqrt{3}m!}(h-k)f^{(m)}(a), \tag{2.153}$$

where the fact that has been used that the derivative $f^{(m)}(a)$ of order m of $f(u)$ at $u = a$ is related to the expansion coefficient in Eq. (1.89), with $u_0 = a$, according to Eq. (1.93), with $u_0 = a$. The relation (2.153) can also be obtained by successive derivations of Eq. (2.151).

If a function $f(u)$ is expanded in positive and negative powers of $u - u_j$, where u_j are fourcomplex constants, j being an index, the integral of f on a closed loop Γ is determined by the terms in the expansion of f which are of the form $a_j/(u - u_j)$,

$$f(u) = \cdots + \sum_j \frac{a_j}{u - u_j} + \cdots. \tag{2.154}$$

In Eq. (2.154), u_j is the pole and a_j is the residue relative to the pole u_j. Then the integral of f on a closed loop Γ is

$$\oint_\Gamma f(u)du = \frac{2\pi}{\sqrt{3}}(h-k)\sum_j \text{int}(u_{j\Pi}, \Gamma_\Pi)a_j, \tag{2.155}$$

where the functional $\text{int}(M, C)$, defined for a point M and a closed curve C in a two-dimensional plane, is given by

$$\text{int}(M, C) = \begin{cases} 1 & \text{if } M \text{ is an interior point of } C, \\ 0 & \text{if } M \text{ is exterior to } C, \end{cases} \tag{2.156}$$

and $u_{j\Pi}, \Gamma_\Pi$ are the projections of the point u_j and of the curve Γ on the nodal plane Π, as shown in Fig. 2.9.

2.9 Factorization of tricomplex polynomials

A polynomial of degree m of the tricomplex variable $u = x + hy + kz$ has the form

$$P_m(u) = u^m + a_1 u^{m-1} + \cdots + a_{m-1}u + a_m, \tag{2.157}$$

where the constants are in general tricomplex numbers, $a_l = p_l + hq_l + kr_l$, $l = 1, \cdots, m$. In order to write the polynomial $P_m(u)$ as a product of

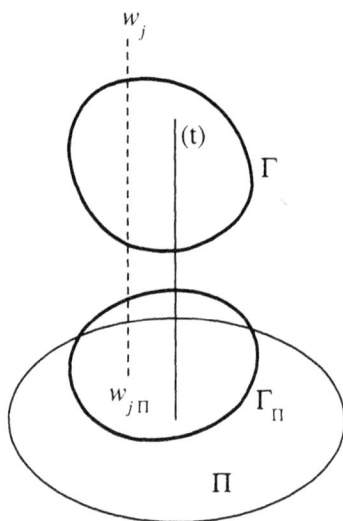

Figure 2.9: Integration path Γ, pole u_j and their projections $\Gamma_\Pi, u_{j\Pi}$ on the nodal plane Π.

factors, the variable u and the constants a_l will be written in the form which explicits the transverse and longitudinal components,

$$P_m(u) = \sum_{l=0}^{m}(a_{l1}e_1 + \tilde{a}_{l1}\tilde{e}_1)(v_1e_1 + \tilde{v}_1\tilde{e}_1)^{m-l} + e_+\sum_{l=0}^{m}a_{l+}v_+^{m-l}, \quad (2.158)$$

where the constants have been defined previously in Eq. (2.132). Due to the properties in Eq. (2.128), the transverse part of the polynomial $P_m(u)$ can be written as a product of linear factors of the form

$$\sum_{l=0}^{m}(a_{l1}e_1+\tilde{a}_{l1}\tilde{e}_1)(v_1e_1+\tilde{v}_1\tilde{e}_1)^{m-l} = \prod_{l=1}^{m}[(v_1-v_{l1})e_1+(\tilde{v}_1-\tilde{v}_{l1})\tilde{e}_1], \quad (2.159)$$

where the quantities $v_{l1}, \tilde{v}_{l1}$ are real numbers. The longitudinal part of $P_m(u)$, Eq. (2.158), can be written as a product of linear or quadratic factors with real coefficients, or as a product of linear factors which, if imaginary, appear always in complex conjugate pairs. Using the latter form for the simplicity of notations, the longitudinal part can be written as

$$\sum_{l=0}^{m}a_{l+}v_+^{m-l} = \prod_{l=1}^{m}(v_+ - v_{l+}), \quad (2.160)$$

where the quantities v_{l+} appear always in complex conjugate pairs. Due to the orthogonality of the transverse and longitudinal components, Eq. (2.128), the polynomial $P_m(u)$ can be written as a product of factors of the form

$$P_m(u) = \prod_{l=1}^{m} [(v_1 - v_{l1})e_1 + (\tilde{v}_1 - \tilde{v}_{l1})\tilde{e}_1 + (v_+ - v_{l+})e_+]. \qquad (2.161)$$

These relations can be written with the aid of Eqs. (2.125) as

$$P_m(u) = \prod_{l=1}^{m} (u - u_l), \qquad (2.162)$$

where

$$u_l = v_{l1}e_1 + \tilde{v}_{l1}\tilde{e}_1 + v_{l+}e_+. \qquad (2.163)$$

The roots v_{l+} and the roots $v_{l1}e_1 + \tilde{v}_{l1}\tilde{e}_1$ defined in Eq. (2.159) may be ordered arbitrarily. This means that Eq. (2.163) gives sets of m roots $u_1, ..., u_m$ of the polynomial $P_m(u)$, corresponding to the various ways in which the roots $v_{l+}, v_{l1}e_1 + \tilde{v}_{l1}\tilde{e}_1$ are ordered according to l in each group. Thus, while the tricomplex components in Eq. (2.158) taken separately have unique factorizations, the polynomial $P_m(q)$ can be written in many different ways as a product of linear factors.

If $P(u) = u^2 - 1$, the degree is $m = 2$, the coefficients of the polynomial are $a_1 = 0, a_2 = -1$, the coefficients defined in Eq. (2.132) are $a_{21} = -1, a_{22} = 0, a_{2u} = -1$. The expression of $P(u)$, Eq. (2.158), is $(e_1 v_1 + \tilde{e}_1 \tilde{v}_1)^2 - e_1 + e_+(v_+^2 - 1)$. The factorizations in Eqs. (2.159) and (2.160) are $(e_1 v_1 + \tilde{e}_1 \tilde{v}_1)^2 - e_1 = [e_1(v_1 + 1) + \tilde{e}_1 \tilde{v}_1][e_1(v_1 - 1) + \tilde{e}_1 \tilde{v}_1]$ and $v_+^2 - 1 = (v_+ + 1)(v_+ - 1)$. The factorization of $P(u)$, Eq. (2.162), is $P(u) = (u - u_1)(u - u_2)$, where according to Eq. (2.163) the roots are $u_1 = \pm e_1 \pm e_+, u_2 = -u_1$. If e_1 and e_+ are expressed with the aid of Eq. (2.127) in terms of h and k, the factorizations of $P(u)$ are obtained as

$$u^2 - 1 = (u + 1)(u - 1), \qquad (2.164)$$

or as

$$u^2 - 1 = \left(u + \frac{1 - 2h - 2k}{3}\right)\left(u - \frac{1 - 2h - 2k}{3}\right). \qquad (2.165)$$

It can be checked that $(\pm e_1 \pm e_+)^2 = e_1 + e_+ = 1$.

2.10   Representation of tricomplex numbers by irreducible matrices

If the matrix in Eq. (2.34) representing the tricomplex number u is called U, and

$$T = \begin{pmatrix} \sqrt{\frac{2}{3}} & -\frac{1}{\sqrt{6}} & -\frac{1}{\sqrt{6}} \\ 0 & \frac{1}{\sqrt{2}} & -\frac{1}{\sqrt{2}} \\ \frac{1}{\sqrt{3}} & \frac{1}{\sqrt{3}} & \frac{1}{\sqrt{3}} \end{pmatrix}, \tag{2.166}$$

which is the matrix appearing in Eq. (2.17), it can be checked that

$$TUT^{-1} = \begin{pmatrix} x - \frac{y+z}{2} & \frac{\sqrt{3}}{2}(y-z) & 0 \\ -\frac{\sqrt{3}}{2}(y-z) & x - \frac{y+z}{2} & 0 \\ 0 & 0 & x+y+z \end{pmatrix}. \tag{2.167}$$

The relations for the variables $x - (y+z)/2$, $(\sqrt{3}/2)(y-z)$ and $x+y+z$ for the multiplication of tricomplex numbers have been written in Eqs. (2.26), (2.28) and (2.29). The matrices TUT^{-1} provide an irreducible representation [7] of the tricomplex numbers $u = x + hy + kz$, in terms of matrices with real coefficients.

Chapter 3

Commutative Complex Numbers in Four Dimensions

Systems of hypercomplex numbers in 4 dimensions of the form $u = x + \alpha y + \beta z + \gamma t$ are described in this chapter, where the variables x, y, z and t are real numbers, for which the multiplication is both associative and commutative. The product of two fourcomplex numbers is equal to zero if both numbers are equal to zero, or if the numbers belong to certain four-dimensional hyperplanes as discussed further in this chapter. The fourcomplex numbers have exponential and trigonometric forms, and the concepts of analytic fourcomplex function, contour integration and residue can be defined. Expressions are given for the elementary functions of four-complex variable. The functions $f(u)$ of fourcomplex variable defined by power series, have derivatives $\lim_{u \to u_0}[f(u) - f(u_0)]/(u - u_0)$ independent of the direction of approach of u to u_0. If the fourcomplex function $f(u)$ of the fourcomplex variable u is written in terms of the real functions $P(x, y, z, t), Q(x, y, z, t), R(x, y, z, t), S(x, y, z, t)$, then relations of equality exist between partial derivatives of the functions P, Q, R, S. The integral $\int_A^B f(u)du$ of a fourcomplex function between two points A, B is independent of the path connecting A, B.

Four distinct types of hypercomplex numbers are studied, as discussed further. In Sec. 3.1, the multiplication rules for the complex units α, β and γ are $\alpha^2 = -1, \beta^2 = -1, \gamma^2 = 1, \alpha\beta = \beta\alpha = -\gamma, \alpha\gamma = \gamma\alpha = \beta, \beta\gamma = \gamma\beta = \alpha$. The exponential form of a fourcomplex number is $u = \rho \exp\left[\gamma \ln \tan \psi + (1/2)\alpha(\phi + \chi) + (1/2)\beta(\phi - \chi)\right]$, where the amplitude is $\rho^4 = \left[(x + t)^2 + (y + z)^2\right]\left[(x - t)^2 + (y - z)^2\right]$, ϕ, χ are azimuthal angles,

$0 \leq \phi < 2\pi, 0 \leq \chi < 2\pi$, and ψ is a planar angle, $0 < \psi \leq \pi/2$. The trigonometric form of a fourcomplex number is $u = d[\cos(\psi - \pi/4) + \gamma \sin(\psi - \pi/4)]$ $\exp[(1/2)\alpha(\phi + \chi) + (1/2)\beta(\phi - \chi)]$, where $d^2 = x^2 + y^2 + z^2 + t^2$. The amplitude ρ and $\tan \psi$ are multiplicative and the angles ϕ, χ are additive upon the multiplication of fourcomplex numbers. Since there are two cyclic variables, ϕ and χ, these fourcomplex numbers are called circular fourcomplex numbers. If $f(u)$ is an analytic fourcomplex function, then $\oint_\Gamma f(u)du/(u-u_0) = \pi[(\alpha+\beta) \text{ int}(u_{0\xi v}, \Gamma_{\xi v}) + (\alpha-\beta) \text{ int}(u_{0\tau\varsigma}, \Gamma_{\tau\varsigma})] f(u_0)$, where the functional int takes the values 0 or 1 depending on the relation between the projections of the point u_0 and of the curve Γ on certain planes. A polynomial can be written as a product of linear or quadratic factors, although the factorization may not be unique.

In Sec. 3.2, the multiplication rules for the complex units α, β and γ are $\alpha^2 = 1$, $\beta^2 = 1$, $\gamma^2 = 1, \alpha\beta = \beta\alpha = \gamma$, $\alpha\gamma = \gamma\alpha = \beta$, $\beta\gamma = \gamma\beta = \alpha$. The exponential form of a fourcomplex number, which can be defined for $s = x + y + z + t > 0$, $s' = x - y + z - t > 0$, $s'' = x + y - z - t > 0$, $s''' = x - y - z + t > 0$, is $u = \mu \exp(\alpha y_1 + \beta z_1 + \gamma t_1)$, where the amplitude is $\mu = (ss's''s''')^{1/4}$ and the arguments are $y_1 = (1/4) \ln(ss''/s's''')$, $z_1 = (1/4) \ln(ss'''/s's'')$, $t_1 = (1/4) \ln(ss'/s''s''')$. Since there is no cyclic variable, these fourcomplex numbers are called hyperbolic fourcomplex numbers. The amplitude μ is multiplicative and the arguments y_1, z_1, t_1 are additive upon the multiplication of fourcomplex numbers. A polynomial can be written as a product of linear or quadratic factors, although the factorization may not be unique.

In Sec. 3.3, the multiplication rules for the complex units α, β and γ are $\alpha^2 = \beta$, $\beta^2 = -1$, $\gamma^2 = -\beta, \alpha\beta = \beta\alpha = \gamma$, $\alpha\gamma = \gamma\alpha = -1$, $\beta\gamma = \gamma\beta = -\alpha$. The exponential function of a fourcomplex number can be expanded in terms of the four-dimensional cosexponential functions $f_{40}(x) = 1 - x^4/4! + x^8/8! - \cdots$, $f_{41}(x) = x - x^5/5! + x^9/9! - \cdots$, $f_{42}(x) = x^2/2! - x^6/6! + x^{10}/10! - \cdots$, $f_{43}(x) = x^3/3! - x^7/7! + x^{11}/11! - \cdots$. Expressions are obtained for the four-dimensional cosexponential functions in terms of elementary functions. The exponential form of a fourcomplex number is $u = \rho \exp \{(1/2)(\alpha - \gamma) \ln \tan \psi + (1/2)[\beta + (\alpha + \gamma)/\sqrt{2}]\phi$ $-(1/2)[\beta - (\alpha + \gamma)/\sqrt{2}]\chi\}$, where the amplitude is $\rho^4 = \left\{ \left[x + (y - t)/\sqrt{2} \right]^2 + \left[z + (y + t)/\sqrt{2} \right]^2 \right\} \left\{ \left[x - (y - t)/\sqrt{2} \right]^2 + \left[z - (y + t)/\sqrt{2} \right]^2 \right\}$, ϕ, χ are azimuthal angles, $0 \leq \phi < 2\pi, 0 \leq \chi < 2\pi$, and ψ is a planar angle, $0 \leq \psi \leq \pi/2$. The trigonometric form of a fourcomplex number is $u = d \left[\cos (\psi - \pi/4) + (1/\sqrt{2})(\alpha - \gamma) \sin (\psi - \pi/4) \right] \exp \left\{ (1/2)[\beta + (\alpha + \gamma)/\sqrt{2}]\phi \right.$

$-(1/2)[\beta - (\alpha + \gamma)/\sqrt{2}]\chi\}$, where $d^2 = x^2 + y^2 + z^2 + t^2$. The amplitude ρ and $\tan\psi$ are multiplicative and the angles ϕ, χ are additive upon the multiplication of fourcomplex numbers. There are two cyclic variables, ϕ and χ, so that these fourcomplex numbers are also of a circular type. In order to distinguish them from the circular hypercomplex numbers, these are called planar fourcomplex numbers. If $f(u)$ is an analytic fourcomplex function, then $\oint_\Gamma f(u)du/(u - u_0) = \pi\left[\left(\beta + (\alpha + \gamma)/\sqrt{2}\right) \operatorname{int}(u_{0\xi v}, \Gamma_{\xi v})\right.$ $\left. + \left(\beta - (\alpha + \gamma)/\sqrt{2}\right) \operatorname{int}(u_{0\tau\zeta}, \Gamma_{\tau\zeta})\right] f(u_0)$, where the functional int takes the values 0 or 1 depending on the relation between the projections of the point u_0 and of the curve Γ on certain planes. A polynomial can be written as a product of linear or quadratic factors, although the factorization may not be unique. The fourcomplex numbers described in this chapter are a particular case for $n = 4$ of the planar hypercomplex numbers in n dimensions discussed in Sec. 6.2.

In Sec. 3.4, the multiplication rules for the complex units α, β and γ are $\alpha^2 = \beta$, $\beta^2 = 1$, $\gamma^2 = \beta, \alpha\beta = \beta\alpha = \gamma$, $\alpha\gamma = \gamma\alpha = 1$, $\beta\gamma = \gamma\beta = \alpha$. The product of two fourcomplex numbers is equal to zero if both numbers are equal to zero, or if the numbers belong to certain four-dimensional hyperplanes described further in this section. The exponential function of a fourcomplex number can be expanded in terms of the four-dimensional cosexponential functions $g_{40}(x) = 1 + x^4/4! + x^8/8! + \cdots$, $g_{41}(x) = x + x^5/5! + x^9/9! + \cdots$, $g_{42}(x) = x^2/2! + x^6/6! + x^{10}/10! + \cdots$, $g_{43}(x) = x^3/3! + x^7/7! + x^{11}/11! + \cdots$. Addition theorems and other relations are obtained for these four-dimensional cosexponential functions. The exponential form of a fourcomplex number, which can be defined for $x + y + z + t > 0, x - y + z - t > 0$, is $u = \rho\exp\left[(1/4)(\alpha + \beta + \gamma)\ln(\sqrt{2}/\tan\theta_+) - (1/4)(\alpha - \beta + \gamma)\ln(\sqrt{2}/\tan\theta_-)\right.$ $\left. + (\alpha - \gamma)\phi/2\right]$, where $\rho = (\mu_+\mu_-)^{1/2}$, $\mu_+^2 = (x - z)^2 + (y - t)^2$, $\mu_-^2 = (x + z)^2 - (y + t)^2$, $e_+ = (1 + \alpha + \beta + \gamma)/4, e_- = (1 - \alpha + \beta - \gamma)/4, e_1 = (1 - \beta)/2, \tilde{e}_1 = (\alpha - \gamma)/2$, the polar angles are $\tan\theta_+ = \sqrt{2}\mu_+/v_+, \tan\theta_- = \sqrt{2}\mu_+/v_-$, $0 \le \theta_+ \le \pi, 0 \le \theta_- \le \pi$, and the azimuthal angle ϕ is defined by the relations $x - y = \mu_+\cos\phi$, $z - t = \mu_+\sin\phi$, $0 \le \phi < 2\pi$. The trigonometric form of the fourcomplex number u is $u = d\sqrt{2}$ $(1 + 1/\tan^2\theta_+ + 1/\tan^2\theta_-)^{-1/2}\left\{e_1 + e_+\sqrt{2}/\tan\theta_+ + e_-\sqrt{2}/\tan\theta_-\right\}$ $\exp[\tilde{e}_1\phi]$. The amplitude ρ and $\tan\theta_+/\sqrt{2}, \tan\theta_-/\sqrt{2}$, are multiplicative, and the azimuthal angle ϕ is additive upon the multiplication of fourcomplex numbers. There is only one cyclic variable, ϕ, and there are two axes v_+, v_- which play an important role in the description of these numbers, so that these hypercomplex numbers are called polar fourcomplex numbers. If $f(u)$ is an analytic fourcomplex function, then $\oint_\Gamma f(u)du/(u - u_0) =$

$\pi(\beta - \gamma) \, \text{int}(u_{0\xi v}, \Gamma_{\xi v}) f(u_0)$, where the functional int takes the values 0 or 1 depending on the relation between the projections of the point u_0 and of the curve Γ on certain planes. A polynomial can be written as a product of linear or quadratic factors, although the factorization may not be unique. The fourcomplex numbers described in this section are a particular case for $n = 4$ of the polar hypercomplex numbers in n dimensions discussed in Sec. 6.1.

3.1 Circular complex numbers in four dimensions

3.1.1 Operations with circular fourcomplex numbers

A circular fourcomplex number is determined by its four components (x, y, z, t). The sum of the circular fourcomplex numbers (x, y, z, t) and (x', y', z', t') is the circular fourcomplex number $(x + x', y + y', z + z', t + t')$. The product of the circular fourcomplex numbers (x, y, z, t) and (x', y', z', t') is defined in this section to be the circular fourcomplex number $(xx' - yy' - zz' + tt', xy' + yx' + zt' + tz', xz' + zx' + yt' + ty', xt' + tx' - yz' - zy')$.

Circular fourcomplex numbers and their operations can be represented by writing the circular fourcomplex number (x, y, z, t) as $u = x + \alpha y + \beta z + \gamma t$, where α, β and γ are bases for which the multiplication rules are

$$\alpha^2 = -1, \ \beta^2 = -1, \ \gamma^2 = 1,$$
$$\alpha\beta = \beta\alpha = -\gamma, \ \alpha\gamma = \gamma\alpha = \beta, \ \beta\gamma = \gamma\beta = \alpha. \tag{3.1}$$

Two circular fourcomplex numbers $u = x + \alpha y + \beta z + \gamma t, u' = x' + \alpha y' + \beta z' + \gamma t'$ are equal, $u = u'$, if and only if $x = x', y = y', z = z', t = t'$. If $u = x + \alpha y + \beta z + \gamma t, u' = x' + \alpha y' + \beta z' + \gamma t'$ are circular fourcomplex numbers, the sum $u + u'$ and the product uu' defined above can be obtained by applying the usual algebraic rules to the sum $(x + \alpha y + \beta z + \gamma t) + (x' + \alpha y' + \beta z' + \gamma t')$ and to the product $(x + \alpha y + \beta z + \gamma t)(x' + \alpha y' + \beta z' + \gamma t')$, and grouping of the resulting terms,

$$u + u' = x + x' + \alpha(y + y') + \beta(z + z') + \gamma(t + t'), \tag{3.2}$$

$$uu' = xx' - yy' - zz' + tt' + \alpha(xy' + yx' + zt' + tz')$$
$$+ \beta(xz' + zx' + yt' + ty') + \gamma(xt' + tx' - yz' - zy'). \tag{3.3}$$

If u, u', u'' are circular fourcomplex numbers, the multiplication is associative

$$(uu')u'' = u(u'u'') \tag{3.4}$$

and commutative

$$uu' = u'u, \tag{3.5}$$

as can be checked through direct calculation. The circular fourcomplex zero is $0 + \alpha \cdot 0 + \beta \cdot 0 + \gamma \cdot 0$, denoted simply 0, and the circular fourcomplex unity is $1 + \alpha \cdot 0 + \beta \cdot 0 + \gamma \cdot 0$, denoted simply 1.

The inverse of the circular fourcomplex number $u = x + \alpha y + \beta z + \gamma t$ is a circular fourcomplex number $u' = x' + \alpha y' + \beta z' + \gamma t'$ having the property that

$$uu' = 1. \tag{3.6}$$

Written on components, the condition, Eq. (3.6), is

$$\begin{aligned}
xx' - yy' - zz' + tt' &= 1, \\
yx' + xy' + tz' + zt' &= 0, \\
zx' + ty' + xz' + yt' &= 0, \\
tx' - zy' - yz' + xt' &= 0.
\end{aligned} \tag{3.7}$$

The system (3.7) has the solution

$$x' = \frac{x(x^2 + y^2 + z^2 - t^2) - 2yzt}{\rho^4}, \tag{3.8}$$

$$y' = \frac{y(-x^2 - y^2 + z^2 - t^2) + 2xzt}{\rho^4}, \tag{3.9}$$

$$z' = \frac{z(-x^2 + y^2 - z^2 - t^2) + 2xyt}{\rho^4}, \tag{3.10}$$

$$t' = \frac{t(-x^2 + y^2 + z^2 + t^2) - 2xyz}{\rho^4}, \tag{3.11}$$

provided that $\rho \neq 0$, where

$$\begin{aligned}
\rho^4 &= x^4 + y^4 + z^4 + t^4 + 2(x^2y^2 + x^2z^2 - x^2t^2 - y^2z^2 + y^2t^2 + z^2t^2) \\
&\quad - 8xyzt.
\end{aligned} \tag{3.12}$$

The quantity ρ will be called amplitude of the circular fourcomplex number $x + \alpha y + \beta z + \gamma t$. Since

$$\rho^4 = \rho_+^2 \rho_-^2, \tag{3.13}$$

where

$$\rho_+^2 = (x + t)^2 + (y + z)^2, \quad \rho_-^2 = (x - t)^2 + (y - z)^2, \tag{3.14}$$

a circular fourcomplex number $u = x + \alpha y + \beta z + \gamma t$ has an inverse, unless

$$x + t = 0, \quad y + z = 0, \tag{3.15}$$

or

$$x - t = 0, \quad y - z = 0. \tag{3.16}$$

Because of conditions (3.15)-(3.16) these 2-dimensional surfaces will be called nodal hyperplanes. It can be shown that if $uu' = 0$ then either $u = 0$, or $u' = 0$, or one of the circular fourcomplex numbers is of the form $x + \alpha y + \beta y + \gamma x$ and the other of the form $x' + \alpha y' - \beta y' - \gamma x'$.

3.1.2   Geometric representation of circular fourcomplex numbers

The circular fourcomplex number $x + \alpha y + \beta z + \gamma t$ can be represented by the point A of coordinates (x, y, z, t). If O is the origin of the four-dimensional space x, y, z, t, the distance from A to the origin O can be taken as

$$d^2 = x^2 + y^2 + z^2 + t^2. \tag{3.17}$$

The distance d will be called modulus of the circular fourcomplex number $x + \alpha y + \beta z + \gamma t$, $d = |u|$. The orientation in the four-dimensional space of the line OA can be specified with the aid of three angles ϕ, χ, ψ defined with respect to the rotated system of axes

$$\xi = \frac{x+t}{\sqrt{2}}, \quad \tau = \frac{x-t}{\sqrt{2}}, \quad \upsilon = \frac{y+z}{\sqrt{2}}, \quad \zeta = \frac{y-z}{\sqrt{2}}. \tag{3.18}$$

The variables $\xi, \upsilon, \tau, \zeta$ will be called canonical circular fourcomplex variables. The use of the rotated axes $\xi, \upsilon, \tau, \zeta$ for the definition of the angles ϕ, χ, ψ is convenient for the expression of the circular fourcomplex numbers in exponential and trigonometric forms, as it will be discussed further. The angle ϕ is the angle between the projection of A in the plane ξ, υ and the $O\xi$ axis, $0 \leq \phi < 2\pi$, χ is the angle between the projection of A in the plane τ, ζ and the $O\tau$ axis, $0 \leq \chi < 2\pi$, and ψ is the angle between the line OA and the plane $\tau O\zeta$, $0 \leq \psi \leq \pi/2$, as shown in Fig. 3.1. The angles ϕ and χ will be called azimuthal angles, the angle ψ will be called planar angle . The fact that $0 \leq \psi \leq \pi/2$ means that ψ has the same sign on both faces of the two-dimensional hyperplane $\upsilon O\zeta$. The components of the point A in terms of the distance d and the angles ϕ, χ, ψ are thus

$$\frac{x+t}{\sqrt{2}} = d \cos\phi \sin\psi, \tag{3.19}$$

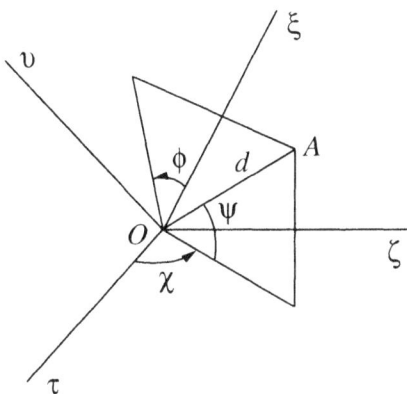

Figure 3.1: Azimuthal angles ϕ, χ and planar angle ψ of the fourcomplex number $x + \alpha y + \beta z + \gamma t$, represented by the point A, situated at a distance d from the origin O.

$$\frac{x - t}{\sqrt{2}} = d \cos \chi \cos \psi, \tag{3.20}$$

$$\frac{y + z}{\sqrt{2}} = d \sin \phi \sin \psi, \tag{3.21}$$

$$\frac{y - z}{\sqrt{2}} = d \sin \chi \cos \psi. \tag{3.22}$$

It can be checked that $\rho_+ = \sqrt{2} d \sin \psi, \rho_- = \sqrt{2} d \cos \psi$. The coordinates x, y, z, t in terms of the variables d, ϕ, χ, ψ are

$$x = \frac{d}{\sqrt{2}} (\cos \psi \cos \chi + \sin \psi \cos \phi), \tag{3.23}$$

$$y = \frac{d}{\sqrt{2}} (\cos \psi \sin \chi + \sin \psi \sin \phi), \tag{3.24}$$

$$z = \frac{d}{\sqrt{2}} (- \cos \psi \sin \chi + \sin \psi \sin \phi), \tag{3.25}$$

$$t = \frac{d}{\sqrt{2}} (- \cos \psi \cos \chi + \sin \psi \cos \phi). \tag{3.26}$$

The angles ϕ, χ, ψ can be expressed in terms of the coordinates x, y, z, t as

$$\sin \phi = (y + z)/\rho_+, \quad \cos \phi = (x + t)/\rho_+, \tag{3.27}$$

$$\sin \chi = (y - z)/\rho_-, \quad \cos \chi = (x - t)/\rho_-, \tag{3.28}$$

$$\tan \psi = \rho_+/\rho_-. \tag{3.29}$$

The nodal hyperplanes are $\xi O v$, for which $\tau = 0, \zeta = 0$, and $\tau O \zeta$, for which $\xi = 0, v = 0$. For points in the nodal hyperplane $\xi O v$ the planar angle is $\psi = \pi/2$, for points in the nodal hyperplane $\tau O \zeta$ the planar angle is $\psi = 0$.

It can be shown that if $u_1 = x_1 + \alpha y_1 + \beta z_1 + \gamma t_1, u_2 = x_2 + \alpha y_2 + \beta z_2 + \gamma t_2$ are circular fourcomplex numbers of amplitudes and angles $\rho_1, \phi_1, \chi_1, \psi_1$ and respectively $\rho_2, \phi_2, \chi_2, \psi_2$, then the amplitude ρ and the angles ϕ, χ, ψ of the product circular fourcomplex number $u_1 u_2$ are

$$\rho = \rho_1 \rho_2, \tag{3.30}$$

$$\phi = \phi_1 + \phi_2, \quad \chi = \chi_1 + \chi_2, \quad \tan \psi = \tan \psi_1 \tan \psi_2. \tag{3.31}$$

The relations (3.30)-(3.31) are consequences of the definitions (3.12)-(3.14), (3.27)-(3.29) and of the identities

$$[(x_1 x_2 - y_1 y_2 - z_1 z_2 + t_1 t_2) + (x_1 t_2 + t_1 x_2 - y_1 z_2 - z_1 y_2)]^2$$
$$+ [(x_1 y_2 + y_1 x_2 + z_1 t_2 + t_1 z_2) + (x_1 z_2 + z_1 x_2 + y_1 t_2 + t_1 y_2)]^2$$
$$= [(x_1 + t_1)^2 + (y_1 + z_1)^2][(x_2 + t_2)^2 + (y_2 + z_2)^2], \tag{3.32}$$

$$[(x_1 x_2 - y_1 y_2 - z_1 z_2 + t_1 t_2) - (x_1 t_2 + t_1 x_2 - y_1 z_2 - z_1 y_2)]^2$$
$$+ [(x_1 y_2 + y_1 x_2 + z_1 t_2 + t_1 z_2) - (x_1 z_2 + z_1 x_2 + y_1 t_2 + t_1 y_2)]^2$$
$$= [(x_1 - t_1)^2 + (y_1 - z_1)^2][(x_2 - t_2)^2 + (y_2 - z_2)^2], \tag{3.33}$$

$$(x_1 x_2 - y_1 y_2 - z_1 z_2 + t_1 t_2) + (x_1 t_2 + t_1 x_2 - y_1 z_2 - z_1 y_2)$$
$$= (x_1 + t_1)(x_2 + t_2) - (y_1 + z_1)(y_2 + z_2), \tag{3.34}$$

$$(x_1 x_2 - y_1 y_2 - z_1 z_2 + t_1 t_2) - (x_1 t_2 + t_1 x_2 - y_1 z_2 - z_1 y_2)$$
$$= (x_1 - t_1)(x_2 - t_2) - (y_1 - z_1)(y_2 - z_2), \tag{3.35}$$

$$(x_1 y_2 + y_1 x_2 + z_1 t_2 + t_1 z_2) + (x_1 z_2 + z_1 x_2 + y_1 t_2 + t_1 y_2)$$
$$= (y_1 + z_1)(x_2 + t_2) + (y_2 + z_2)(x_2 + t_2), \tag{3.36}$$

$$(x_1y_2 + y_1x_2 + z_1t_2 + t_1z_2) - (x_1z_2 + z_1x_2 + y_1t_2 + t_1y_2)$$
$$= (y_1 - z_1)(x_2 - t_2) + (y_2 - z_2)(x_2 - t_2). \tag{3.37}$$

The identities (3.32) and (3.33) can also be written as

$$\rho_+^2 = \rho_{1+}\rho_{2+}, \tag{3.38}$$

$$\rho_-^2 = \rho_{1-}\rho_{2-}, \tag{3.39}$$

where

$$\rho_{j+}^2 = (x_j + t_j)^2 + (y_j + z_j)^2, \ \rho_{j-}^2 = (x_j - t_j)^2 + (y_j - z_j)^2, \ j = 1, 2. \tag{3.40}$$

The fact that the amplitude of the product is equal to the product of the amplitudes, as written in Eq. (3.30), can be demonstrated also by using a representation of the multiplication of the circular fourcomplex numbers by matrices, in which the circular fourcomplex number $u = x + \alpha y + \beta z + \gamma t$ is represented by the matrix

$$A = \begin{pmatrix} x & y & z & t \\ -y & x & t & -z \\ -z & t & x & -y \\ t & z & y & x \end{pmatrix}. \tag{3.41}$$

The product $u = x + \alpha y + \beta z + \gamma t$ of the circular fourcomplex numbers $u_1 = x_1 + \alpha y_1 + \beta z_1 + \gamma t_1, u_2 = x_2 + \alpha y_2 + \beta z_2 + \gamma t_2$, can be represented by the matrix multiplication

$$A = A_1 A_2. \tag{3.42}$$

It can be checked that the determinant $\det(A)$ of the matrix A is

$$\det A = \rho^4. \tag{3.43}$$

The identity (3.30) is then a consequence of the fact the determinant of the product of matrices is equal to the product of the determinants of the factor matrices.

3.1.3 The exponential and trigonometric forms of circular fourcomplex numbers

The exponential function of a hypercomplex variable u and the addition theorem for the exponential function have been written in Eqs. (1.35)-(1.36). If $u = x + \alpha y + \beta z + \gamma t$, then $\exp u$ can be calculated as $\exp u = \exp x \cdot \exp(\alpha y) \cdot \exp(\beta z) \cdot \exp(\gamma t)$. According to Eq. (3.1),

$$\alpha^{2m} = (-1)^m, \alpha^{2m+1} = (-1)^m \alpha, \beta^{2m} = (-1)^m, \beta^{2m+1} = (-1)^m \beta,$$
$$\gamma^m = 1, \tag{3.44}$$

where m is a natural number, so that $\exp(\alpha y)$, $\exp(\beta z)$ and $\exp(\gamma t)$ can be written as

$$\exp(\alpha y) = \cos y + \alpha \sin y, \ \exp(\beta z) = \cos z + \beta \sin z, \tag{3.45}$$

and

$$\exp(\gamma t) = \cosh t + \gamma \sinh t. \tag{3.46}$$

From Eqs. (3.45)-(3.46) it can be inferred that

$$
\begin{aligned}
(\cos y + \alpha \sin y)^m &= \cos my + \alpha \sin my, \\
(\cos z + \beta \sin z)^m &= \cos mz + \beta \sin mz, \\
(\cosh t + \gamma \sinh t)^m &= \cosh mt + \gamma \sinh mt.
\end{aligned}
\tag{3.47}
$$

Any circular fourcomplex number $u = x + \alpha y + \beta z + \gamma t$ can be writen in the form

$$x + \alpha y + \beta z + \gamma t = e^{x_1 + \alpha y_1 + \beta z_1 + \gamma t_1}. \tag{3.48}$$

The expressions of x_1, y_1, z_1, t_1 as functions of x, y, z, t can be obtained by developing $e^{\alpha y_1}, e^{\beta z_1}$ and $e^{\gamma t_1}$ with the aid of Eqs. (3.45) and (3.46), by multiplying these expressions and separating the hypercomplex components,

$$x = e^{x_1}(\cos y_1 \cos z_1 \cosh t_1 - \sin y_1 \sin z_1 \sinh t_1), \tag{3.49}$$

$$y = e^{x_1}(\sin y_1 \cos z_1 \cosh t_1 + \cos y_1 \sin z_1 \sinh t_1), \tag{3.50}$$

$$z = e^{x_1}(\cos y_1 \sin z_1 \cosh t_1 + \sin y_1 \cos z_1 \sinh t_1), \tag{3.51}$$

$$t = e^{x_1}(-\sin y_1 \sin z_1 \cosh t_1 + \cos y_1 \cos z_1 \sinh t_1). \tag{3.52}$$

From Eqs. (3.49)-(3.52) it can be shown by direct calculation that

$$x^2 + y^2 + z^2 + t^2 = e^{2x_1} \cosh 2t_1, \tag{3.53}$$

$$2(xt + yz) = e^{2x_1} \sinh 2t_1, \tag{3.54}$$

so that

$$e^{4x_1} = (x^2 + y^2 + z^2 + t^2)^2 - 4(xt + yz)^2. \tag{3.55}$$

By comparing the expression in the right-hand side of Eq. (3.55) with the expression of ρ, Eq. (3.13), it can be seen that

$$e^{x_1} = \rho. \tag{3.56}$$

The variable t_1 is then given by

$$\cosh 2t_1 = \frac{d^2}{\rho^2}, \ \sinh 2t_1 = \frac{2(xt + yz)}{\rho^2}. \tag{3.57}$$

From the fact that $\rho^4 = d^4 - 4(xt + yz)^2$ it follows that $d^2/\rho^2 \geq 1$, so that Eq. (3.57) has always a real solution, and $t_1 = 0$ for $xt + yz = 0$. It can be shown similarly that

$$\cos 2y_1 = \frac{x^2 - y^2 + z^2 - t^2}{\rho^2}, \ \sin 2y_1 = \frac{2(xy - zt)}{\rho^2}, \tag{3.58}$$

$$\cos 2z_1 = \frac{x^2 + y^2 - z^2 - t^2}{\rho^2}, \ \sin 2z_1 = \frac{2(xz - yt)}{\rho^2}. \tag{3.59}$$

It can be shown that $(x^2 - y^2 + z^2 - t^2)^2 \leq \rho^4$, the equality taking place for $xy = zt$, and $(x^2 + y^2 - z^2 - t^2)^2 \leq \rho^4$, the equality taking place for $xz = yt$, so that Eqs. (3.58) and Eqs. (3.59) have always real solutions.

The variables

$$y_1 = \frac{1}{2} \arcsin \frac{2(xy - zt)}{\rho^2}, \ z_1 = \frac{1}{2} \arcsin \frac{2(xz - yt)}{\rho^2},$$
$$t_1 = \frac{1}{2} \text{argsinh} \frac{2(xt + yz)}{\rho^2} \tag{3.60}$$

are additive upon the multiplication of circular fourcomplex numbers, as can be seen from the identities

$$(xx' - yy' - zz' + tt')(xy' + yx' + zt' + tz')$$
$$-(xz' + zx' + yt' + ty')(xt' - yz' - zy' + tx')$$
$$= (xy - zt)(x'^2 - y'^2 + z'^2 - t'^2) + (x^2 - y^2 + z^2 - t^2)(x'y' - z't'), \tag{3.61}$$

$$(xx' - yy' - zz' + tt')(xz' + zx' + yt' + ty')$$
$$-(xy' + yx' + zt' + tz')(xt' - yz' - zy' + tx')$$
$$= (xz - yt)(x'^2 + y'^2 - z'^2 - t'^2) + (x^2 + y^2 - z^2 - t^2)(x'z' - y't'), \tag{3.62}$$

$$(xx' - yy' - zz' + tt')(xt' - yz' - zy' + tx')$$
$$+(xy' + yx' + zt' + tz')(xz' + zx' + yt' + ty')$$
$$= (xt + yz)(x'^2 + y'^2 + z'^2 + t'^2) + (x^2 + y^2 + z^2 + t^2)(x't' + y'z'). \tag{3.63}$$

The expressions appearing in Eqs. (3.57)-(3.59) can be calculated in terms of the angles ϕ, χ, ψ with the aid of Eqs. (3.23)-(3.26) as

$$\frac{d^2}{\rho^2} = \frac{1}{\sin 2\psi}, \quad \frac{2(xt+yz)}{\rho^2} = -\frac{1}{\tan 2\psi}, \tag{3.64}$$

$$\frac{x^2 - y^2 + z^2 - t^2}{\rho^2} = \cos(\phi + \chi), \quad \frac{2(xy - zt)}{\rho^2} = \sin(\phi + \chi), \tag{3.65}$$

$$\frac{x^2 + y^2 - z^2 - t^2}{\rho^2} = \cos(\phi - \chi), \quad \frac{2(xz - yt)}{\rho^2} = \sin(\phi - \chi). \tag{3.66}$$

Then from Eqs. (3.57)-(3.59) and (3.64)-(3.66) it results that

$$y_1 = \frac{\phi + \chi}{2}, \quad z_1 = \frac{\phi - \chi}{2}, \quad t_1 = \frac{1}{2} \ln \tan \psi, \tag{3.67}$$

so that the circular fourcomplex number u, Eq. (3.48), can be written as

$$u = \rho \exp \left[\alpha \frac{\phi + \chi}{2} + \beta \frac{\phi - \chi}{2} + \gamma \frac{1}{2} \ln \tan \psi \right]. \tag{3.68}$$

In Eq. (3.68) the circular fourcomplex number $u = x + \alpha y + \beta z + \gamma t$ is written as the product of the amplitude ρ and of an exponential function, and therefore this form of u will be called the exponential form of the circular fourcomplex number. It can be checked that

$$\exp \left(\frac{\alpha + \beta}{2} \phi \right) = \frac{1 - \gamma}{2} + \frac{1 + \gamma}{2} \cos \phi + \frac{\alpha + \beta}{2} \sin \phi, \tag{3.69}$$

$$\exp \left(\frac{\alpha - \beta}{2} \chi \right) = \frac{1 + \gamma}{2} + \frac{1 - \gamma}{2} \cos \chi + \frac{\alpha - \beta}{2} \sin \phi, \tag{3.70}$$

which shows that $e^{(\alpha+\beta)\phi/2}$ and $e^{(\alpha-\beta)\chi/2}$ are periodic functions of ϕ and respectively χ, with period 2π.

The relations between the variables y_1, z_1, t_1 and the angles ϕ, χ, ψ can be obtained alternatively by substituting in Eqs. (3.49)-(3.52) the expression $e^{x_1} = d/(\cosh 2t_1)^{1/2}$, Eq. (3.53), and summing and subtracting of the relations,

$$\frac{x + t}{\sqrt{2}} = d \cos(y_1 + z_1) \sin(\eta + \pi/4), \tag{3.71}$$

$$\frac{x - t}{\sqrt{2}} = d \cos(y_1 - z_1) \cos(\eta + \pi/4), \tag{3.72}$$

$$\frac{y+z}{\sqrt{2}} = d\sin(y_1 + z_1)\sin(\eta + \pi/4), \tag{3.73}$$

$$\frac{y-z}{\sqrt{2}} = d\sin(y_1 - z_1)\cos(\eta + \pi/4), \tag{3.74}$$

where the variable η is defined by the relations

$$\frac{\cosh t_1}{(\cosh 2t_1)^{1/2}} = \cos\eta, \quad \frac{\sinh t_1}{(\cosh 2t_1)^{1/2}} = \sin\eta, \tag{3.75}$$

and when $-\infty < t_1 < \infty$, the range of the variable η is $-\pi/4 \le \eta \le \pi/4$. The comparison of Eqs. (3.19)-(3.22) and (3.71)-(3.74) shows that

$$\phi = y_1 + z_1, \ \chi = y_1 - z_1, \ \psi = \eta + \pi/4. \tag{3.76}$$

It can be shown with the aid of Eq. (3.46) that

$$\exp\left(\frac{1}{2}\gamma \ln \tan \psi\right) = \frac{1}{(\sin 2\psi)^{1/2}}\left[\cos(\psi - \pi/4) + \gamma \sin(\psi - \pi/4)\right]. \tag{3.77}$$

The circular fourcomplex number u, Eq. (3.68), can then be written equivalently as

$$u = d\{\cos(\psi - \pi/4) + \gamma \sin(\psi - \pi/4)\}\exp\left[\alpha\frac{\phi + \chi}{2} + \beta\frac{\phi - \chi}{2}\right]. \tag{3.78}$$

In Eq. (3.78), the circular fourcomplex number $u = x + \alpha y + \beta z + \gamma t$ is written as the product of the modulus d and of factors depending on the geometric angles ϕ, χ and ψ, and this form will be called the trigonometric form of the circular fourcomplex number.

If u_1, u_2 are circular fourcomplex numbers of moduli and angles $d_1, \phi_1,$ χ_1, ψ_1 and respectively $d_2, \phi_2, \chi_2, \psi_2$, the product of the planar factors can be calculated to be

$$[\cos(\psi_1 - \pi/4) + \gamma \sin(\psi_1 - \pi/4)][\cos(\psi_2 - \pi/4) + \gamma \sin(\psi_2 - \pi/4)]$$
$$= [\cos(\psi_1 - \psi_2) - \gamma \cos(\psi_1 + \psi_2)]. \tag{3.79}$$

The right-hand side of Eq. (3.79) can be written as

$$\cos(\psi_1 - \psi_2) - \gamma \cos(\psi_1 + \psi_2)$$
$$= [2(\cos^2\psi_1 \cos^2\psi_2 + \sin^2\psi_1 \sin^2\psi_2)]^{1/2}[\cos(\psi - \pi/4)$$
$$+ \gamma \sin(\psi - \pi/4)], \tag{3.80}$$

where the angle ψ, determined by the condition that

$$\tan(\psi - \pi/4) = -\cos(\psi_1 + \psi_2)/\cos(\psi_1 - \psi_2), \tag{3.81}$$

is given by $\tan \psi = \tan \psi_1 \tan \psi_2$, which is consistent with Eq. (3.31). It can be checked that the modulus d of the product $u_1 u_2$ is

$$d = \sqrt{2} d_1 d_2 \left(\cos^2 \psi_1 \cos^2 \psi_2 + \sin^2 \psi_1 \sin^2 \psi_2 \right)^{1/2} . \tag{3.82}$$

3.1.4   Elementary functions of a circular fourcomplex variable

The logarithm u_1 of the circular fourcomplex number u, $u_1 = \ln u$, can be defined as the solution of the equation

$$u = e^{u_1}, \tag{3.83}$$

written explicitly previously in Eq. (3.48), for u_1 as a function of u. From Eq. (3.68) it results that

$$\ln u = \ln \rho + \frac{1}{2} \gamma \ln \tan \psi + \alpha \frac{\phi + \chi}{2} + \beta \frac{\phi - \chi}{2}. \tag{3.84}$$

It can be inferred from Eqs. (3.30) and (3.31) that

$$\ln(uu') = \ln u + \ln u', \tag{3.85}$$

up to multiples of $\pi(\alpha + \beta)$ and $\pi(\alpha - \beta)$.

The power function u^n can be defined for real values of n as

$$u^m = e^{m \ln u}. \tag{3.86}$$

The power function is multivalued unless n is an integer. For integer n, it can be inferred from Eq. (3.85) that

$$(uu')^m = u^n \, u'^m. \tag{3.87}$$

If, for example, $m = 2$, it can be checked with the aid of Eq. (3.78) that Eq. (3.86) gives indeed $(x + \alpha y + \beta z + \gamma t)^2 = x^2 - y^2 - z^2 + t^2 + 2\alpha(xy + zt) + 2\beta(xz + yt) + 2\gamma(2xt - yz)$.

The trigonometric functions of the hypercomplex variable u and the addition theorems for these functions have been written in Eqs. (1.57)-(1.60). The cosine and sine functions of the hypercomplex variables $\alpha y, \beta z$ and γt can be expressed as

$$\cos \alpha y = \cosh y, \ \sin \alpha y = \alpha \sinh y, \tag{3.88}$$

$$\cos \beta y = \cosh y, \ \sin \beta y = \beta \sinh y, \tag{3.89}$$

$$\cos \gamma y = \cos y, \ \sin \gamma y = \gamma \sin y. \tag{3.90}$$

The cosine and sine functions of a circular fourcomplex number $x + \alpha y + \beta z + \gamma t$ can then be expressed in terms of elementary functions with the aid of the addition theorems Eqs. (1.59), (1.60) and of the expressions in Eqs. (3.88)-(3.90).

The hyperbolic functions of the hypercomplex variable u and the addition theorems for these functions have been written in Eqs. (1.62)-(1.65). The hyperbolic cosine and sine functions of the hypercomplex variables $\alpha y, \beta z$ and γt can be expressed as

$$\cosh \alpha y = \cos y, \ \sinh \alpha y = \alpha \sin y, \tag{3.91}$$

$$\cosh \beta y = \cos y, \ \sinh \beta y = \beta \sin y, \tag{3.92}$$

$$\cosh \gamma y = \cosh y, \ \sinh \gamma y = \gamma \sinh y. \tag{3.93}$$

The hyperbolic cosine and sine functions of a circular fourcomplex number $x + \alpha y + \beta z + \gamma t$ can then be expressed in terms of elementary functions with the aid of the addition theorems Eqs. (1.64), (1.65) and of the expressions in Eqs. (3.91)-(3.93).

3.1.5 Power series of circular fourcomplex variables

A circular fourcomplex series is an infinite sum of the form

$$a_0 + a_1 + a_2 + \cdots + a_n + \cdots, \tag{3.94}$$

where the coefficients a_n are circular fourcomplex numbers. The convergence of the series (3.94) can be defined in terms of the convergence of its 4 real components. The convergence of a circular fourcomplex series can however be studied using circular fourcomplex variables. The main criterion for absolute convergence remains the comparison theorem, but this requires a number of inequalities which will be discussed further.

The modulus of a circular fourcomplex number $u = x + \alpha y + \beta z + \gamma t$ can be defined as

$$|u| = (x^2 + y^2 + z^2 + t^2)^{1/2}, \tag{3.95}$$

so that, according to Eq. (3.17), $d = |u|$. Since $|x| \leq |u|, |y| \leq |u|, |z| \leq |u|, |t| \leq |u|$, a property of absolute convergence established via a comparison theorem based on the modulus of the series (3.94) will ensure the absolute convergence of each real component of that series.

The modulus of the sum $u_1 + u_2$ of the circular fourcomplex numbers u_1, u_2 fulfils the inequality

$$||u_1| - |u_2|| \leq |u_1 + u_2| \leq |u_1| + |u_2|. \tag{3.96}$$

For the product the relation is

$$|u_1 u_2| \leq \sqrt{2}|u_1||u_2|, \tag{3.97}$$

which replaces the relation of equality extant for regular complex numbers. The equality in Eq. (3.97) takes place for $x_1 = t_1, y_1 = z_1, x_2 = t_2, y_2 = z_2$ or $x_1 = -t_1, y_1 = -z_1, x_2 = -t_2, y_2 = -z_2$. In Eq. (3.82), this corresponds to $\psi_1 = 0, \psi_2 = 0$ or $\psi_1 = \pi/2, \psi_2 = \pi/2$. The modulus of a product, which has the property that $0 \leq |u_1 u_2|$, becomes equal to zero for $x_1 = t_1, y_1 = z_1, x_2 = -t_2, y_2 = -z_2$ or $x_1 = -t_1, y_1 = -z_1, x_2 = t_2, y_2 = z_2$, as discussed after Eq. (3.16). In Eq. (3.82), the latter situation corresponds to $\psi_1 = 0, \psi_2 = \pi/2$ or $\psi_1 = 0, \psi_2 = \pi/2$.

It can be shown that

$$x^2 + y^2 + z^2 + t^2 \leq |u^2| \leq \sqrt{2}(x^2 + y^2 + z^2 + t^2). \tag{3.98}$$

The left relation in Eq. (3.98) becomes an equality, $x^2 + y^2 + z^2 + t^2 = |u^2|$, for $xt + yz = 0$. This condition corresponds to $\psi_1 = \psi_2 = \pi/4$ in Eq. (3.82). The inequality in Eq. (3.97) implies that

$$|u^l| \leq 2^{(l-1)/2}|u|^l. \tag{3.99}$$

From Eqs. (3.97) and (3.99) it results that

$$|au^l| \leq 2^{l/2}|a||u|^l. \tag{3.100}$$

A power series of the circular fourcomplex variable u is a series of the form

$$a_0 + a_1 u + a_2 u^2 + \cdots + a_l u^l + \cdots. \tag{3.101}$$

Since

$$\left| \sum_{l=0}^{\infty} a_l u^l \right| \leq \sum_{l=0}^{\infty} 2^{l/2}|a_l||u|^l, \tag{3.102}$$

a sufficient condition for the absolute convergence of this series is that

$$\lim_{l \to \infty} \frac{\sqrt{2}|a_{l+1}||u|}{|a_l|} < 1. \tag{3.103}$$

Thus the series is absolutely convergent for

$$|u| < c, \tag{3.104}$$

where

$$c = \lim_{l \to \infty} \frac{|a_l|}{\sqrt{2}|a_{l+1}|}. \tag{3.105}$$

The convergence of the series (3.101) can be also studied with the aid of the transformation

$$x + \alpha y + \beta z + \gamma t = \sqrt{2}(e_1 \xi + \tilde{e}_1 \upsilon + e_2 \tau + \tilde{e}_2 \zeta), \tag{3.106}$$

where $\xi, \upsilon, \tau, \zeta$ have been defined in Eq. (3.18), and

$$e_1 = \frac{1+\gamma}{2}, \quad \tilde{e}_1 = \frac{\alpha+\beta}{2}, \quad e_2 = \frac{1-\gamma}{2}, \quad \tilde{e}_2 = \frac{\alpha-\beta}{2}. \tag{3.107}$$

The ensemble $e_1, \tilde{e}_1, e_2, \tilde{e}_2$ will be called the canonical circular fourcomplex base, and Eq. (3.106) gives the canonical form of the circular fourcomplex number. It can be checked that

$$e_1^2 = e_1, \quad \tilde{e}_1^2 = -e_1, \quad e_1\tilde{e}_1 = \tilde{e}_1, \quad e_2^2 = e_2, \quad \tilde{e}_2^2 = -e_2, \quad e_2\tilde{e}_2 = \tilde{e}_2,$$
$$e_1 e_2 = 0, \quad \tilde{e}_1\tilde{e}_2 = 0, \quad e_1\tilde{e}_2 = 0, \quad e_2\tilde{e}_1 = 0. \tag{3.108}$$

The moduli of the bases in Eq. (3.107) are

$$|e_1| = \frac{1}{\sqrt{2}}, \quad |\tilde{e}_1| = \frac{1}{\sqrt{2}}, \quad |e_2| = \frac{1}{\sqrt{2}}, \quad |\tilde{e}_2| = \frac{1}{\sqrt{2}}, \tag{3.109}$$

and it can be checked that

$$|x + \alpha y + \beta z + \gamma t|^2 = \xi^2 + \upsilon^2 + \tau^2 + \zeta^2. \tag{3.110}$$

If $u = u'u''$, the components $\xi, \upsilon, \tau, \zeta$ are related, according to Eqs. (3.34)-(3.37) by

$$\xi = \sqrt{2}(\xi'\xi'' - \upsilon'\upsilon''), \quad \upsilon = \sqrt{2}(\xi'\upsilon'' + \upsilon'\xi''), \quad \tau = \sqrt{2}(\tau'\tau'' - \zeta'\zeta''),$$
$$\zeta = \sqrt{2}(\tau'\zeta'' + \zeta'\tau''), \tag{3.111}$$

which show that, upon multiplication, the components ξ, υ and τ, ζ obey, up to a normalization constant, the same rules as the real and imaginary components of usual, two-dimensional complex numbers.

If the coefficients in Eq. (3.101) are

$$a_l = a_{l0} + \alpha a_{l1} + \beta a_{l2} + \gamma a_{l3}, \tag{3.112}$$

and

$$A_{l1} = a_{l0} + a_{l3}, \quad \tilde{A}_{l1} = a_{l1} + a_{l2}, \quad A_{l2} = a_{l0} - a_{l3}, \quad \tilde{A}_{l2} = a_{l1} - a_{l2}, \tag{3.113}$$

the series (3.101) can be written as

$$\sum_{l=0}^{\infty} 2^{l/2} \left[(e_1 A_{l1} + \tilde{e}_1 \tilde{A}_{l1})(e_1 \xi + \tilde{e}_1 v)^l + (e_2 A_{l2} + \tilde{e}_2 \tilde{A}_{l2})(e_2 \tau + \tilde{e}_2 \zeta)^l \right].$$

$$(3.114)$$

Thus, the series in Eqs. (3.101) and (3.114) are absolutely convergent for

$$\rho_+ < c_1, \ \rho_- < c_2, \tag{3.115}$$

where

$$c_1 = \lim_{l \to \infty} \frac{\left[A_{l1}^2 + \tilde{A}_{l1}^2 \right]^{1/2}}{\sqrt{2} \left[A_{l+1,1}^2 + \tilde{A}_{l+1,1}^2 \right]^{1/2}}, \quad c_2 = \lim_{l \to \infty} \frac{\left[A_{l2}^2 + \tilde{A}_{l2}^2 \right]^{1/2}}{\sqrt{2} \left[A_{l+1,2}^2 + \tilde{A}_{l+1,2}^2 \right]^{1/2}}.$$

$$(3.116)$$

It can be shown that $c = (1/\sqrt{2})\min(c_1, c_2)$, where min designates the smallest of the numbers c_1, c_2. Using the expression of $|u|$ in Eq. (3.110), it can be seen that the spherical region of convergence defined in Eqs. (3.104), (3.105) is included in the cylindrical region of convergence defined in Eqs. (3.115) and (3.116).

3.1.6 Analytic functions of circular fourcomplex variables

The analytic functions of the hypercomplex variable u and the series expansion of functions have been discussed in Eqs. (1.85)-(1.93). If the fourcomplex function $f(u)$ of the fourcomplex variable u can be expressed in terms of the real functions $P(x, y, z, t)$, $Q(x, y, z, t)$, $R(x, y, z, t)$, $S(x, y, z, t)$ of real variables x, y, z, t as

$$f(u) = P(x,y,z,t) + \alpha Q(x,y,z,t) + \beta R(x,y,z,t) + \gamma S(x,y,z,t), \tag{3.117}$$

then relations of equality exist between partial derivatives of the functions P, Q, R, S. These relations can be obtained by writing the derivative of the function f as

$$\lim_{u \to u_0} \frac{1}{\Delta x + \alpha \Delta y + \beta \Delta z + \gamma \Delta t} \left[\frac{\partial P}{\partial x} \Delta x + \frac{\partial P}{\partial y} \Delta y + \frac{\partial P}{\partial z} \Delta z + \frac{\partial P}{\partial t} \Delta t \right.$$

$$+ \alpha \left(\frac{\partial Q}{\partial x} \Delta x + \frac{\partial Q}{\partial y} \Delta y + \frac{\partial Q}{\partial z} \Delta z + \frac{\partial Q}{\partial t} \Delta t \right)$$

$$+ \beta \left(\frac{\partial R}{\partial x} \Delta x + \frac{\partial R}{\partial y} \Delta y + \frac{\partial R}{\partial z} \Delta z + \frac{\partial R}{\partial t} \Delta t \right)$$

$$\left. + \gamma \left(\frac{\partial S}{\partial x} \Delta x + \frac{\partial S}{\partial y} \Delta y + \frac{\partial S}{\partial z} \Delta z + \frac{\partial S}{\partial t} \Delta t \right) \right], \tag{3.118}$$

where the difference appearing in Eq. (1.86) is $u - u_0 = \Delta x + \alpha \Delta y + \beta \Delta z + \gamma \Delta t$.

The relations between the partial derivatives of the functions P, Q, R, S are obtained by setting succesively in Eq. (3.118) $\Delta x \to 0, \Delta y = \Delta z = \Delta t = 0$; then $\Delta y \to 0, \Delta x = \Delta z = \Delta t = 0$; then $\Delta z \to 0, \Delta x = \Delta y = \Delta t = 0$; and finally $\Delta t \to 0, \Delta x = \Delta y = \Delta z = 0$. The relations are

$$\frac{\partial P}{\partial x} = \frac{\partial Q}{\partial y} = \frac{\partial R}{\partial z} = \frac{\partial S}{\partial t}, \tag{3.119}$$

$$\frac{\partial Q}{\partial x} = -\frac{\partial P}{\partial y} = -\frac{\partial S}{\partial z} = \frac{\partial R}{\partial t}, \tag{3.120}$$

$$\frac{\partial R}{\partial x} = -\frac{\partial S}{\partial y} = -\frac{\partial P}{\partial z} = \frac{\partial Q}{\partial t}, \tag{3.121}$$

$$\frac{\partial S}{\partial x} = \frac{\partial R}{\partial y} = \frac{\partial Q}{\partial z} = \frac{\partial P}{\partial t}. \tag{3.122}$$

The relations (3.119)-(3.122) are analogous to the Riemann relations for the real and imaginary components of a complex function. It can be shown from Eqs. (3.119)-(3.122) that the component P is a solution of the equations

$$\frac{\partial^2 P}{\partial x^2} + \frac{\partial^2 P}{\partial y^2} = 0, \quad \frac{\partial^2 P}{\partial x^2} + \frac{\partial^2 P}{\partial z^2} = 0, \quad \frac{\partial^2 P}{\partial y^2} + \frac{\partial^2 P}{\partial t^2} = 0,$$
$$\frac{\partial^2 P}{\partial z^2} + \frac{\partial^2 P}{\partial t^2} = 0, \tag{3.123}$$

$$\frac{\partial^2 P}{\partial x^2} - \frac{\partial^2 P}{\partial t^2} = 0, \quad \frac{\partial^2 P}{\partial y^2} - \frac{\partial^2 P}{\partial z^2} = 0, \tag{3.124}$$

and the components Q, R, S are solutions of similar equations.

As can be seen from Eqs. (3.123)-(3.124), the components P, Q, R, S of an analytic function of circular fourcomplex variable are harmonic with respect to the pairs of variables $x, y; x, z; y, t$ and z, t, and are solutions of the wave equation with respect to the pairs of variables x, t and y, z. The components P, Q, R, S are also solutions of the mixed-derivative equations

$$\frac{\partial^2 P}{\partial x \partial y} = \frac{\partial^2 P}{\partial z \partial t}, \quad \frac{\partial^2 P}{\partial x \partial z} = \frac{\partial^2 P}{\partial y \partial t}, \quad \frac{\partial^2 P}{\partial x \partial t} = -\frac{\partial^2 P}{\partial y \partial z}, \tag{3.125}$$

and the components Q, R, S are solutions of similar equations.

3.1.7 Integrals of functions of circular fourcomplex variables

The singularities of circular fourcomplex functions arise from terms of the form $1/(u-u_0)^m$, with $m > 0$. Functions containing such terms are singular not only at $u = u_0$, but also at all points of the two-dimensional hyperplanes passing through u_0 and which are parallel to the nodal hyperplanes.

The integral of a circular fourcomplex function between two points A, B along a path situated in a region free of singularities is independent of path, which means that the integral of an analytic function along a loop situated in a region free from singularities is zero,

$$\oint_\Gamma f(u)du = 0, \tag{3.126}$$

where it is supposed that a surface Σ spanning the closed loop Γ is not intersected by any of the two-dimensional hyperplanes associated with the singularities of the function $f(u)$. Using the expression, Eq. (3.117), for $f(u)$ and the fact that $du = dx + \alpha dy + \beta dz + \gamma dt$, the explicit form of the integral in Eq. (3.126) is

$$\oint_\Gamma f(u)du = \oint_\Gamma [(Pdx - Qdy - Rdz + Sdt)$$
$$+\alpha(Qdx + Pdy + Sdz + Rdt) + \beta(Rdx + Sdy + Pdz + Qdt)$$
$$+\gamma(Sdx - Rdy - Qdz + Pdt)]. \tag{3.127}$$

If the functions P, Q, R, S are regular on a surface Σ spanning the loop Γ, the integral along the loop Γ can be transformed with the aid of the theorem of Stokes in an integral over the surface Σ of terms of the form $\partial P/\partial y + \partial Q/\partial x$, $\partial P/\partial z + \partial R/\partial x$, $\partial P/\partial t - \partial S/\partial x$, $\partial Q/\partial z - \partial R/\partial y$, $\partial Q/\partial t + \partial S/\partial y$, $\partial R/\partial t + \partial S/\partial z$ and of similar terms arising from the α, β and γ components, which are equal to zero by Eqs. (3.119)-(3.122), and this proves Eq. (3.126).

The integral of the function $(u - u_0)^m$ on a closed loop Γ is equal to zero for m a positive or negative integer not equal to -1,

$$\oint_\Gamma (u - u_0)^m du = 0, \ \ m \text{ integer}, \ m \neq -1. \tag{3.128}$$

This is due to the fact that $\int (u - u_0)^m du = (u - u_0)^{m+1}/(m + 1)$, and to the fact that the function $(u - u_0)^{m+1}$ is singlevalued for m an integer.

The integral $\oint du/(u - u_0)$ can be calculated using the exponential form (3.68),

$$u - u_0 = \rho \exp\left(\alpha\frac{\phi + \chi}{2} + \beta\frac{\phi - \chi}{2} + \gamma \ln \tan \psi\right), \tag{3.129}$$

so that

$$\frac{du}{u - u_0} = \frac{d\rho}{\rho} + \frac{\alpha + \beta}{2} d\phi + \frac{\alpha - \beta}{2} d\chi + \gamma d \ln \tan \psi. \tag{3.130}$$

Since ρ and $\ln \tan \psi$ are singlevalued variables, it follows that $\oint_\Gamma d\rho/\rho = 0$, $\oint_\Gamma d \ln \tan \psi = 0$. On the other hand, ϕ and χ are cyclic variables, so that they may give a contribution to the integral around the closed loop Γ. Thus, if C_+ is a circle of radius r parallel to the $\xi O v$ plane, and the projection of the center of this circle on the $\xi O v$ plane coincides with the projection of the point u_0 on this plane, the points of the circle C_+ are described according to Eqs. (3.18)-(3.22) by the equations

$$\xi = \xi_0 + r \sin \psi \cos \phi, \quad v = v_0 + r \sin \psi \sin \phi, \quad \tau = \tau_0 + r \cos \psi \cos \chi,$$

$$\zeta = \zeta_0 + r \cos \psi \sin \chi, \tag{3.131}$$

for constant values of χ and ψ, $\psi \neq 0, \pi/2$, where $u_0 = x_0 + \alpha y_0 + \beta z_0 + \gamma t_0$, and $\xi_0, v_0, \tau_0, \zeta_0$ are calculated from x_0, y_0, z_0, t_0 according to Eqs. (3.18). Then

$$\oint_{C_+} \frac{du}{u - u_0} = \pi(\alpha + \beta). \tag{3.132}$$

If C_- is a circle of radius r parallel to the $\tau O \zeta$ plane, and the projection of the center of this circle on the $\tau O \zeta$ plane coincides with the projection of the point u_0 on this plane, the points of the circle C_- are described by the same Eqs. (3.131) but for constant values of ϕ and ψ, $\psi \neq 0, \pi/2$. Then

$$\oint_{C_-} \frac{du}{u - u_0} = \pi(\alpha - \beta). \tag{3.133}$$

The expression of $\oint_\Gamma du/(u - u_0)$ can be written as a single equation with the aid of a functional $\text{int}(M, C)$ defined for a point M and a closed curve C in a two-dimensional plane, such that

$$\text{int}(M, C) = \begin{cases} 1 \text{ if } M \text{ is an interior point of } C, \\ 0 \text{ if } M \text{ is exterior to } C. \end{cases} \tag{3.134}$$

With this notation the result of the integration along a closed path Γ can be written as

$$\oint_\Gamma \frac{du}{u - u_0} = \pi(\alpha + \beta) \, \text{int}(u_{0\xi v}, \Gamma_{\xi v}) + \pi(\alpha - \beta) \, \text{int}(u_{0\tau\zeta}, \Gamma_{\tau\zeta}), \tag{3.135}$$

where $u_{0\xi v}, u_{0\tau\zeta}$ and $\Gamma_{\xi v}, \Gamma_{\tau\zeta}$ are respectively the projections of the point u_0 and of the loop Γ on the planes ξv and $\tau \zeta$.

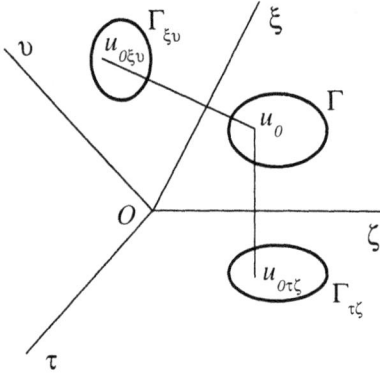

Figure 3.2: Integration path Γ and the pole u_0, and their projections $\Gamma_{\xi v}, \Gamma_{\tau \zeta}$ and $u_{0\xi v}, u_{0\tau \zeta}$ on the planes ξv and respectively $\tau \zeta$.

If $f(u)$ is an analytic circular fourcomplex function which can be expanded in a series as written in Eq. (1.89), and the expansion holds on the curve Γ and on a surface spanning Γ, then from Eqs. (3.128) and (3.135) it follows that

$$\oint_\Gamma \frac{f(u)du}{u - u_0} = \pi[(\alpha + \beta) \text{ int}(u_{0\xi v}, \Gamma_{\xi v}) + (\alpha - \beta) \text{ int}(u_{0\tau \zeta}, \Gamma_{\tau \zeta})] \, f(u_0),$$

$$(3.136)$$

where $\Gamma_{\xi v}, \Gamma_{\tau \zeta}$ are the projections of the curve Γ on the planes ξv and respectively $\tau \zeta$, as shown in Fig. 3.2. Substituting in the right-hand side of Eq. (3.136) the expression of $f(u)$ in terms of the real components P, Q, R, S, Eq. (3.117), yields

$$\oint_\Gamma \frac{f(u)du}{u - u_0} = \pi[-(1 + \gamma)(Q + R) + (\alpha + \beta)(P + S)] \text{ int}(u_{0\xi v}, \Gamma_{\xi v})$$
$$+ \pi[-(1 - \gamma)(Q - R) + (\alpha - \beta)(P - S)] \text{ int}(u_{0\tau \zeta}, \Gamma_{\tau \zeta}), \quad (3.137)$$

where P, Q, R, S are the values of the components of f at $u = u_0$.

If $f(u)$ can be expanded as written in Eq. (1.89) on Γ and on a surface spanning Γ, then from Eqs. (3.128) and (3.135) it also results that

$$\oint_\Gamma \frac{f(u)du}{(u - u_0)^{m+1}} = \frac{\pi}{m!}[(\alpha + \beta) \text{ int}(u_{0\xi v}, \Gamma_{\xi v})$$
$$+ (\alpha - \beta) \text{ int}(u_{0\tau \zeta}, \Gamma_{\tau \zeta})] \, f^{(m)}(u_0), \quad (3.138)$$

where it has been used the fact that the derivative $f^{(m)}(u_0)$ of order m of $f(u)$ at $u = u_0$ is related to the expansion coefficient in Eq. (1.89) according to Eq. (1.93).

If a function $f(u)$ is expanded in positive and negative powers of $u - u_j$, where u_j are circular fourcomplex constants, j being an index, the integral of f on a closed loop Γ is determined by the terms in the expansion of f which are of the form $a_j/(u - u_j)$,

$$f(u) = \cdots + \sum_j \frac{a_j}{u - u_j} + \cdots \tag{3.139}$$

Then the integral of f on a closed loop Γ is

$$\oint_\Gamma f(u)du = \pi(\alpha + \beta) \sum_j \text{int}(u_{j\xi v}, \Gamma_{\xi v})a_j$$
$$+ \pi(\alpha - \beta) \sum_j \text{int}(u_{j\tau\zeta}, \Gamma_{\tau\zeta})a_j. \tag{3.140}$$

3.1.8 Factorization of circular fourcomplex polynomials

A polynomial of degree m of the circular fourcomplex variable $u = x + \alpha y + \beta z + \gamma t$ has the form

$$P_m(u) = u^m + a_1 u^{m-1} + \cdots + a_{m-1}u + a_m, \tag{3.141}$$

where the constants are in general circular fourcomplex numbers.

It can be shown that any circular fourcomplex polynomial has a circular fourcomplex root, whence it follows that a polynomial of degree m can be written as a product of m linear factors of the form $u - u_j$, where the circular fourcomplex numbers u_j are the roots of the polynomials, although the factorization may not be unique,

$$P_m(u) = \prod_{j=1}^{m}(u - u_j). \tag{3.142}$$

The fact that any circular fourcomplex polynomial has a root can be shown by considering the transformation of a fourdimensional sphere with the center at the origin by the function u^m. The points of the hypersphere of radius d are of the form written in Eq. (3.78), with d constant and $0 \leq \phi < 2\pi, 0 \leq \chi < 2\pi, 0 \leq \psi \leq \pi/2$. The point u^m is

$$u^m = d^m \exp\left(\alpha m \frac{\phi + \chi}{2} + \beta m \frac{\phi - \chi}{2}\right)$$
$$[\cos(\psi - \pi/4) + \gamma \sin(\psi - \pi/4)]^m. \tag{3.143}$$

It can be shown with the aid of Eq. (3.82) that

$$\left| u \exp\left(\alpha \frac{\phi + \chi}{2} + \beta \frac{\phi - \chi}{2} \right) \right| = |u|, \tag{3.144}$$

so that

$$\left| [\cos(\psi - \pi/4) + \gamma \sin(\psi - \pi/4)]^m \exp\left(\alpha m \frac{\phi + \chi}{2} + \beta m \frac{\phi - \chi}{2} \right) \right|$$
$$= \left| [\cos(\psi - \pi/4) + \gamma \sin(\psi - \pi/4)]^m \right|. \tag{3.145}$$

The right-hand side of Eq. (3.145) is

$$|(\cos \epsilon + \gamma \sin \epsilon)^m|^2 = \sum_{k=0}^{m} C_{2m}^{2k} \cos^{2m-2k} \epsilon \sin^{2k} \epsilon, \tag{3.146}$$

where $\epsilon = \psi - \pi/4$, and since $C_{2m}^{2k} \geq C_m^k$, it can be concluded that

$$|(\cos \epsilon + \gamma \sin \epsilon)^m|^2 \geq 1. \tag{3.147}$$

Then

$$d^m \leq |u^m| \leq 2^{(m-1)/2} d^m, \tag{3.148}$$

which shows that the image of a four-dimensional sphere via the transformation operated by the function u^m is a finite hypersurface.

If $u' = u^m$, and

$$u' = d'[\cos(\psi' - \pi/4) + \gamma \sin(\psi' - \pi/4)] \exp\left(\alpha \frac{\phi' + \chi'}{2} + \beta \frac{\phi' - \chi'}{2} \right), \tag{3.149}$$

then

$$\phi' = m\phi, \ \chi' = m\chi, \ \tan \psi' = \tan^m \psi. \tag{3.150}$$

Since for any values of the angles ϕ', χ', ψ' there is a set of solutions ϕ, χ, ψ of Eqs. (3.150), and since the image of the hypersphere is a finite hypersurface, it follows that the image of the four-dimensional sphere via the function u^m is also a closed hypersurface. A continuous hypersurface is called closed when any ray issued from the origin intersects that surface at least once in the finite part of the space.

A transformation of the four-dimensional space by the polynomial $P_m(u)$ will be considered further. By this transformation, a hypersphere of radius d having the center at the origin is changed into a certain finite closed surface, as discussed previously. The transformation of the four-dimensional space by the polynomial $P_m(u)$ associates to the point $u = 0$ the point $f(0) = a_m$, and the image of a hypersphere of very large radius d can be represented with good approximation by the image of that hypersphere by the function u^m. The origin of the axes is an inner point of the latter image. If the radius of the hypersphere is now reduced continuously from the initial very large values to zero, the image hypersphere encloses initially the origin, but the image shrinks to a_m when the radius approaches the value zero. Thus, the origin is initially inside the image hypersurface, and it lies outside the image hypersurface when the radius of the hypersphere tends to zero. Then since the image hypersurface is closed, the image surface must intersect at some stage the origin of the axes, which means that there is a point u_1 such that $f(u_1) = 0$. The factorization in Eq. (3.142) can then be obtained by iterations.

The roots of the polynomial P_m can be obtained by the following method. If the constants in Eq. (3.141) are $a_l = a_{l0} + \alpha a_{l1} + \beta a_{l2} + \gamma a_{l3}$, and with the notations of Eq. (3.113), the polynomial $P_m(u)$ can be written as

$$P_m = \sum_{l=0}^{m} 2^{(m-l)/2} (e_1 A_{l1} + \tilde{e}_1 \tilde{A}_{l1})(e_1 \xi + \tilde{e}_1 v)^{m-l}$$

$$+ \sum_{l=0}^{m} 2^{(m-l)/2} (e_2 A_{l2} + \tilde{e}_2 \tilde{A}_{l2})(e_2 \tau + \tilde{e}_2 \zeta)^{m-l}, \qquad (3.151)$$

where the constants $A_{lk}, \tilde{A}_{lk}, k = 1, 2$, are real numbers. Each of the polynomials of degree m in $e_1 \xi + \tilde{e}_1 v, e_2 \tau + \tilde{e}_2 \zeta$ in Eq. (3.151) can always be written as a product of linear factors of the form $e_1(\xi - \xi_p) + \tilde{e}_1(v - v_p)$ and respectively $e_2(\tau - \tau_p) + \tilde{e}_2(\zeta - \zeta_p)$, where the constants $\xi_p, v_p, \tau_p, \zeta_p$ are real,

$$\sum_{l=0}^{m} 2^{(m-l)/2} (e_1 A_{l1} + \tilde{e}_1 \tilde{A}_{l1})(e_1 \xi + \tilde{e}_1 v)^{m-l}$$

$$= \prod_{p=1}^{m} 2^{m/2} \{ e_1(\xi - \xi_p) + \tilde{e}_1(v - v_p) \}, \qquad (3.152)$$

$$\sum_{l=0}^{m} 2^{(m-l)/2} (e_2 A_{l2} + \tilde{e}_2 \tilde{A}_{l2})(e_2 \tau + \tilde{e}_2 \zeta)^{m-l}$$

$$= \prod_{p=1}^{m} 2^{m/2} \{ e_2(\tau - \tau_p) + \tilde{e}_2(\zeta - \zeta_p) \}. \qquad (3.153)$$

Due to the relations (3.108), the polynomial $P_m(u)$ can be written as a product of factors of the form

$$P_m(u) = \prod_{p=1}^{m} 2^{m/2} \left\{ e_1(\xi - \xi_p) + \tilde{e}_1(v - v_p) + e_2(\tau - \tau_p) + \tilde{e}_2(\zeta - \zeta_p) \right\}.$$

$$(3.154)$$

This relation can be written with the aid of Eq. (3.106) in the form (3.142), where

$$u_p = \sqrt{2}(e_1\xi_p + \tilde{e}_1 v_p + e_2\tau_p + \tilde{e}_2\zeta_p).$$ $$(3.155)$$

The roots $e_1\xi_p + \tilde{e}_1 v_p$ and $e_2\tau_p + \tilde{e}_2\zeta_p$ defined in Eqs. (3.152) and respectively (3.153) may be ordered arbitrarily. This means that Eq. (3.155) gives sets of m roots $u_1, ..., u_m$ of the polynomial $P_m(u)$, corresponding to the various ways in which the roots $e_1\xi_p + \tilde{e}_1 v_p$ and $e_2\tau_p + \tilde{e}_2\zeta_p$ are ordered according to p for each polynomial. Thus, while the hypercomplex components in Eqs. (3.152), Eqs. (3.153) taken separately have unique factorizations, the polynomial $P_m(u)$ can be written in many different ways as a product of linear factors. The result of the circular fourcomplex integration, Eq. (3.140), is however unique.

If, for example, $P(u) = u^2 + 1$, the possible factorizations are $P = (u - \tilde{e}_1 - \tilde{e}_2)(u + \tilde{e}_1 + \tilde{e}_2)$ and $P = (u - \tilde{e}_1 + \tilde{e}_2)(u + \tilde{e}_1 - \tilde{e}_2)$, which can also be written as $u^2 + 1 = (u - \alpha)(u + \alpha)$ or as $u^2 + 1 = (u - \beta)(u + \beta)$. The result of the circular fourcomplex integration, Eq. (3.140), is however unique. It can be checked that $(\pm\tilde{e}_1 \pm \tilde{e}_2)^2 = -e_1 - e_2 = -1$.

3.1.9 Representation of circular fourcomplex numbers by irreducible matrices

If T is the unitary matrix,

$$T = \begin{pmatrix} \frac{1}{\sqrt{2}} & 0 & 0 & \frac{1}{\sqrt{2}} \\ 0 & \frac{1}{\sqrt{2}} & \frac{1}{\sqrt{2}} & 0 \\ \frac{1}{\sqrt{2}} & 0 & 0 & -\frac{1}{\sqrt{2}} \\ 0 & \frac{1}{\sqrt{2}} & -\frac{1}{\sqrt{2}} & 0 \end{pmatrix},$$

$$(3.156)$$

it can be shown that the matrix TUT^{-1} has the form

$$TUT^{-1} = \begin{pmatrix} V_1 & 0 \\ 0 & V_2 \end{pmatrix}.$$ $$(3.157)$$

where U is the matrix in Eq. (3.41) used to represent the circular fourcomplex number u. In Eq. (3.157), V_1, V_2 are the matrices

$$V_1 = \begin{pmatrix} x+t & y+z \\ -y-z & x+t \end{pmatrix}, \quad V_2 = \begin{pmatrix} x-t & y-z \\ -y+z & x-t \end{pmatrix}.$$ (3.158)

In Eq. (3.157), the symbols 0 denote the matrix

$$\begin{pmatrix} 0 & 0 \\ 0 & 0 \end{pmatrix}.$$ (3.159)

The relations between the variables $x+t, y+z, x-t, y-z$ for the multiplication of circular fourcomplex numbers have been written in Eqs. (3.34)-(3.37). The matrix TUT^{-1} provides an irreducible representation [7] of the circular fourcomplex number u in terms of matrices with real coefficients.

3.2 Hyperbolic complex numbers in four dimensions

3.2.1 Operations with hyperbolic fourcomplex numbers

A hyperbolic fourcomplex number is determined by its four components (x, y, z, t). The sum of the hyperbolic fourcomplex numbers (x, y, z, t) and (x', y', z', t') is the hyperbolic fourcomplex number $(x+x', y+y', z+z', t+t')$. The product of the hyperbolic fourcomplex numbers (x, y, z, t) and (x', y', z', t') is defined in this section to be the hyperbolic fourcomplex number $(xx' + yy' + zz' + tt', xy' + yx' + zt' + tz', xz' + zx' + yt' + ty', xt' + tx' + yz' + zy')$.

Hyperbolic fourcomplex numbers and their operations can be represented by writing the hyperbolic fourcomplex number (x, y, z, t) as $u = x + \alpha y + \beta z + \gamma t$, where α, β and γ are bases for which the multiplication rules are

$$\alpha^2 = 1, \ \beta^2 = 1, \ \gamma^2 = 1, \alpha\beta = \beta\alpha = \gamma, \ \alpha\gamma = \gamma\alpha = \beta, \ \beta\gamma = \gamma\beta = \alpha.$$ (3.160)

Two hyperbolic fourcomplex numbers $u = x + \alpha y + \beta z + \gamma t, u' = x' + \alpha y' + \beta z' + \gamma t'$ are equal, $u = u'$, if and only if $x = x', y = y', z = z', t = t'$. If $u = x + \alpha y + \beta z + \gamma t, u' = x' + \alpha y' + \beta z' + \gamma t'$ are hyperbolic fourcomplex numbers, the sum $u + u'$ and the product uu' defined above can be obtained by applying the usual algebraic rules to the sum $(x + \alpha y + \beta z + \gamma t) + (x' +$

$\alpha y' + \beta z' + \gamma t')$ and to the product $(x + \alpha y + \beta z + \gamma t)(x' + \alpha y' + \beta z' + \gamma t')$, and grouping of the resulting terms.

$$u + u' = x + x' + \alpha(y + y') + \beta(z + z') + \gamma(t + t'), \tag{3.161}$$

$$uu' = xx' + yy' + zz' + tt' + \alpha(xy' + yx' + zt' + tz')$$
$$+ \beta(xz' + zx' + yt' + ty') + \gamma(xt' + tx' + yz' + zy'). \tag{3.162}$$

If u, u', u'' are hyperbolic fourcomplex numbers, the multiplication is associative

$$(uu')u'' = u(u'u'') \tag{3.163}$$

and commutative

$$uu' = u'u, \tag{3.164}$$

as can be checked through direct calculation. The hyperbolic fourcomplex zero is $0 + \alpha \cdot 0 + \beta \cdot 0 + \gamma \cdot 0$, denoted simply 0, and the hyperbolic fourcomplex unity is $1 + \alpha \cdot 0 + \beta \cdot 0 + \gamma \cdot 0$, denoted simply 1.

The inverse of the hyperbolic fourcomplex number $u = x + \alpha y + \beta z + \gamma t$ is a hyperbolic fourcomplex number $u' = x' + \alpha y' + \beta z' + \gamma t'$ having the property that

$$uu' = 1. \tag{3.165}$$

Written on components, the condition, Eq. (3.165), is

$$\begin{aligned}
xx' + yy' + zz' + tt' &= 1, \\
yx' + xy' + tz' + zt' &= 0, \\
zx' + ty' + xz' + yt' &= 0, \\
tx' + zy' + yz' + xt' &= 0.
\end{aligned} \tag{3.166}$$

The system (3.166) has the solution

$$x' = \frac{x(x^2 - y^2 - z^2 - t^2) + 2yzt}{\nu}, \tag{3.167}$$

$$y' = \frac{y(-x^2 + y^2 - z^2 - t^2) + 2xzt}{\nu}. \tag{3.168}$$

$$z' = \frac{z(-x^2 - y^2 + z^2 - t^2) + 2xyt}{\nu}, \tag{3.169}$$

$$t' = \frac{t(-x^2 - y^2 - z^2 + t^2) + 2xyz}{\nu}, \tag{3.170}$$

provided that $\nu \neq 0$, where

$$\nu = x^4 + y^4 + z^4 + t^4 - 2(x^2y^2 + x^2z^2 + x^2t^2 + y^2z^2 + y^2t^2 + z^2t^2)$$
$$+ 8xyzt. \tag{3.171}$$

The quantity ν can be written as

$$\nu = ss's''s''', \tag{3.172}$$

where

$$s = x+y+z+t, s' = x-y+z-t, s'' = x+y-z-t, s''' = x-y-z+t. \tag{3.173}$$

The variables s, s', s'', s''' will be called canonical hyperbolic fourcomplex variables.

Then a hyperbolic fourcomplex number $u = x + \alpha y + \beta z + \gamma t$ has an inverse, unless

$$s = 0, \text{ or } s' = 0, \text{ or } s'' = 0, \text{ or } s''' = 0. \tag{3.174}$$

For arbitrary values of the variables x, y, z, t, the quantity ν can be positive or negative. If $\nu \geq 0$, the quantity $\mu = \nu^{1/4}$ will be called amplitude of the hyperbolic fourcomplex number $x + \alpha y + \beta z + \gamma t$. The normals of the hyperplanes in Eq. (3.174) are orthogonal to each other. Because of conditions (3.174) these hyperplanes will be also called the nodal hyperplanes. It can be shown that if $uu' = 0$ then either $u = 0$, or $u' = 0$, or q, q' belong to pairs of orthogonal hypersurfaces as described further. Thus, divisors of zero exist if one of the hyperbolic fourcomplex numbers u, u' belongs to one of the nodal hyperplanes and the other hyperbolic fourcomplex number belongs to the straight line through the origin which is normal to that hyperplane,

$$x + y + z + t = 0, \text{ and } x' = y' = z' = t', \tag{3.175}$$

or

$$x - y + z - t = 0, \text{ and } x' = -y' = z' = -t', \tag{3.176}$$

or

$$x + y - z - t = 0, \text{ and } x' = y' = -z' = -t', \tag{3.177}$$

or

$$x - y - z + t = 0, \text{ and } x' = -y' = -z' = t'. \tag{3.178}$$

Divisors of zero also exist if the hyperbolic fourcomplex numbers u, u' belong to different members of the pairs of two-dimensional hypersurfaces listed further,

$$x + y = 0, \; z + t = 0 \text{ and } x' - y' = 0, \; z' - t' = 0, \tag{3.179}$$

or

$$x + z = 0, \; y + t = 0 \text{ and } x' - z' = 0, \; y' - t' = 0, \tag{3.180}$$

or

$$y + z = 0, \; x + t = 0 \text{ and } y' - z' = 0, \; x' - t' = 0. \tag{3.181}$$

3.2.2 Geometric representation of hyperbolic fourcomplex numbers

The hyperbolic fourcomplex number $x + \alpha y + \beta z + \gamma t$ can be represented by the point A of coordinates (x, y, z, t). If O is the origin of the four-dimensional space x, y, z, t, the distance from A to the origin O can be taken as

$$d^2 = x^2 + y^2 + z^2 + t^2. \tag{3.182}$$

The distance d will be called modulus of the hyperbolic fourcomplex number $x + \alpha y + \beta z + \gamma t$, $d = |u|$.

If $u = x+\alpha y+\beta z+\gamma t, u_1 = x_1+\alpha y_1+\beta z_1+\gamma t_1, u_2 = x_2+\alpha y_2+\beta z_2+\gamma t_2$, and $u = u_1 u_2$, and if

$$s_j = x_j + y_j + z_j + t_j, \; s'_j = x_j - y_j + z_j - t_j, \; s''_j = x_j + y_j - z_j - t_j,$$
$$s'''_j = x_j - y_j - z_j + t_j, \tag{3.183}$$

for $j = 1, 2$, it can be shown that

$$s = s_1 s_2, \; s' = s'_1 s'_2, \; s'' = s''_1 s''_2, \; s''' = s'''_1 s'''_2. \tag{3.184}$$

The relations (3.184) are a consequence of the identities

$$(x_1 x_2 + y_1 y_2 + z_1 z_2 + t_1 t_2) + (x_1 y_2 + y_1 x_2 + z_1 t_2 + t_1 z_2)$$
$$+ (x_1 z_2 + z_1 x_2 + y_1 t_2 + t_1 y_2) + (x_1 t_2 + t_1 x_2 + y_1 z_2 + z_1 y_2)$$
$$= (x_1 + y_1 + z_1 + t_1)(x_2 + y_2 + z_2 + t_2), \tag{3.185}$$

$$(x_1 x_2 + y_1 y_2 + z_1 z_2 + t_1 t_2) - (x_1 y_2 + y_1 x_2 + z_1 t_2 + t_1 z_2)$$
$$+ (x_1 z_2 + z_1 x_2 + y_1 t_2 + t_1 y_2) - (x_1 t_2 + t_1 x_2 + y_1 z_2 + z_1 y_2)$$
$$= (x_1 - y_1 + z_1 - t_1)(x_2 - y_2 + z_2 - t_2), \tag{3.186}$$

$$(x_1x_2 + y_1y_2 + z_1z_2 + t_1t_2) + (x_1y_2 + y_1x_2 + z_1t_2 + t_1z_2)$$
$$- (x_1z_2 + z_1x_2 + y_1t_2 + t_1y_2) - (x_1t_2 + t_1x_2 + y_1z_2 + z_1y_2)$$
$$= (x_1 + y_1 - z_1 - t_1)(x_2 + y_2 - z_2 - t_2), \qquad (3.187)$$

$$(x_1x_2 + y_1y_2 + z_1z_2 + t_1t_2) - (x_1y_2 + y_1x_2 + z_1t_2 + t_1z_2)$$
$$- (x_1z_2 + z_1x_2 + y_1t_2 + t_1y_2) + (x_1t_2 + t_1x_2 + y_1z_2 + z_1y_2)$$
$$= (x_1 - y_1 - z_1 + t_1)(x_2 - y_2 - z_2 + t_2). \qquad (3.188)$$

A consequence of the relations (3.184) is that if $u = u_1u_2$, then

$$\nu = \nu_1\nu_2, \qquad (3.189)$$

where

$$\nu_j = s_j s'_j s''_j s'''_j, j = 1, 2. \qquad (3.190)$$

The hyperbolic fourcomplex numbers

$$e = \frac{1 + \alpha + \beta + \gamma}{4}, \; e' = \frac{1 - \alpha + \beta - \gamma}{4}, \; e'' = \frac{1 + \alpha - \beta - \gamma}{4},$$
$$e''' = \frac{1 - \alpha - \beta + \gamma}{4} \qquad (3.191)$$

are orthogonal,

$$ee' = 0, \; ee'' = 0, \; ee''' = 0, \; e'e'' = 0, \; e'e''' = 0, \; e''e''' = 0, \qquad (3.192)$$

and have also the property that

$$e^2 = e, \; e'^2 = e', \; e''^2 = e'', \; e'''^2 = e'''. \qquad (3.193)$$

The hyperbolic fourcomplex number $u = x + \alpha y + \beta z + \gamma t$ can be written as

$$x + \alpha y + \beta z + \gamma t = (x + y + z + t)e + (x - y + z - t)e'$$
$$+ (x + y - z - t)e'' + (x - y - z + t)e''', \qquad (3.194)$$

or, by using Eq. (3.173),

$$u = se + s'e' + s''e'' + s'''e'''. \qquad (3.195)$$

The ensemble e, e', e'', e''' will be called the canonical hyperbolic fourcomplex base, and Eq. (3.195) gives the canonical form of the hyperbolic fourcomplex number. Thus, if $u_j = s_j e + s'_j e' + s''_j e'' + s'''_j e'''$, $j = 1, 2$, and $u = u_1u_2$, then the multiplication of the hyperbolic fourcomplex numbers is expressed by the relations (3.184). The moduli of the bases e, e', e'', e''' are

$$|e| = \frac{1}{2}, \ |e'| = \frac{1}{2}, \ |e''| = \frac{1}{2}, \ |e'''| = \frac{1}{2}. \tag{3.196}$$

The distance d, Eq. (3.182), is given by

$$d^2 = \frac{1}{4}\left(s^2 + s'^2 + s''^2 + s'''^2\right). \tag{3.197}$$

The relation (3.197) shows that the variables s, s', s'', s''' can be written as

$$s = 2d\cos\psi\cos\phi, \ \ s' = 2d\cos\psi\sin\phi, \ \ s'' = 2d\sin\psi\cos\chi,$$
$$s''' = 2d\sin\psi\sin\chi, \tag{3.198}$$

where ϕ is the azimuthal angle in the s, s' plane, $0 \le \phi < 2\pi$, χ is the azimuthal angle in the s'', s''' plane, $0 \le \chi < 2\pi$, and ψ is the angle between the line OA and the plane s, s', $0 \le \psi \le \pi/2$. The variables x, y, z, t can be expressed in terms of the distance d and the angles ϕ, χ, ψ as

$$\begin{aligned}
x &= (d/2)(\cos\psi\cos\phi + \cos\psi\sin\phi + \sin\psi\cos\chi + \sin\psi\sin\chi),\\
y &= (d/2)(\cos\psi\cos\phi - \cos\psi\sin\phi + \sin\psi\cos\chi - \sin\psi\sin\chi),\\
z &= (d/2)(\cos\psi\cos\phi + \cos\psi\sin\phi - \sin\psi\cos\chi - \sin\psi\sin\chi),\\
t &= (d/2)(\cos\psi\cos\phi - \cos\psi\sin\phi - \sin\psi\cos\chi + \sin\psi\sin\chi).
\end{aligned} \tag{3.199}$$

If $u = u_1 u_2$, and the hypercomplex numbers u_1, u_2 are described by the variables $d_1, \phi_1, \chi_1, \psi_1$ and respectively $d_2, \phi_2, \chi_2, \psi_2$, then from Eq. (3.198) it results that

$$\tan\phi = \tan\phi_1 \tan\phi_2, \ \ \tan\chi = \tan\chi_1 \tan\chi_2,$$
$$\frac{\tan^2\psi\sin 2\chi}{\sin 2\phi} = \frac{\tan^2\psi_1\sin 2\chi_1}{\sin 2\phi_1}\frac{\tan^2\psi_2\sin 2\chi_2}{\sin 2\phi_2}. \tag{3.200}$$

The relation (3.189) for the product of hyperbolic fourcomplex numbers can be demonstrated also by using a representation of the multiplication of the hyperbolic fourcomplex numbers by matrices, in which the hyperbolic fourcomplex number $u = x + \alpha y + \beta z + \gamma t$ is represented by the matrix

$$A = \begin{pmatrix} x & y & z & t \\ y & x & t & z \\ z & t & x & y \\ t & z & y & x \end{pmatrix}. \tag{3.201}$$

The product $u = x + \alpha y + \beta z + \gamma t$ of the hyperbolic fourcomplex numbers $u_1 = x_1 + \alpha y_1 + \beta z_1 + \gamma t_1, u_2 = x_2 + \alpha y_2 + \beta z_2 + \gamma t_2$, can be represented by the matrix multiplication

$$A = A_1 A_2. \tag{3.202}$$

It can be checked that the determinant $\det(A)$ of the matrix A is

$$\det A = \nu. \tag{3.203}$$

The identity (3.189) is then a consequence of the fact the determinant of the product of matrices is equal to the product of the determinants of the factor matrices.

3.2.3 Exponential form of a hyperbolic fourcomplex number

The exponential function of a hypercomplex variable u and the addition theorem for the exponential function have been written in Eqs. (1.35)-(1.36). If $u = x + \alpha y + \beta z + \gamma t$, then $\exp u$ can be calculated as $\exp u = \exp x \cdot \exp(\alpha y) \cdot \exp(\beta z) \cdot \exp(\gamma t)$. According to Eqs. (3.160),

$$\alpha^{2m} = 1, \alpha^{2m+1} = \alpha, \beta^{2m} = 1, \beta^{2m+1} = \beta, \gamma^{2m} = 1, \gamma^{2m+1} = \gamma, \tag{3.204}$$

where m is a natural number, so that $\exp(\alpha y)$, $\exp(\beta z)$ and $\exp(\gamma t)$ can be written as

$$\exp(\alpha y) = \cosh y + \alpha \sinh y, \ \exp(\beta z) = \cosh y + \beta \sinh z,$$
$$\exp(\gamma t) = \cosh t + \gamma \sinh t. \tag{3.205}$$

From Eqs. (3.205) it can be inferred that

$$(\cosh t + \alpha \sinh t)^m = \cosh mt + \alpha \sinh mt,$$
$$(\cosh t + \beta \sinh t)^m = \cosh mt + \beta \sinh mt,$$
$$(\cosh t + \gamma \sinh t)^m = \cosh mt + \gamma \sinh mt. \tag{3.206}$$

The hyperbolic fourcomplex numbers $u = x + \alpha y + \beta z + \gamma t$ for which $s = x + y + z + t > 0$, $s' = x - y + z - t > 0$, $s'' = x + y - z - t > 0$, $s''' = x - y - z + t > 0$ can be written in the form

$$x + \alpha y + \beta z + \gamma t = e^{x_1 + \alpha y_1 + \beta z_1 + \gamma t_1}. \tag{3.207}$$

The conditions $s = x + y + z + t > 0$, $s' = x - y + z - t > 0$, $s'' = x + y - z - t > 0$, $s''' = x - y - z + t > 0$ correspond in Eq. (3.198) to a range of angles $0 < \phi < \pi/2, 0 < \chi < \pi/2, 0 < \psi \le \pi/2$. The expressions of x_1, y_1, z_1, t_1 as functions of x, y, z, t can be obtained by developing $e^{\alpha y_1}, e^{\beta z_1}$ and $e^{\gamma t_1}$ with the aid of Eqs. (3.205), by multiplying these expressions and separating the hypercomplex components,

$$x = e^{x_1}(\cosh y_1 \cosh z_1 \cosh t_1 + \sinh y_1 \sinh z_1 \sinh t_1), \qquad (3.208)$$

$$y = e^{x_1}(\sinh y_1 \cosh z_1 \cosh t_1 + \cosh y_1 \sinh z_1 \sinh t_1), \qquad (3.209)$$

$$z = e^{x_1}(\cosh y_1 \sinh z_1 \cosh t_1 + \sinh y_1 \cosh z_1 \sinh t_1), \qquad (3.210)$$

$$t = e^{x_1}(\sinh y_1 \sinh z_1 \cosh t_1 + \cosh y_1 \cosh z_1 \sinh t_1). \qquad (3.211)$$

It can be shown from Eqs. (3.208)-(3.211) that

$$x_1 = \frac{1}{4}\ln(ss's''s'''), \ y_1 = \frac{1}{4}\ln\frac{ss''}{s's'''}, \ z_1 = \frac{1}{4}\ln\frac{ss'}{s''s'''}, \ t_1 = \frac{1}{4}\ln\frac{ss'''}{s's''}. \tag{3.212}$$

The exponential form of the hyperbolic fourcomplex number u can be written as

$$u = \mu \exp\left(\frac{1}{4}\alpha\ln\frac{ss''}{s's'''} + \frac{1}{4}\beta\ln\frac{ss'}{s''s'''} + \frac{1}{4}\gamma\ln\frac{ss'''}{s's''}\right), \tag{3.213}$$

where

$$\mu = (ss's''s''')^{1/4}. \tag{3.214}$$

The exponential form of the hyperbolic fourcomplex number u can be written with the aid of the relations (3.198) as

$$u = \mu \exp\left(\frac{1}{4}\alpha\ln\frac{1}{\tan\phi\tan\chi} + \frac{1}{4}\beta\ln\frac{\sin 2\phi}{\tan^2\psi\sin 2\chi} + \frac{1}{4}\gamma\ln\frac{\tan\chi}{\tan\phi}\right). \tag{3.215}$$

The amplitude μ can be expressed in terms of the distance d with the aid of Eqs. (3.198) as

$$\mu = d\sin^{1/2}2\psi\sin^{1/4}2\phi\sin^{1/4}2\chi. \tag{3.216}$$

The hypercomplex number can be written as

$$u = d\sin^{1/2}2\psi\sin^{1/4}2\phi\sin^{1/4}2\chi\exp\left(\frac{1}{4}\alpha\ln\frac{1}{\tan\phi\tan\chi}\right.$$
$$\left. +\frac{1}{4}\beta\ln\frac{\sin 2\phi}{\tan^2\psi\sin 2\chi} + \frac{1}{4}\gamma\ln\frac{\tan\chi}{\tan\phi}\right), \tag{3.217}$$

which is the trigonometric form of the hypercomplex number u.

3.2.4 Elementary functions of a hyperbolic fourcomplex variable

The logarithm u_1 of the hyperbolic fourcomplex number u, $u_1 = \ln u$, can be defined for $s > 0, s' > 0, s'' > 0, s''' > 0$ as the solution of the equation

$$u = e^{u_1}, \tag{3.218}$$

for u_1 as a function of u. From Eq. (3.213) it results that

$$\ln u = \frac{1}{4} \ln \mu + + \frac{1}{4} \alpha \ln \frac{ss''}{s's'''} + \frac{1}{4} \beta \ln \frac{ss'}{s''s'''} + \frac{1}{4} \gamma \ln \frac{ss'''}{s's''}. \tag{3.219}$$

Using the expression in Eq. (3.215), the logarithm can be written as

$$\ln u = \frac{1}{4} \ln \mu + \frac{1}{4} \alpha \ln \frac{1}{\tan \phi \tan \chi} + \frac{1}{4} \beta \ln \frac{\sin 2\phi}{\tan^2 \psi \sin 2\chi} + \frac{1}{4} \gamma \ln \frac{\tan \chi}{\tan \phi}. \tag{3.220}$$

It can be inferred from Eqs. (3.219) and (3.184) that

$$\ln(u_1 u_2) = \ln u_1 + \ln u_2. \tag{3.221}$$

The explicit form of Eq. (3.219) is

$$\ln(x + \alpha y + \beta z + \gamma t) = \frac{1}{4}(1 + \alpha + \beta + \gamma) \ln(x + y + z + t)$$

$$+ \frac{1}{4}(1 - \alpha + \beta - \gamma) \ln(x - y + z - t)$$

$$+ \frac{1}{4}(1 + \alpha - \beta - \gamma) \ln(x + y - z - t)$$

$$+ \frac{1}{4}(1 - \alpha - \beta + \gamma) \ln(x - y - z + t). \tag{3.222}$$

The relation (3.222) can be written with the aid of Eq. (3.191) as

$$\ln u = e \ln s + e' \ln s' + e'' \ln s'' + e''' \ln s'''. \tag{3.223}$$

The power function u^n can be defined for $s > 0, s' > 0, s'' > 0, s''' > 0$ and real values of n as

$$u^n = e^{n \ln u}. \tag{3.224}$$

It can be inferred from Eqs. (3.224) and (3.221) that

$$(u_1 u_2)^n = u_1^n u_2^n. \tag{3.225}$$

Using the expression (3.222) for $\ln u$ and the relations (3.192) and (3.193) it can be shown that

$$(x + \alpha y + \beta z + \gamma t)^n = \frac{1}{4}(1 + \alpha + \beta + \gamma)(x + y + z + t)^n$$

$$+\frac{1}{4}(1 - \alpha + \beta - \gamma)(x - y + z - t)^n$$

$$+\frac{1}{4}(1 + \alpha - \beta - \gamma)(x + y - z - t)^n$$

$$+\frac{1}{4}(1 - \alpha - \beta + \gamma)(x - y - z + t)^n. \qquad (3.226)$$

For integer n, the relation (3.226) is valid for any x, y, z, t. The relation (3.226) for $n = -1$ is

$$\frac{1}{x + \alpha y + \beta z + \gamma t} = \frac{1}{4}\left(\frac{1 + \alpha + \beta + \gamma}{x + y + z + t} + \frac{1 - \alpha + \beta - \gamma}{x - y + z - t}\right.$$

$$\left.+\frac{1 + \alpha - \beta - \gamma}{x + y - z - t} + \frac{1 - \alpha - \beta + \gamma}{x - y - z + t}\right). \qquad (3.227)$$

The trigonometric functions of the hypercomplex variable u and the addition theorems for these functions have been written in Eqs. (1.57)-(1.60). The cosine and sine functions of the hypercomplex variables $\alpha y, \beta z$ and γt can be expressed as

$$\cos \alpha y = \cos y, \ \sin \alpha y = \alpha \sin y, \qquad (3.228)$$

$$\cos \beta y = \cos y, \ \sin \beta y = \beta \sin y, \qquad (3.229)$$

$$\cos \gamma y = \cos y, \ \sin \gamma y = \gamma \sin y. \qquad (3.230)$$

The cosine and sine functions of a hyperbolic fourcomplex number $x + \alpha y + \beta z + \gamma t$ can then be expressed in terms of elementary functions with the aid of the addition theorems Eqs. (1.59), (1.60) and of the expressions in Eqs. (3.228)-(3.230).

The hyperbolic functions of the hypercomplex variable u and the addition theorems for these functions have been written in Eqs. (1.62)-(1.65). The hyperbolic cosine and sine functions of the hypercomplex variables $\alpha y, \beta z$ and γt can be expressed as

$$\cosh \alpha y = \cosh y, \ \sinh \alpha y = \alpha \sinh y, \qquad (3.231)$$

$$\cosh \beta y = \cosh y, \ \sinh \beta y = \beta \sinh y, \qquad (3.232)$$

$$\cosh \gamma y = \cosh y, \ \sinh \gamma y = \gamma \sinh y. \qquad (3.233)$$

The hyperbolic cosine and sine functions of a hyperbolic fourcomplex number $x + \alpha y + \beta z + \gamma t$ can then be expressed in terms of elementary functions with the aid of the addition theorems Eqs. (1.64), (1.65) and of the expressions in Eqs. (3.231)-(3.233).

3.2.5 Power series of hyperbolic fourcomplex variables

A hyperbolic fourcomplex series is an infinite sum of the form

$$a_0 + a_1 + a_2 + \cdots + a_l + \cdots, \tag{3.234}$$

where the coefficients a_l are hyperbolic fourcomplex numbers. The convergence of the series (3.234) can be defined in terms of the convergence of its 4 real components. The convergence of a hyperbolic fourcomplex series can however be studied using hyperbolic fourcomplex variables. The main criterion for absolute convergence remains the comparison theorem, but this requires a number of inequalities which will be discussed further.

The modulus of a hyperbolic fourcomplex number $u = x + \alpha y + \beta z + \gamma t$ can be defined as

$$|u| = (x^2 + y^2 + z^2 + t^2)^{1/2}, \tag{3.235}$$

so that according to Eq. (3.182) $d = |u|$. Since $|x| \leq |u|, |y| \leq |u|, |z| \leq |u|, |t| \leq |u|$, a property of absolute convergence established via a comparison theorem based on the modulus of the series (3.234) will ensure the absolute convergence of each real component of that series.

The modulus of the sum $u_1 + u_2$ of the hyperbolic fourcomplex numbers u_1, u_2 fulfils the inequality

$$||u_1| - |u_2|| \leq |u_1 + u_2| \leq |u_1| + |u_2|. \tag{3.236}$$

For the product the relation is

$$|u_1 u_2| \leq 2|u_1||u_2|, \tag{3.237}$$

which replaces the relation of equality extant for regular complex numbers. The equality in Eq. (3.237) takes place for $x_1^2 = y_1^2 = z_1^2 = t_1^2$ and $x_2/x_1 = y_2/y_1 = z_2/z_1 = t_2/t_1$. In particular

$$|u^2| \leq 2(x^2 + y^2 + z^2 + t^2). \tag{3.238}$$

The inequality in Eq. (3.237) implies that

$$|u^l| \leq 2^{l-1}|u|^l. \tag{3.239}$$

From Eqs. (3.237) and (3.239) it results that

$$|au^l| \leq 2^l |a||u|^l. \tag{3.240}$$

A power series of the hyperbolic fourcomplex variable u is a series of the form

$$a_0 + a_1 u + a_2 u^2 + \cdots + a_l u^l + \cdots. \tag{3.241}$$

Since

$$\left| \sum_{l=0}^{\infty} a_l u^l \right| \le \sum_{l=0}^{\infty} 2^l |a_l| |u|^l, \tag{3.242}$$

a sufficient condition for the absolute convergence of this series is that

$$\lim_{l \to \infty} \frac{2|a_{l+1}||u|}{|a_l|} < 1. \tag{3.243}$$

Thus the series is absolutely convergent for

$$|u| < c_0, \tag{3.244}$$

where

$$c_0 = \lim_{l \to \infty} \frac{|a_l|}{2|a_{l+1}|}. \tag{3.245}$$

The convergence of the series (3.241) can be also studied with the aid of the formula (3.226) which, for integer values of l, is valid for any x, y, z, t. If $a_l = a_{lx} + \alpha a_{ly} + \beta a_{lz} + \gamma a_{lt}$, and

$$A_l = a_{lx} + a_{ly} + a_{lz} + a_{lt}, \tag{3.246}$$
$$A_l' = a_{lx} - a_{ly} + a_{lz} - a_{lt}, \tag{3.247}$$
$$A_l'' = a_{lx} + a_{ly} - a_{lz} - a_{lt}, \tag{3.248}$$
$$A_l''' = a_{lx} - a_{ly} - a_{lz} + a_{lt}, \tag{3.249}$$

it can be shown with the aid of relations (3.192) and (3.193) that

$$a_l e = A_l e, \ a_l e' = A_l' e', \ a_l e'' = A_l'' e'', \ a_l e''' = A_l''' e''', \tag{3.250}$$

so that the expression of the series (3.241) becomes

$$\sum_{l=0}^{\infty} \left(A_l s^l e + A_l' s'^l e' + A_l'' s''^l e'' + A_l''' s'''^l e''' \right), \tag{3.251}$$

where the quantities s, s', s'', s''' have been defined in Eq. (3.173). The sufficient conditions for the absolute convergence of the series in Eq. (3.251) are that

$$\lim_{l \to \infty} \frac{|A_{l+1}||s|}{|A_l|} < 1, \lim_{l \to \infty} \frac{|A_{l+1}'||s'|}{|A_l'|} < 1, \lim_{l \to \infty} \frac{|A_{l+1}''||s''|}{|A_l''|} < 1,$$
$$\lim_{l \to \infty} \frac{|A_{l+1}'''||s'''|}{|A_l'''|} < 1, \tag{3.252}$$

Thus the series in Eq. (3.251) is absolutely convergent for

$$|x + y + z + t| < c, \ |x - y + z - t| < c', \ |x + y - z - t| < c'',$$
$$|x - y - z + t| < c''', \tag{3.253}$$

where

$$c = \lim_{l \to \infty} \frac{|A_l|}{|A_{l+1}|}, \ c' = \lim_{l \to \infty} \frac{|A_l'|}{|A_{l+1}'|}, \ c'' = \lim_{l \to \infty} \frac{|A_l''|}{|A_{l+1}''|}, \ c''' = \lim_{l \to \infty} \frac{|A_l'''|}{|A_{l+1}'''|}. \tag{3.254}$$

The relations (3.253) show that the region of convergence of the series (3.251) is a four-dimensional parallelepiped. It can be shown that $c_0 = (1/2)\min(c, c', c'', c''')$, where min designates the smallest of the numbers c, c', c'', c'''. Using Eq. (3.197), it can be seen that the circular region of convergence defined in Eqs. (3.244), (3.245) is included in the parallelogram defined in Eqs. (3.253) and (3.254).

3.2.6 Analytic functions of hyperbolic fourcomplex variables

The fourcomplex function $f(u)$ of the fourcomplex variable u has been expressed in Eq. (3.117) in terms of the real functions $P(x, y, z, t), Q(x, y, z, t)$, $R(x, y, z, t), S(x, y, z, t)$ of real variables x, y, z, t. The relations between the partial derivatives of the functions P, Q, R, S are obtained by setting succesively in Eq. (3.118) $\Delta x \to 0, \Delta y = \Delta z = \Delta t = 0$; then $\Delta y \to 0, \Delta x = \Delta z = \Delta t = 0$; then $\Delta z \to 0, \Delta x = \Delta y = \Delta t = 0$; and finally $\Delta t \to 0, \Delta x = \Delta y = \Delta z = 0$. The relations are

$$\frac{\partial P}{\partial x} = \frac{\partial Q}{\partial y} = \frac{\partial R}{\partial z} = \frac{\partial S}{\partial t}, \tag{3.255}$$

$$\frac{\partial Q}{\partial x} = \frac{\partial P}{\partial y} = \frac{\partial S}{\partial z} = \frac{\partial R}{\partial t}, \tag{3.256}$$

$$\frac{\partial R}{\partial x} = \frac{\partial S}{\partial y} = \frac{\partial P}{\partial z} = \frac{\partial Q}{\partial t}, \tag{3.257}$$

$$\frac{\partial S}{\partial x} = \frac{\partial R}{\partial y} = \frac{\partial Q}{\partial z} = \frac{\partial P}{\partial t}. \tag{3.258}$$

The relations (3.255)-(3.258) are analogous to the Riemann relations for the real and imaginary components of a complex function. It can be shown from Eqs. (3.255)-(3.258) that the component P is a solution of the equations

$$\frac{\partial^2 P}{\partial x^2} - \frac{\partial^2 P}{\partial y^2} = 0, \quad \frac{\partial^2 P}{\partial x^2} - \frac{\partial^2 P}{\partial z^2} = 0, \quad \frac{\partial^2 P}{\partial y^2} - \frac{\partial^2 P}{\partial t^2} = 0,$$

$$\frac{\partial^2 P}{\partial z^2} - \frac{\partial^2 P}{\partial t^2} = 0, \quad \frac{\partial^2 P}{\partial x^2} - \frac{\partial^2 P}{\partial t^2} = 0, \quad \frac{\partial^2 P}{\partial y^2} - \frac{\partial^2 P}{\partial z^2} = 0, \quad (3.259)$$

and the components Q, R, S are solutions of similar equations. As can be seen from Eqs. (3.259), the components P, Q, R, S of an analytic function of hyperbolic fourcomplex variable are solutions of the wave equation with respect to pairs of the variables x, y, z, t. The component P is also a solution of the mixed-derivative equations

$$\frac{\partial^2 P}{\partial x \partial y} = \frac{\partial^2 P}{\partial z \partial t}, \quad \frac{\partial^2 P}{\partial x \partial z} = \frac{\partial^2 P}{\partial y \partial t}, \quad \frac{\partial^2 P}{\partial x \partial t} = \frac{\partial^2 P}{\partial y \partial z}, \quad (3.260)$$

and the components Q, R, S are solutions of similar equations.

3.2.7   Integrals of functions of hyperbolic fourcomplex variables

The singularities of hyperbolic fourcomplex functions arise from terms of the form $1/(u - u_0)^m$, with $m > 0$. Functions containing such terms are singular not only at $u = u_0$, but also at all points of the two-dimensional hyperplanes passing through u_0 and which are parallel to the nodal hyperplanes.

The integral of a hyperbolic fourcomplex function between two points A, B along a path situated in a region free of singularities is independent of path, which means that the integral of an analytic function along a loop situated in a region free from singularities is zero,

$$\oint_\Gamma f(u) du = 0, \quad (3.261)$$

where it is supposed that a surface Σ spanning the closed loop Γ is not intersected by any of the two-dimensional hyperplanes associated with the singularities of the function $f(u)$. Using the expression, Eq. (3.117), for $f(u)$ and the fact that $du = dx + \alpha dy + \beta dz + \gamma dt$, the explicit form of the integral in Eq. (3.261) is

$$\oint_\Gamma f(u) du = \oint_\Gamma [(P dx + Q dy + R dz + S dt)$$
$$+ \alpha(Q dx + P dy + S dz + R dt) + \beta(R dx + S dy + P dz + Q dt)$$
$$+ \gamma(S dx + R dy + Q dz + P dt)]. \quad (3.262)$$

If the functions P, Q, R, S are regular on a surface Σ spanning the loop Γ, the integral along the loop Γ can be transformed with the aid of the theorem of Stokes in an integral over the surface Σ of terms of the form $\partial P/\partial y - \partial Q/\partial x$, $\partial P/\partial z - \partial R/\partial x$, $\partial P/\partial t - \partial S/\partial x$, $\partial Q/\partial z - \partial R/\partial y$, $\partial Q/\partial t - \partial S/\partial y$, $\partial R/\partial t - \partial S/\partial z$ and of similar terms arising from the α, β and γ components, which are equal to zero by Eqs. (3.255)-(3.258), and this proves Eq. (3.261).

The exponential form of the hyperbolic fourcomplex numbers, Eq. (3.215), contains no cyclic variable, and therefore the concept of residue is not applicable to the hyperbolic fourcomplex numbers defined in Eqs. (3.160).

3.2.8 Factorization of hyperbolic fourcomplex polynomials

A polynomial of degree m of the hyperbolic fourcomplex variable $u = x + \alpha y + \beta z + \gamma t$ has the form

$$P_m(u) = u^m + a_1 u^{m-1} + \cdots + a_{m-1} u + a_m, \tag{3.263}$$

where the constants are in general hyperbolic fourcomplex numbers. If $a_m = a_{mx} + \alpha a_{my} + \beta a_{mz} + \gamma a_{mt}$, and with the notations of Eqs. (3.173) and (3.246)-(3.249) applied for $l = 0, 1, \cdots, m$, the polynomial $P_m(u)$ can be written as

$$
\begin{aligned}
P_m = {} & \left[s^m + A_1 s^{m-1} + \cdots + A_{m-1} s + A_m \right] e \\
& + \left[s'^m + A_1' s'^{m-1} + \cdots + A_{m-1}' s' + A_m' \right] e' \\
& + \left[s''^m + A_1'' s''^{m-1} + \cdots + A_{m-1}'' s'' + A_m'' \right] e'' \\
& + \left[s'''^m + A_1''' s'''^{m-1} + \cdots + A_{m-1}''' s''' + A_m''' \right] e'''. \tag{3.264}
\end{aligned}
$$

Each of the polynomials of degree m with real coefficients in Eq. (3.264) can be written as a product of linear or quadratic factors with real coefficients, or as a product of linear factors which, if imaginary, appear always in complex conjugate pairs. Using the latter form for the simplicity of notations, the polynomial P_m can be written as

$$P_m = \prod_{l=1}^{m}(s - s_l)e + \prod_{l=1}^{m}(s' - s_l')e' + \prod_{l=1}^{m}(s'' - s_l'')e'' + \prod_{l=1}^{m}(s''' - s_l''')e''', \tag{3.265}$$

where the quantities s_l appear always in complex conjugate pairs, and the same is true for the quantities s_l', for the quantities s_l'', and for the quantities s_l'''. Due to the properties in Eqs. (3.192) and (3.193), the polynomial $P_m(u)$ can be written as a product of factors of the form

$$P_m(u) = \prod_{l=1}^{m} \left[(s - s_l)e + (s' - s_l')e' + (s'' - s_l'')e'' + (s''' - s_l''')e''' \right].$$

(3.266)

These relations can be written with the aid of Eq. (3.194) as

$$P_m(u) = \prod_{p=1}^{m} (u - u_p),$$

(3.267)

where

$$u_p = s_p e + s_p' e' + s_p'' e'' + s_p''' e'''.$$

(3.268)

The roots s_p, s_p', s_p'', s_p''' of the corresponding polynomials in Eq. (3.265) may be ordered arbitrarily. This means that Eq. (3.268) gives sets of m roots $u_1, ..., u_m$ of the polynomial $P_m(u)$, corresponding to the various ways in which the roots s_p, s_p', s_p'', s_p''' are ordered according to p in each group. Thus, while the hypercomplex components in Eq. (3.264) taken separately have unique factorizations, the polynomial $P_m(u)$ can be written in many different ways as a product of linear factors.

If $P(u) = u^2 - 1$, the factorization in Eq. (3.267) is $u^2 - 1 = (u - u_1)(u - u_2)$, where $u_1 = \pm e \pm e' \pm e'' \pm e'''$, $u_2 = -u_1$, so that there are 8 distinct factorizations of $u^2 - 1$,

$$
\begin{aligned}
u^2 - 1 &= (u - e - e' - e'' - e''')(u + e + e' + e'' + e'''), \\
u^2 - 1 &= (u - e - e' - e'' + e''')(u + e + e' + e'' - e'''), \\
u^2 - 1 &= (u - e - e' + e'' - e''')(u + e + e' - e'' + e'''), \\
u^2 - 1 &= (u - e + e' - e'' - e''')(u + e - e' + e'' + e'''), \\
u^2 - 1 &= (u - e - e' + e'' + e''')(u + e + e' - e'' - e'''), \\
u^2 - 1 &= (u - e + e' - e'' + e''')(u + e - e' + e'' - e'''), \\
u^2 - 1 &= (u - e + e' + e'' - e''')(u + e - e' - e'' + e'''), \\
u^2 - 1 &= (u - e + e' + e'' + e''')(u + e - e' - e'' - e''').
\end{aligned}
$$

(3.269)

It can be checked that $\{\pm e \pm e' \pm e'' \pm e'''\}^2 = e + e' + e'' + e''' = 1$.

3.2.9 Representation of hyperbolic fourcomplex numbers by irreducible matrices

If T is the unitary matrix,

$$T = \begin{pmatrix} \frac{1}{2} & \frac{1}{2} & \frac{1}{2} & \frac{1}{2} \\ \frac{1}{2} & -\frac{1}{2} & \frac{1}{2} & -\frac{1}{2} \\ \frac{1}{2} & \frac{1}{2} & -\frac{1}{2} & -\frac{1}{2} \\ \frac{1}{2} & -\frac{1}{2} & -\frac{1}{2} & \frac{1}{2} \end{pmatrix},$$

(3.270)

it can be shown that the matrix TUT^{-1} has the form

$$TUT^{-1}$$
$$= \begin{pmatrix} x+y+z+t & 0 & 0 & 0 \\ 0 & x-y+z-t & 0 & 0 \\ 0 & 0 & x+y-z-t & 0 \\ 0 & 0 & 0 & x-y-z+t \end{pmatrix},$$

(3.271)

where U is the matrix in Eq. (3.201) used to represent the hyperbolic fourcomplex number u. The relations between the variables $x+y+z+t, x-y+z-t, x+y-z-t, x-y-z+t$ for the multiplication of hyperbolic fourcomplex numbers have been written in Eqs. (3.185)-(3.188). The matrix TUT^{-1} provides an irreducible representation [7] of the hyperbolic fourcomplex number u in terms of matrices with real coefficients.

3.3 Planar complex numbers in four dimensions

3.3.1 Operations with planar fourcomplex numbers

A planar fourcomplex number is determined by its four components (x, y, z, t). The sum of the planar fourcomplex numbers (x, y, z, t) and (x', y', z', t') is the planar fourcomplex number $(x+x', y+y', z+z', t+t')$. The product of the planar fourcomplex numbers (x, y, z, t) and (x', y', z', t') is defined in this section to be the planar fourcomplex number $(xx' - yt' - zz' - ty', xy' + yx' - zt' - tz', xz' + yy' + zx' - tt', xt' + yz' + zy' + tx')$.

Planar fourcomplex numbers and their operations can be represented by writing the planar fourcomplex number (x, y, z, t) as $u = x + \alpha y + \beta z + \gamma t$, where α, β and γ are bases for which the multiplication rules are

$$\alpha^2 = \beta, \ \beta^2 = -1, \ \gamma^2 = -\beta,$$
$$\alpha\beta = \beta\alpha = \gamma, \ \alpha\gamma = \gamma\alpha = -1, \ \beta\gamma = \gamma\beta = -\alpha.$$

(3.272)

Two planar fourcomplex numbers $u = x + \alpha y + \beta z + \gamma t, u' = x' + \alpha y' + \beta z' + \gamma t'$ are equal, $u = u'$, if and only if $x = x', y = y', z = z', t = t'$. If $u = x + \alpha y + \beta z + \gamma t, u' = x' + \alpha y' + \beta z' + \gamma t'$ are planar fourcomplex numbers, the sum $u + u'$ and the product uu' defined above can be obtained by applying the usual algebraic rules to the sum $(x + \alpha y + \beta z + \gamma t) + (x' + \alpha y' + \beta z' + \gamma t')$ and to the product $(x + \alpha y + \beta z + \gamma t)(x' + \alpha y' + \beta z' + \gamma t')$, and grouping of the resulting terms,

$$u + u' = x + x' + \alpha(y + y') + \beta(z + z') + \gamma(t + t'),$$

(3.273)

$$uu' = xx' - yt' - zz' - ty' + \alpha(xy' + yx' - zt' - tz')$$
$$+\beta(xz' + yy' + zx' - tt') + \gamma(xt' + yz' + zy' + tx'). \qquad (3.274)$$

If u, u', u'' are planar fourcomplex numbers, the multiplication is associative

$$(uu')u'' = u(u'u'') \qquad (3.275)$$

and commutative

$$uu' = u'u, \qquad (3.276)$$

as can be checked through direct calculation. The planar fourcomplex zero is $0 + \alpha \cdot 0 + \beta \cdot 0 + \gamma \cdot 0$, denoted simply 0, and the planar fourcomplex unity is $1 + \alpha \cdot 0 + \beta \cdot 0 + \gamma \cdot 0$, denoted simply 1.

The inverse of the planar fourcomplex number $u = x + \alpha y + \beta z + \gamma t$ is a planar fourcomplex number $u' = x' + \alpha y' + \beta z' + \gamma t'$ having the property that

$$uu' = 1. \qquad (3.277)$$

Written on components, the condition, Eq. (3.277), is

$$\begin{aligned} xx' - ty' - zz' - yt' &= 1, \\ yx' + xy' - tz' - zt' &= 0, \\ zx' + yy' + xz' - tt' &= 0, \\ tx' + zy' + yz' + xt' &= 0. \end{aligned} \qquad (3.278)$$

The system (3.278) has the solution

$$x' = \frac{x(x^2 + z^2) - z(y^2 - t^2) + 2xyt}{\rho^4}, \qquad (3.279)$$

$$y' = -\frac{y(x^2 - z^2) + t(y^2 + t^2) + 2xzt}{\rho^4}, \qquad (3.280)$$

$$z' = \frac{-z(x^2 + z^2) + x(y^2 - t^2) + 2zyt}{\rho^4}, \qquad (3.281)$$

$$t' = -\frac{t(x^2 - z^2) + y(y^2 + t^2) - 2xyz}{\rho^4}, \qquad (3.282)$$

provided that $\rho \neq 0$, where

$$\rho^4 = x^4 + z^4 + y^4 + t^4$$
$$+2x^2z^2 + 2y^2t^2 + 4x^2yt - 4xy^2z + 4xzt^2 - 4yz^2t. \qquad (3.283)$$

The quantity ρ will be called amplitude of the planar fourcomplex number $x + \alpha y + \beta z + \gamma t$. Since

$$\rho^4 = \rho_+^2 \rho_-^2,\qquad(3.284)$$

where

$$\rho_+^2 = \left(x + \frac{y-t}{\sqrt{2}}\right)^2 + \left(z + \frac{y+t}{\sqrt{2}}\right)^2,$$
$$\rho_-^2 = \left(x - \frac{y-t}{\sqrt{2}}\right)^2 + \left(z - \frac{y+t}{\sqrt{2}}\right)^2,\qquad(3.285)$$

a planar fourcomplex number $u = x + \alpha y + \beta z + \gamma t$ has an inverse, unless

$$x + \frac{y-t}{\sqrt{2}} = 0,\ z + \frac{y+t}{\sqrt{2}} = 0,\qquad(3.286)$$

or

$$x - \frac{y-t}{\sqrt{2}} = 0,\ z - \frac{y+t}{\sqrt{2}} = 0.\qquad(3.287)$$

Because of conditions (3.286)-(3.287) these 2-dimensional hypersurfaces will be called nodal hyperplanes. It can be shown that if $uu' = 0$ then either $u = 0$, or $u' = 0$, or one of the planar fourcomplex numbers is of the form $x + \alpha(x + z)/\sqrt{2} + \beta z - \gamma(x - z)/\sqrt{2}$ and the other of the form $x' - \alpha(x' + z')/\sqrt{2} + \beta z' + \gamma(x' - z')/\sqrt{2}$.

3.3.2 Geometric representation of planar fourcomplex numbers

The planar fourcomplex number $x + \alpha y + \beta z + \gamma t$ can be represented by the point A of coordinates (x, y, z, t). If O is the origin of the four-dimensional space x, y, z, t, the distance from A to the origin O can be taken as

$$d^2 = x^2 + y^2 + z^2 + t^2.\qquad(3.288)$$

The distance d will be called modulus of the planar fourcomplex number $x + \alpha y + \beta z + \gamma t$, $d = |u|$. The orientation in the four-dimensional space of the line OA can be specified with the aid of three angles ϕ, χ, ψ defined with respect to the rotated system of axes

$$\xi = \frac{x}{\sqrt{2}} + \frac{y-t}{2},\ \tau = \frac{x}{\sqrt{2}} - \frac{y-t}{2},\ \upsilon = \frac{z}{\sqrt{2}} + \frac{y+t}{2},$$
$$\zeta = -\frac{z}{\sqrt{2}} + \frac{y+t}{2}.\qquad(3.289)$$

The variables $\xi, \upsilon, \tau, \zeta$ will be called canonical planar fourcomplex variables. The use of the rotated axes $\xi, \upsilon, \tau, \zeta$ for the definition of the angles ϕ, χ, ψ is convenient for the expression of the planar fourcomplex numbers in exponential and trigonometric forms, as it will be discussed further. The angle ϕ is the angle between the projection of A in the plane ξ, υ and the $O\xi$ axis, $0 \leq \phi < 2\pi$, χ is the angle between the projection of A in the plane τ, ζ and the $O\tau$ axis, $0 \leq \chi < 2\pi$, and ψ is the angle between the line OA and the plane $\tau O\zeta$, $0 \leq \psi \leq \pi/2$, as shown in Fig. 3.1. The definition of the variables in this section is different from the definition used for the circular fourcomplex numbers, because the definition of the rotated axes in Eq. (3.289) is different from the definition of the rotated circular axes, Eq. (3.18). The angles ϕ and χ will be called azimuthal angles, the angle ψ will be called planar angle. The fact that $0 \leq \psi \leq \pi/2$ means that ψ has the same sign on both faces of the two-dimensional hyperplane $\upsilon O\zeta$. The components of the point A in terms of the distance d and the angles ϕ, χ, ψ are thus

$$\frac{x}{\sqrt{2}} + \frac{y-t}{2} = d\cos\phi\sin\psi, \tag{3.290}$$

$$\frac{x}{\sqrt{2}} - \frac{y-t}{2} = d\cos\chi\cos\psi, \tag{3.291}$$

$$\frac{z}{\sqrt{2}} + \frac{y+t}{2} = d\sin\phi\sin\psi, \tag{3.292}$$

$$-\frac{z}{\sqrt{2}} + \frac{y+t}{2} = d\sin\chi\cos\psi. \tag{3.293}$$

It can be checked that $\rho_+ = \sqrt{2}d\sin\psi, \rho_- = \sqrt{2}d\cos\psi$. The coordinates x, y, z, t in terms of the variables d, ϕ, χ, ψ are

$$x = \frac{d}{\sqrt{2}}(\cos\phi\sin\psi + \cos\chi\cos\psi), \tag{3.294}$$

$$y = \frac{d}{\sqrt{2}}[\sin(\phi + \pi/4)\sin\psi + \sin(\chi - \pi/4)\cos\psi], \tag{3.295}$$

$$z = \frac{d}{\sqrt{2}}(\sin\phi\sin\psi - \sin\chi\cos\psi), \tag{3.296}$$

$$t = \frac{d}{\sqrt{2}}[-\cos(\phi + \pi/4)\sin\psi + \cos(\chi - \pi/4)\cos\psi]. \tag{3.297}$$

The angles ϕ, χ, ψ can be expressed in terms of the coordinates x, y, z, t as

$$\sin\phi = \frac{z + (y + t)/\sqrt{2}}{\rho_+}, \quad \cos\phi = \frac{x + (y - t)/\sqrt{2}}{\rho_+}, \tag{3.298}$$

$$\sin\chi = \frac{-z + (y + t)/\sqrt{2}}{\rho_-}, \quad \cos\chi = \frac{x - (y - t)/\sqrt{2}}{\rho_-}, \tag{3.299}$$

$$\tan\psi = \rho_+/\rho_-. \tag{3.300}$$

The nodal hyperplanes are $\xi O v$, for which $\tau = 0, \zeta = 0$, and $\tau O \zeta$, for which $\xi = 0, v = 0$. For points in the nodal hyperplane $\xi O v$ the planar angle is $\psi = \pi/2$, for points in the nodal hyperplane $\tau O \zeta$ the planar angle is $\psi = 0$.

It can be shown that if $u_1 = x_1 + \alpha y_1 + \beta z_1 + \gamma t_1$, $u_2 = x_2 + \alpha y_2 + \beta z_2 + \gamma t_2$ are planar fourcomplex numbers of amplitudes and angles $\rho_1, \phi_1, \chi_1, \psi_1$ and respectively $\rho_2, \phi_2, \chi_2, \psi_2$, then the amplitude ρ and the angles ϕ, χ, ψ of the product planar fourcomplex number $u_1 u_2$ are

$$\rho = \rho_1 \rho_2, \tag{3.301}$$

$$\phi = \phi_1 + \phi_2, \quad \chi = \chi_1 + \chi_2, \quad \tan\psi = \tan\psi_1 \tan\psi_2. \tag{3.302}$$

The relations (3.301)-(3.302) are consequences of the definitions (3.283)-(3.285), (3.298)-(3.300) and of the identities

$$\left[(x_1 x_2 - z_1 z_2 - y_1 t_2 - t_1 y_2) + \frac{(x_1 y_2 + y_1 x_2 - z_1 t_2 - t_1 z_2)}{\sqrt{2}} \right.$$
$$\left. - \frac{(x_1 t_2 + t_1 x_2 + z_1 y_2 + y_1 z_2)}{\sqrt{2}} \right]^2$$
$$+ \left[(x_1 z_2 + z_1 x_2 + y_1 y_2 - t_1 t_2) + \frac{(x_1 y_2 + y_1 x_2 - z_1 t_2 - t_1 z_2)}{\sqrt{2}} \right.$$
$$\left. + \frac{(x_1 t_2 + t_1 x_2 + z_1 y_2 + y_1 z_2)}{\sqrt{2}} \right]^2$$
$$= \left[\left(x_1 + \frac{y_1 - t_1}{\sqrt{2}} \right)^2 + \left(z_1 + \frac{y_1 + t_1}{\sqrt{2}} \right)^2 \right]$$
$$\left[\left(x_2 + \frac{y_2 - t_2}{\sqrt{2}} \right)^2 + \left(z_2 + \frac{y_2 + t_2}{\sqrt{2}} \right)^2 \right], \tag{3.303}$$

$$\left[(x_1 x_2 - z_1 z_2 - y_1 t_2 - t_1 y_2) - \frac{(x_1 y_2 + y_1 x_2 - z_1 t_2 - t_1 z_2)}{\sqrt{2}} \right.$$

$$\left. - \frac{(x_1 t_2 + t_1 x_2 + z_1 y_2 + y_1 z_2)}{\sqrt{2}} \right]^2$$

$$+ \left[(x_1 z_2 + z_1 x_2 + y_1 y_2 - t_1 t_2) - \frac{(x_1 y_2 + y_1 x_2 - z_1 t_2 - t_1 z_2)}{\sqrt{2}} \right.$$

$$\left. + \frac{(x_1 t_2 + t_1 x_2 + z_1 y_2 + y_1 z_2)}{\sqrt{2}} \right]^2$$

$$= \left[\left(x_1 - \frac{y_1 - t_1}{\sqrt{2}} \right)^2 + \left(z_1 - \frac{y_1 + t_1}{\sqrt{2}} \right)^2 \right]$$

$$\left[\left(x_2 - \frac{y_2 - t_2}{\sqrt{2}} \right)^2 + \left(z_2 - \frac{y_2 + t_2}{\sqrt{2}} \right)^2 \right], \tag{3.304}$$

$$(x_1 x_2 - z_1 z_2 - y_1 t_2 - t_1 y_2)$$

$$+ \frac{(x_1 y_2 + y_1 x_2 - z_1 t_2 - t_1 z_2) - (x_1 t_2 + t_1 x_2 + z_1 y_2 + y_1 z_2)}{\sqrt{2}}$$

$$= \left(x_1 + \frac{y_1 - t_1}{\sqrt{2}} \right) \left(x_2 + \frac{y_2 - t_2}{\sqrt{2}} \right)$$

$$- \left(z_1 + \frac{y_1 + t_1}{\sqrt{2}} \right) \left(z_2 + \frac{y_2 + t_2}{\sqrt{2}} \right), \tag{3.305}$$

$$(x_1 z_2 + z_1 x_2 + y_1 y_2 - t_1 t_2)$$

$$+ \frac{(x_1 y_2 + y_1 x_2 - z_1 t_2 - t_1 z_2) + (x_1 t_2 + t_1 x_2 + z_1 y_2 + y_1 z_2)}{\sqrt{2}}$$

$$= \left(z_1 + \frac{y_1 + t_1}{\sqrt{2}} \right) \left(x_2 + \frac{y_2 - t_2}{\sqrt{2}} \right)$$

$$+ \left(x_1 + \frac{y_1 - t_1}{\sqrt{2}} \right) \left(z_2 + \frac{y_2 + t_2}{\sqrt{2}} \right), \tag{3.306}$$

$$(x_1 x_2 - z_1 z_2 - y_1 t_2 - t_1 y_2)$$

$$- \frac{(x_1 y_2 + y_1 x_2 - z_1 t_2 - t_1 z_2) - (x_1 t_2 + t_1 x_2 + z_1 y_2 + y_1 z_2)}{\sqrt{2}}$$

$$= \left(x_1 - \frac{y_1 - t_1}{\sqrt{2}} \right) \left(x_2 - \frac{y_2 - t_2}{\sqrt{2}} \right)$$

$$- \left(-z_1 + \frac{y_1 + t_1}{\sqrt{2}} \right) \left(-z_2 + \frac{y_2 + t_2}{\sqrt{2}} \right), \tag{3.307}$$

$$-(x_1 z_2 + z_1 x_2 + y_1 y_2 - t_1 t_2)$$

$$+ \frac{(x_1 y_2 + y_1 x_2 - z_1 t_2 - t_1 z_2) + (x_1 t_2 + t_1 x_2 + z_1 y_2 + y_1 z_2)}{\sqrt{2}}$$

$$= \left(-z_1 + \frac{y_1 + t_1}{\sqrt{2}} \right) \left(x_2 - \frac{y_2 - t_2}{\sqrt{2}} \right)$$
$$+ \left(x_1 - \frac{y_1 - t_1}{\sqrt{2}} \right) \left(-z_2 + \frac{y_2 + t_2}{\sqrt{2}} \right). \tag{3.308}$$

The identities (3.303) and (3.304) can also be written as

$$\rho_+^2 = \rho_{1+}\rho_{2+}, \tag{3.309}$$

$$\rho_-^2 = \rho_{1-}\rho_{2-}, \tag{3.310}$$

where

$$\rho_{j+}^2 = \left(x_j + \frac{y_j - t_j}{\sqrt{2}} \right)^2 + \left(z_j + \frac{y_j + t_j}{\sqrt{2}} \right)^2,$$
$$\rho_{j-}^2 = \left(x_j - \frac{y_j - t_j}{\sqrt{2}} \right)^2 + \left(z - \frac{y_j + t_j}{\sqrt{2}} \right)^2, \tag{3.311}$$

for $j = 1, 2$.

The fact that the amplitude of the product is equal to the product of the amplitudes, as written in Eq. (3.301), can be demonstrated also by using a representation of the multiplication of the planar fourcomplex numbers by matrices, in which the planar fourcomplex number $u = x + \alpha y + \beta z + \gamma t$ is represented by the matrix

$$A = \begin{pmatrix} x & y & z & t \\ -t & x & y & z \\ -z & -t & x & y \\ -y & -z & -t & x \end{pmatrix}. \tag{3.312}$$

The product $u = x + \alpha y + \beta z + \gamma t$ of the planar fourcomplex numbers $u_1 = x_1 + \alpha y_1 + \beta z_1 + \gamma t_1, u_2 = x_2 + \alpha y_2 + \beta z_2 + \gamma t_2$, can be represented by the matrix multiplication

$$A = A_1 A_2. \tag{3.313}$$

It can be checked that the determinant $\det(A)$ of the matrix A is

$$\det A = \rho^4. \tag{3.314}$$

The identity (3.301) is then a consequence of the fact the determinant of the product of matrices is equal to the product of the determinants of the factor matrices.

3.3.3   The planar fourdimensional cosexponential functions

The exponential function of a hypercomplex variable u and the addition theorem for the exponential function have been written in Eqs. (1.35)-(1.36). If $u = x + \alpha y + \beta z + \gamma t$, then $\exp u$ can be calculated as $\exp u = \exp x \cdot \exp(\alpha y) \cdot \exp(\beta z) \cdot \exp(\gamma t)$. According to Eq. (3.272),

$$
\begin{aligned}
&\alpha^{8m} = 1, \alpha^{8m+1} = \alpha, \alpha^{8m+2} = \beta, \alpha^{8m+3} = \gamma, \\
&\alpha^{8m+4} = -1, \alpha^{8m+5} = -\alpha, \alpha^{8m+6} = -\beta, \alpha^{8m+7} = -\gamma, \\
&\beta^{4m} = 1, \beta^{4m+1} = \beta, \beta^{4m+2} = -1, \beta^{4m+3} = -\beta, \\
&\gamma^{8m} = 1, \gamma^{8m+1} = \gamma, \gamma^{8m+2} = -\beta, \gamma^{8m+3} = \alpha, \\
&\gamma^{8m+4} = -1, \gamma^{8m+5} = -\gamma, \gamma^{8m+6} = \beta, \gamma^{8m+7} = -\alpha,
\end{aligned}
\tag{3.315}
$$

where n is a natural number, so that $\exp(\alpha y)$, $\exp(\beta z)$ and $\exp(\gamma t)$ can be written as

$$
\exp(\beta z) = \cos z + \beta \sin z,
\tag{3.316}
$$

and

$$
\exp(\alpha y) = f_{40}(y) + \alpha f_{41}(y) + \beta f_{42}(y) + \gamma f_{43}(y),
\tag{3.317}
$$

$$
\exp(\gamma t) = f_{40}(t) + \gamma f_{41}(t) - \beta f_{42}(t) + \alpha f_{43}(t),
\tag{3.318}
$$

where the four-dimensional cosexponential functions $f_{40}, f_{41}, f_{42}, f_{43}$ are defined by the series

$$
f_{40}(x) = 1 - x^4/4! + x^8/8! - \cdots,
\tag{3.319}
$$

$$
f_{41}(x) = x - x^5/5! + x^9/9! - \cdots,
\tag{3.320}
$$

$$
f_{42}(x) = x^2/2! - x^6/6! + x^{10}/10! - \cdots,
\tag{3.321}
$$

$$
f_{43}(x) = x^3/3! - x^7/7! + x^{11}/11! - \cdots.
\tag{3.322}
$$

The functions f_{40}, f_{42} are even, the functions f_{41}, f_{43} are odd,

$$
\begin{aligned}
&f_{40}(-u) = f_{40}(u), \ f_{42}(-u) = f_{42}(u), \ f_{41}(-u) = -f_{41}(u), \\
&f_{43}(-u) = -f_{43}(u).
\end{aligned}
\tag{3.323}
$$

Addition theorems for the four-dimensional cosexponential functions can be obtained from the relation $\exp \alpha(x + y) = \exp \alpha x \cdot \exp \alpha y$, by substituting the expression of the exponentials as given in Eq. (3.317),

$$
\begin{aligned}
f_{40}(x + y) = &f_{40}(x)f_{40}(y) - f_{41}(x)f_{43}(y) - f_{42}(x)f_{42}(y) \\
&- f_{43}(x)f_{41}(y),
\end{aligned}
\tag{3.324}
$$

$$f_{41}(x + y) = f_{40}(x)f_{41}(y) + f_{41}(x)f_{40}(y) - f_{42}(x)f_{43}(y)$$
$$- f_{43}(x)f_{42}(y), \tag{3.325}$$

$$f_{42}(x + y) = f_{40}(x)f_{42}(y) + f_{41}(x)f_{41}(y) + f_{42}(x)f_{40}(y)$$
$$- f_{43}(x)f_{43}(y), \tag{3.326}$$

$$f_{43}(x + y) = f_{40}(x)f_{43}(y) + f_{41}(x)f_{42}(y) + f_{42}(x)f_{41}(y)$$
$$+ f_{43}(x)f_{40}(y). \tag{3.327}$$

For $x = y$ the relations (3.324)-(3.327) take the form

$$f_{40}(2x) = f_{40}^2(x) - f_{42}^2(x) - 2f_{41}(x)f_{43}(x), \tag{3.328}$$

$$f_{41}(2x) = 2f_{40}(x)f_{41}(x) - 2f_{42}(x)f_{43}(x), \tag{3.329}$$

$$f_{42}(2x) = f_{41}^2(x) - f_{43}^2(x) + 2f_{40}(x)f_{42}(x), \tag{3.330}$$

$$f_{43}(2x) = 2f_{40}(x)f_{43}(x) + 2f_{41}(x)f_{42}(x). \tag{3.331}$$

For $x = -y$ the relations (3.324)-(3.327) and (3.323) yield

$$f_{40}^2(x) - f_{42}^2(x) + 2f_{41}(x)f_{43}(x) = 1, \tag{3.332}$$

$$f_{41}^2(x) - f_{43}^2(x) - 2f_{40}(x)f_{42}(x) = 0. \tag{3.333}$$

From Eqs. (3.316)-(3.318) it can be shown that, for m integer,

$$(\cos z + \beta \sin z)^m = \cos mz + \beta \sin mz, \tag{3.334}$$

and

$$[f_{40}(y) + \alpha f_{41}(y) + \beta f_{42}(y) + \gamma f_{43}(y)]^m$$
$$= f_{40}(my) + \alpha f_{41}(my) + \beta f_{42}(my) + \gamma f_{43}(my), \tag{3.335}$$

$$[f_{40}(t) + \gamma f_{41}(t) - \beta f_{42}(t) + \alpha f_{43}(t)]^m$$
$$= f_{40}(mt) + \gamma f_{41}(mt) - \beta f_{42}(mt) + \alpha f_{43}(mt). \tag{3.336}$$

Since

$$(\alpha - \gamma)^{2m} = 2^m, \ (\alpha - \gamma)^{2m+1} = 2^m(\alpha - \gamma), \tag{3.337}$$

it can be shown from the definition of the exponential function, Eq. (1.35) that

$$\exp(\alpha - \gamma)x = \cosh \sqrt{2}x + \frac{\alpha - \gamma}{\sqrt{2}} \sinh \sqrt{2}x. \tag{3.338}$$

Substituting in the relation $\exp(\alpha - \gamma)x = \exp \alpha x \exp(-\gamma x)$ the expression of the exponentials from Eqs. (3.317), (3.318) and (3.338) yields

$$f_{40}^2 + f_{41}^2 + f_{42}^2 + f_{43}^2 = \cosh \sqrt{2}x, \tag{3.339}$$

$$f_{40}f_{41} - f_{40}f_{43} + f_{41}f_{42} + f_{42}f_{43} = \frac{1}{\sqrt{2}}\sinh \sqrt{2}x, \tag{3.340}$$

where $f_{40}, f_{41}, f_{42}, f_{43}$ are functions of x. From relations (3.339) and (3.340) it can be inferred that

$$\left(f_{40} + \frac{f_{41} - f_{43}}{\sqrt{2}}\right)^2 + \left(f_{42} + \frac{f_{41} + f_{43}}{\sqrt{2}}\right)^2 = \exp \sqrt{2}x, \tag{3.341}$$

$$\left(f_{40} - \frac{f_{41} - f_{43}}{\sqrt{2}}\right)^2 + \left(f_{42} - \frac{f_{41} + f_{43}}{\sqrt{2}}\right)^2 = \exp(-\sqrt{2}x), \tag{3.342}$$

which means that

$$\left[\left(f_{40} + \frac{f_{41} - f_{43}}{\sqrt{2}}\right)^2 + \left(f_{42} + \frac{f_{41} + f_{43}}{\sqrt{2}}\right)^2\right]$$
$$\left[\left(f_{40} - \frac{f_{41} - f_{43}}{\sqrt{2}}\right)^2 + \left(f_{42} - \frac{f_{41} + f_{43}}{\sqrt{2}}\right)^2\right] = 1. \tag{3.343}$$

An equivalent form of Eq. (3.343) is

$$f_{40}^4 + f_{41}^4 + f_{42}^4 + f_{43}^4 + 2(f_{40}^2 f_{42}^2 + f_{41}^2 f_{43}^2)$$
$$+ 4(f_{40}^2 f_{41} f_{43} + f_{40} f_{42} f_{43}^2 - f_{40} f_{41}^2 f_{42} - f_{41} f_{42}^2 f_{43}) = 1. \tag{3.344}$$

The form of this relation is similar to the expression in Eq. (3.283). Similarly, since

$$(\alpha + \gamma)^{2m} = (-1)^m 2^m, \quad (\alpha + \gamma)^{2m+1} = (-1)^m 2^m (\alpha + \gamma), \tag{3.345}$$

it can be shown from the definition of the exponential function, Eq. (1.35) that

$$\exp(\alpha + \gamma)x = \cos \sqrt{2}x + \frac{\alpha + \gamma}{\sqrt{2}}\sin \sqrt{2}x. \tag{3.346}$$

Substituting in the relation $\exp(\alpha + \gamma)x = \exp \alpha x \exp \gamma x$ the expression of the exponentials from Eqs. (3.317), (3.318) and (3.346) yields

$$f_{40}^2 - f_{41}^2 + f_{42}^2 - f_{43}^2 = \cos \sqrt{2}x, \tag{3.347}$$

$$f_{40}f_{41} + f_{40}f_{43} - f_{41}f_{42} + f_{42}f_{43} = \frac{1}{\sqrt{2}}\sin \sqrt{2}x, \tag{3.348}$$

where $f_{40}, f_{41}, f_{42}, f_{43}$ are functions of x.

Expressions of the four-dimensional cosexponential functions (3.319)-(3.322) can be obtained using the fact that $[(1 + i)/\sqrt{2}]^4 = -1$, so that

$$f_{40}(x) = \frac{1}{2} \left(\cosh \frac{1+i}{\sqrt{2}} x + \cos \frac{1+i}{\sqrt{2}} x \right), \tag{3.349}$$

$$f_{41}(x) = \frac{1}{\sqrt{2}(1+i)} \left(\sinh \frac{1+i}{\sqrt{2}} x + \sin \frac{1+i}{\sqrt{2}} x \right), \tag{3.350}$$

$$f_{42}(x) = \frac{1}{2i} \left(\cosh \frac{1+i}{\sqrt{2}} x - \cos \frac{1+i}{\sqrt{2}} x \right), \tag{3.351}$$

$$f_{43}(x) = \frac{1}{\sqrt{2}(-1+i)} \left(\sinh \frac{1+i}{\sqrt{2}} x - \sin \frac{1+i}{\sqrt{2}} x \right). \tag{3.352}$$

Using the addition theorems for the functions in the right-hand sides of Eqs. (3.349)-(3.352), the expressions of the four-dimensional cosexponential functions become

$$f_{40}(x) = \cos \frac{x}{\sqrt{2}} \cosh \frac{x}{\sqrt{2}}, \tag{3.353}$$

$$f_{41}(x) = \frac{1}{\sqrt{2}} \left(\sin \frac{x}{\sqrt{2}} \cosh \frac{x}{\sqrt{2}} + \sinh \frac{x}{\sqrt{2}} \cos \frac{x}{\sqrt{2}} \right), \tag{3.354}$$

$$f_{42}(x) = \sin \frac{x}{\sqrt{2}} \sinh \frac{x}{\sqrt{2}}, \tag{3.355}$$

$$f_{43}(x) = \frac{1}{\sqrt{2}} \left(\sin \frac{x}{\sqrt{2}} \cosh \frac{x}{\sqrt{2}} - \sinh \frac{x}{\sqrt{2}} \cos \frac{x}{\sqrt{2}} \right). \tag{3.356}$$

It is remarkable that the series in Eqs. (3.353)-(3.356), in which the terms are either of the form x^{4m}, or x^{4m+1}, or x^{4m+2}, or x^{4m+3} can be expressed in terms of elementary functions whose power series are not subject to such restrictions. The graphs of the four-dimensional cosexponential functions are shown in Fig. 3.3.

It can be checked that the cosexponential functions are solutions of the fourth-order differential equation

$$\frac{d^4\zeta}{du^4} = -\zeta, \tag{3.357}$$

whose solutions are of the form $\zeta(u) = A f_{40}(u) + B f_{41}(u) + C f_{42}(u) + D f_{43}(u)$. It can also be checked that the derivatives of the cosexponential functions are related by

$$\frac{df_{40}}{du} = -f_{43}, \quad \frac{df_{41}}{du} = f_{40}, \quad \frac{df_{42}}{du} = f_{41}, \quad \frac{df_{43}}{du} = f_{42}. \tag{3.358}$$

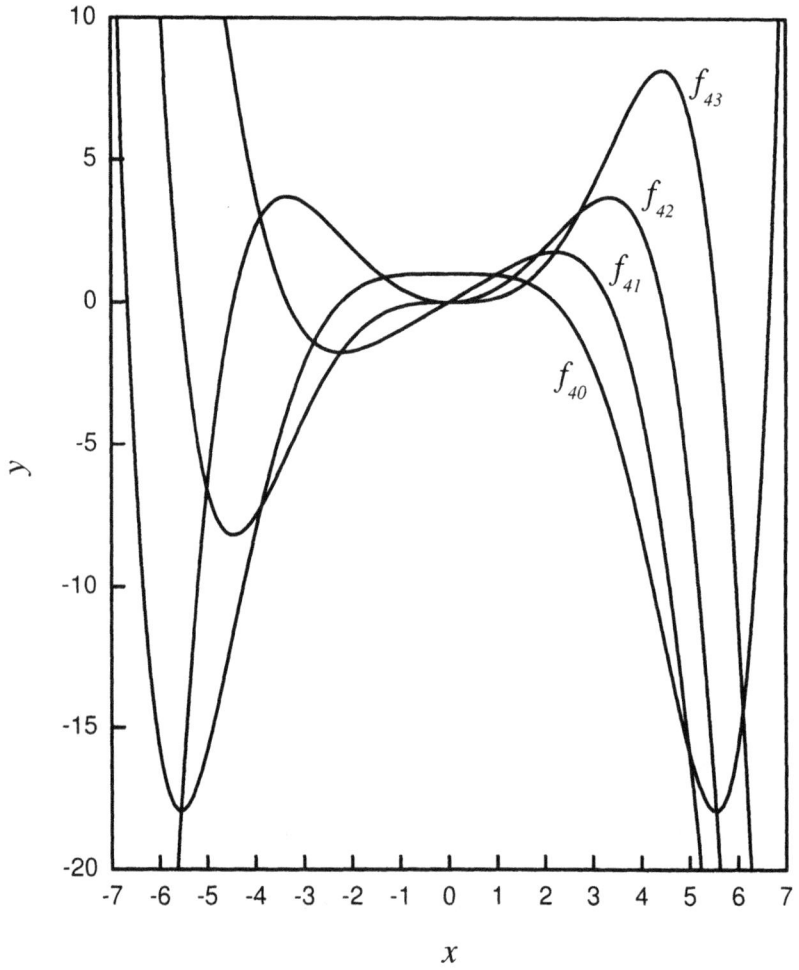

Figure 3.3:   The planar fourdimensional cosexponential functions $f_{40}, f_{41}, f_{42}, f_{43}$.

3.3.4 The exponential and trigonometric forms of planar fourcomplex numbers

Any planar fourcomplex number $u = x + \alpha y + \beta z + \gamma t$ can be writen in the form

$$x + \alpha y + \beta z + \gamma t = e^{x_1 + \alpha y_1 + \beta z_1 + \gamma t_1}. \tag{3.359}$$

The expressions of x_1, y_1, z_1, t_1 as functions of x, y, z, t can be obtained by developing $e^{\alpha y_1}, e^{\beta z_1}$ and $e^{\gamma t_1}$ with the aid of Eqs. (3.316)-(3.318), by multiplying these expressions and separating the hypercomplex components, and then substituting the expressions of the four-dimensional cosexponential functions, Eqs. (3.353)-(3.356),

$$
\begin{aligned}
x = e^{x_1} \Bigg(& \cos z_1 \cos \frac{y_1 + t_1}{\sqrt{2}} \cosh \frac{y_1 - t_1}{\sqrt{2}} \\
& - \sin z_1 \sin \frac{y_1 + t_1}{\sqrt{2}} \sinh \frac{y_1 - t_1}{\sqrt{2}} \Bigg),
\end{aligned}
\tag{3.360}
$$

$$
\begin{aligned}
y = e^{x_1} \Bigg[& \sin\left(z_1 + \frac{\pi}{4}\right) \cos \frac{y_1 + t_1}{\sqrt{2}} \sinh \frac{y_1 - t_1}{\sqrt{2}} \\
& + \cos\left(z_1 + \frac{\pi}{4}\right) \sin \frac{y_1 + t_1}{\sqrt{2}} \cosh \frac{y_1 - t_1}{\sqrt{2}} \Bigg],
\end{aligned}
\tag{3.361}
$$

$$
\begin{aligned}
z = e^{x_1} \Bigg(& \cos z_1 \sin \frac{y_1 + t_1}{\sqrt{2}} \sinh \frac{y_1 - t_1}{\sqrt{2}} \\
& + \sin z_1 \cos \frac{y_1 + t_1}{\sqrt{2}} \cosh \frac{y_1 - t_1}{\sqrt{2}} \Bigg),
\end{aligned}
\tag{3.362}
$$

$$
\begin{aligned}
t = e^{x_1} \Bigg[& -\cos\left(z_1 + \frac{\pi}{4}\right) \cos \frac{y_1 + t_1}{\sqrt{2}} \sinh \frac{y_1 - t_1}{\sqrt{2}} \\
& + \sin\left(z_1 + \frac{\pi}{4}\right) \sin \frac{y_1 + t_1}{\sqrt{2}} \cosh \frac{y_1 - t_1}{\sqrt{2}} \Bigg].
\end{aligned}
\tag{3.363}
$$

The relations (3.360)-(3.363) can be rewritten as

$$x + \frac{y - t}{\sqrt{2}} = e^{x_1} \cos\left(z_1 + \frac{y_1 + t_1}{\sqrt{2}}\right) e^{(y_1 - t_1)/\sqrt{2}}, \tag{3.364}$$

$$z + \frac{y + t}{\sqrt{2}} = e^{x_1} \sin\left(z_1 + \frac{y_1 + t_1}{\sqrt{2}}\right) e^{(y_1 - t_1)/\sqrt{2}}, \tag{3.365}$$

$$x - \frac{y - t}{\sqrt{2}} = e^{x_1} \cos\left(z_1 - \frac{y_1 + t_1}{\sqrt{2}}\right) e^{-(y_1 - t_1)/\sqrt{2}}, \tag{3.366}$$

$$z - \frac{y+t}{\sqrt{2}} = e^{x_1} \sin\left(z_1 - \frac{y_1+t_1}{\sqrt{2}}\right) e^{-(y_1-t_1)/\sqrt{2}}. \tag{3.367}$$

By multiplying the sum of the squares of the first two and of the last two relations (3.364)-(3.367) it results that

$$e^{4x_1} = \rho_+^2 \rho_-^2, \tag{3.368}$$

or

$$e^{x_1} = \rho. \tag{3.369}$$

By summing the squares of all relations (3.364)-(3.367) it results that

$$d^2 = \rho^2 \cosh\left[\sqrt{2}(y_1 - t_1)\right]. \tag{3.370}$$

Then the quantities y_1, z_1, t_1 can be expressed in terms of the angles ϕ, χ, ψ defined in Eqs. (3.290)-(3.293) as

$$z_1 + \frac{y_1+t_1}{\sqrt{2}} = \phi, \tag{3.371}$$

$$-z_1 + \frac{y_1+t_1}{\sqrt{2}} = \chi, \tag{3.372}$$

$$\frac{e^{(y_1-t_1)/\sqrt{2}}}{\sqrt{2}\left[\cosh\sqrt{2}(y_1-t_1)\right]^{1/2}} = \sin\psi, \quad \frac{e^{-(y_1-t_1)/\sqrt{2}}}{\sqrt{2}\left[\cosh\sqrt{2}(y_1-t_1)\right]^{1/2}} = \cos\psi. \tag{3.373}$$

From Eq. (3.373) it results that

$$y_1 - t_1 = \frac{1}{\sqrt{2}}\ln\tan\psi, \tag{3.374}$$

so that

$$y_1 = \frac{\phi+\chi}{2\sqrt{2}} + \frac{1}{2\sqrt{2}}\ln\tan\psi, \ z_1 = \frac{\phi-\chi}{2}, \ t_1 = \frac{\phi+\chi}{2\sqrt{2}} - \frac{1}{2\sqrt{2}}\ln\tan\psi. \tag{3.375}$$

Substituting the expressions of the quantities x_1, y_1, z_1, t_1 in Eq. (3.359) yields

$$u = \rho \exp\left[\frac{1}{2\sqrt{2}}(\alpha-\gamma)\ln\tan\psi + \frac{1}{2}\left(\beta + \frac{\alpha+\gamma}{\sqrt{2}}\right)\phi \right.$$
$$\left. -\frac{1}{2}\left(\beta - \frac{\alpha+\gamma}{\sqrt{2}}\right)\chi\right], \tag{3.376}$$

which will be called the exponential form of the planar fourcomplex number u. It can be checked that

$$\exp\left[\frac{1}{2}\left(\beta + \frac{\alpha+\gamma}{\sqrt{2}}\right)\phi\right] = \frac{1}{2} - \frac{\alpha-\gamma}{2\sqrt{2}} + \left(\frac{1}{2} + \frac{\alpha-\gamma}{2\sqrt{2}}\right)\cos\phi$$
$$+ \left(\frac{\beta}{2} + \frac{\alpha+\gamma}{2\sqrt{2}}\right)\sin\phi, \tag{3.377}$$

$$\exp\left[-\frac{1}{2}\left(\beta - \frac{\alpha+\gamma}{\sqrt{2}}\right)\chi\right] = \frac{1}{2} + \frac{\alpha-\gamma}{2\sqrt{2}} + \left(\frac{1}{2} - \frac{\alpha-\gamma}{2\sqrt{2}}\right)\cos\chi$$
$$- \left(\frac{\beta}{2} - \frac{\alpha+\gamma}{2\sqrt{2}}\right)\sin\chi, \tag{3.378}$$

which shows that $e^{[\beta+(\alpha+\gamma)/\sqrt{2}]\phi/2}$ and $e^{-[\beta-(\alpha+\gamma)/\sqrt{2}]\chi/2}$ are periodic functions of ϕ and respectively χ, with period 2π.

The exponential of the logarithmic term in Eq. (3.376) can be expanded with the aid of the relation (3.338) as

$$\exp\left[\frac{1}{2\sqrt{2}}(\alpha-\gamma)\ln\tan\psi\right]$$
$$= \frac{1}{(\sin 2\psi)^{1/2}}\left[\cos\left(\psi - \frac{\pi}{4}\right) + \frac{\alpha-\gamma}{\sqrt{2}}\sin\left(\psi - \frac{\pi}{4}\right)\right]. \tag{3.379}$$

Since according to Eq. (3.300) $\tan\psi = \rho_+/\rho_-$, then

$$\sin\psi\cos\psi = \frac{\rho_+\rho_-}{\rho_+^2 + \rho_-^2}, \tag{3.380}$$

and it can be checked that

$$\rho_+^2 + \rho_-^2 = 2d^2, \tag{3.381}$$

where d has been defined in Eq. (3.288). Thus

$$\rho^2 = d^2\sin 2\psi, \tag{3.382}$$

so that the planar fourcomplex number u can be written as

$$u = d\left[\cos\left(\psi - \frac{\pi}{4}\right) + \frac{\alpha-\gamma}{\sqrt{2}}\sin\left(\psi - \frac{\pi}{4}\right)\right]\exp\left[\frac{1}{2}\left(\beta + \frac{\alpha+\gamma}{\sqrt{2}}\right)\phi\right.$$
$$\left.- \frac{1}{2}\left(\beta - \frac{\alpha+\gamma}{\sqrt{2}}\right)\chi\right], \tag{3.383}$$

which will be called the trigonometric form of the planar fourcomplex number u.

If u_1, u_2 are planar fourcomplex numbers of moduli and angles $d_1, \phi_1, \chi_1,$ ψ_1 and respectively $d_2, \phi_2, \chi_2, \psi_2$, the product of the factors depending on the planar angles can be calculated to be

$$\left[\cos(\psi_1 - \pi/4) + \frac{\alpha - \gamma}{\sqrt{2}} \sin(\psi_1 - \pi/4)\right]$$
$$\left[\cos(\psi_2 - \pi/4) + \frac{\alpha - \gamma}{\sqrt{2}} \sin(\psi_2 - \pi/4)\right]$$
$$= \left[\cos(\psi_1 - \psi_2) - \frac{\alpha - \gamma}{\sqrt{2}} \cos(\psi_1 + \psi_2)\right]. \tag{3.384}$$

The right-hand side of Eq. (3.384) can be written as

$$\cos(\psi_1 - \psi_2) - \frac{\alpha - \gamma}{\sqrt{2}} \cos(\psi_1 + \psi_2)$$
$$= [2(\cos^2 \psi_1 \cos^2 \psi_2 + \sin^2 \psi_1 \sin^2 \psi_2)]^{1/2}[\cos(\psi - \pi/4)$$
$$+ \frac{\alpha - \gamma}{\sqrt{2}} \sin(\psi - \pi/4)], \tag{3.385}$$

where the angle ψ, determined by the condition that

$$\tan(\psi - \pi/4) = -\cos(\psi_1 + \psi_2)/\cos(\psi_1 - \psi_2), \tag{3.386}$$

is given by $\tan \psi = \tan \psi_1 \tan \psi_2$, which is consistent with Eq. (3.302). The modulus d of the product $u_1 u_2$ is then

$$d = \sqrt{2} d_1 d_2 \left(\cos^2 \psi_1 \cos^2 \psi_2 + \sin^2 \psi_1 \sin^2 \psi_2\right)^{1/2}. \tag{3.387}$$

3.3.5 Elementary functions of planar fourcomplex variables

The logarithm u_1 of the planar fourcomplex number u, $u_1 = \ln u$, can be defined as the solution of the equation

$$u = e^{u_1}, \tag{3.388}$$

written explicitly previously in Eq. (3.359), for u_1 as a function of u. From Eq. (3.376) it results that

$$\ln u = \ln \rho + \frac{1}{2\sqrt{2}}(\alpha - \gamma) \ln \tan \psi + \frac{1}{2}\left(\beta + \frac{\alpha + \gamma}{\sqrt{2}}\right) \phi$$
$$- \frac{1}{2}\left(\beta - \frac{\alpha + \gamma}{\sqrt{2}}\right) \chi, \tag{3.389}$$

which is multivalued because of the presence of the terms proportional to ϕ and χ. It can be inferred from Eqs. (3.301) and (3.302) that

$$\ln(uu') = \ln u + \ln u', \tag{3.390}$$

up to multiples of $\pi[\beta + (\alpha + \gamma)/\sqrt{2}]$ and $\pi[\beta - (\alpha + \gamma)/\sqrt{2}]$.

The power function u^m can be defined for real values of n as

$$u^m = e^{m \ln u}. \tag{3.391}$$

The power function is multivalued unless n is an integer. For integer n, it can be inferred from Eq. (3.390) that

$$(uu')^m = u^m u'^m. \tag{3.392}$$

If, for example, $m = 2$, it can be checked with the aid of Eq. (3.383) that Eq. (3.391) gives indeed $(x + \alpha y + \beta z + \gamma t)^2 = x^2 - z^2 - 2yt + 2\alpha(xy - zt) + \beta(y^2 - t^2 + 2xz) + 2\gamma(xt + yz)$.

The trigonometric functions of the hypercomplex variable u and the addition theorems for these functions have been written in Eqs. (1.57)-(1.60). The cosine and sine functions of the hypercomplex variables $\alpha y, \beta z$ and γt can be expressed as

$$\cos \alpha y = f_{40}(y) - \beta f_{42}(y), \ \sin \alpha y = \alpha f_{41}(y) - \gamma f_{43}(y), \tag{3.393}$$

$$\cos \beta z = \cosh z, \ \sin \beta z = \beta \sinh z, \tag{3.394}$$

$$\cos \gamma t = f_{40}(t) + \beta f_{42}(t), \ \sin \gamma t = \gamma f_{41}(t) - \alpha f_{43}(t). \tag{3.395}$$

The cosine and sine functions of a planar fourcomplex number $x + \alpha y + \beta z + \gamma t$ can then be expressed in terms of elementary functions with the aid of the addition theorems Eqs. (1.59), (1.60) and of the expressions in Eqs. (3.393)-(3.395).

The hyperbolic functions of the hypercomplex variable u and the addition theorems for these functions have been written in Eqs. (1.62)-(1.65). The hyperbolic cosine and sine functions of the hypercomplex variables $\alpha y, \beta z$ and γt can be expressed as

$$\cosh \alpha y = f_{40}(y) + \beta f_{42}(y), \ \sinh \alpha y = \alpha f_{41}(y) + \gamma f_{43}(y), \tag{3.396}$$

$$\cosh \beta z = \cos z, \ \sinh \beta z = \beta \sin z, \tag{3.397}$$

$$\cosh \gamma t = f_{40}(t) - \beta f_{42}(t), \ \sinh \gamma t = \gamma f_{41}(t) + \alpha f_{43}(t). \tag{3.398}$$

The hyperbolic cosine and sine functions of a planar fourcomplex number $x + \alpha y + \beta z + \gamma t$ can then be expressed in terms of elementary functions with the aid of the addition theorems Eqs. (1.64), (1.65) and of the expressions in Eqs. (3.396)-(3.398).

3.3.6   Power series of planar fourcomplex variables

A planar fourcomplex series is an infinite sum of the form

$$a_0 + a_1 + a_2 + \cdots + a_l + \cdots, \tag{3.399}$$

where the coefficients a_l are planar fourcomplex numbers. The convergence of the series (3.399) can be defined in terms of the convergence of its 4 real components. The convergence of a planar fourcomplex series can however be studied using planar fourcomplex variables. The main criterion for absolute convergence remains the comparison theorem, but this requires a number of inequalities which will be discussed further.

The modulus of a planar fourcomplex number $u = x + \alpha y + \beta z + \gamma t$ can be defined as

$$|u| = (x^2 + y^2 + z^2 + t^2)^{1/2}, \tag{3.400}$$

so that, according to Eq. (3.288), $d = |u|$. Since $|x| \le |u|, |y| \le |u|, |z| \le |u|, |t| \le |u|$, a property of absolute convergence established via a comparison theorem based on the modulus of the series (3.399) will ensure the absolute convergence of each real component of that series.

The modulus of the sum $u_1 + u_2$ of the planar fourcomplex numbers u_1, u_2 fulfils the inequality

$$||u_1| - |u_2|| \le |u_1 + u_2| \le |u_1| + |u_2|. \tag{3.401}$$

For the product the relation is

$$|u_1 u_2| \le \sqrt{2}|u_1||u_2|, \tag{3.402}$$

as can be shown from Eq. (3.387). The relation (3.402) replaces the relation of equality extant for regular complex numbers. The equality in Eq. (3.402) takes place for $\cos^2(\psi_1 - \psi_2) = 1$, $\cos^2(\psi_1 + \psi_2) = 1$, which means that $x_1 + (y_1 - t_1)/\sqrt{2} = 0$, $z_1 + (y_1 + t_1)/\sqrt{2} = 0$, $x_2 + (y_2 - t_2)/\sqrt{2} = 0$, $z_2 + (y_2 + t_2)/\sqrt{2} = 0$, or $x_1 - (y_1 - t_1)/\sqrt{2} = 0$, $z_1 - (y_1 + t_1)/\sqrt{2} = 0$, $x_2 - (y_2 - t_2)/\sqrt{2} = 0$, $z_2 - (y_2 + t_2)/\sqrt{2} = 0$. The modulus of the product, which has the property that $0 \le |u_1 u_2|$, becomes equal to zero for $\cos^2(\psi_1 - \psi_2) = 0$, $\cos^2(\psi_1 + \psi_2) = 0$, which means that $x_1 + (y_1 - t_1)/\sqrt{2} = 0$, $z_1 + (y_1 + t_1)/\sqrt{2} = 0$, $x_2 - (y_2 - t_2)/\sqrt{2} = 0$, $z_2 - (y_2 + t_2)/\sqrt{2} = 0$, or $x_1 - (y_1 - t_1)/\sqrt{2} = 0$, $z_1 - (y_1 + t_1)/\sqrt{2} = 0$, $x_2 + (y_2 - t_2)/\sqrt{2} = 0$, $z_2 + (y_2 + t_2)/\sqrt{2} = 0$. as discussed after Eq. (3.287).

It can be shown that

$$x^2 + y^2 + z^2 + t^2 \le |u^2| \le \sqrt{2}(x^2 + y^2 + z^2 + t^2). \tag{3.403}$$

The left relation in Eq. (3.403) becomes an equality for $\sin^2 2\psi = 1$, when $\rho_+ = \rho_-$, which means that $x(y-t) + z(y+t) = 0$. The right relation in Eq. (3.403) becomes an equality for $\sin^2 2\psi = 0$, when $x + (y-t)/\sqrt{2} = 0$, $z + (y+t)/\sqrt{2} = 0$, or $x - (y-t)/\sqrt{2} = 0$, $z - (y+t)/\sqrt{2} = 0$. The inequality in Eq. (3.402) implies that

$$|u^l| \le 2^{(l-1)/2}|u|^l. \tag{3.404}$$

From Eqs. (3.402) and (3.404) it results that

$$|au^l| \le 2^{l/2}|a||u|^l. \tag{3.405}$$

A power series of the planar fourcomplex variable u is a series of the form

$$a_0 + a_1 u + a_2 u^2 + \cdots + a_l u^l + \cdots. \tag{3.406}$$

Since

$$\left| \sum_{l=0}^{\infty} a_l u^l \right| \le \sum_{l=0}^{\infty} 2^{l/2}|a_l||u|^l, \tag{3.407}$$

a sufficient condition for the absolute convergence of this series is that

$$\lim_{l \to \infty} \frac{\sqrt{2}|a_{l+1}||u|}{|a_l|} < 1. \tag{3.408}$$

Thus the series is absolutely convergent for

$$|u| < c, \tag{3.409}$$

where

$$c = \lim_{l \to \infty} \frac{|a_l|}{\sqrt{2}|a_{l+1}|}. \tag{3.410}$$

The convergence of the series (3.406) can be also studied with the aid of the transformation

$$x + \alpha y + \beta z + \gamma t = \sqrt{2}(e_1 \xi + \tilde{e}_1 \upsilon + e_2 \tau + \tilde{e}_2 \zeta), \tag{3.411}$$

where $\xi, \upsilon, \tau, \zeta$ have been defined in Eq. (3.289), and

$$e_1 = \frac{1}{2} + \frac{\alpha - \gamma}{2\sqrt{2}}, \quad \tilde{e}_1 = \frac{\beta}{2} + \frac{\alpha + \gamma}{2\sqrt{2}}, \quad e_2 = \frac{1}{2} - \frac{\alpha - \gamma}{2\sqrt{2}},$$

$$\tilde{e}_2 = -\frac{\beta}{2} + \frac{\alpha + \gamma}{2\sqrt{2}}. \tag{3.412}$$

It can be checked that

$$e_1^2 = e_1, \quad \tilde{e}_1^2 = -e_1, \quad e_1\tilde{e}_1 = \tilde{e}_1, \quad e_2^2 = e_2, \quad \tilde{e}_2^2 = -e_2, \quad e_2\tilde{e}_2 = \tilde{e}_2,$$
$$e_1 e_2 = 0, \quad \tilde{e}_1\tilde{e}_2 = 0, \quad e_1\tilde{e}_2 = 0, \quad e_2\tilde{e}_1 = 0. \tag{3.413}$$

The moduli of the bases in Eq. (3.412) are

$$|e_1| = \frac{1}{\sqrt{2}}, \quad |\tilde{e}_1| = \frac{1}{\sqrt{2}}, \quad |e_2| = \frac{1}{\sqrt{2}}, \quad |\tilde{e}_2| = \frac{1}{\sqrt{2}}, \tag{3.414}$$

and it can be checked that

$$|x + \alpha y + \beta z + \gamma t|^2 = \xi^2 + v^2 + \tau^2 + \zeta^2. \tag{3.415}$$

The ensemble $e_1, \tilde{e}_1, e_2, \tilde{e}_2$ will be called the canonical planar fourcomplex base, and Eq. (3.411) gives the canonical form of the planar fourcomplex number.

If $u = u'u''$, the components ξ, v, τ, ζ are related, according to Eqs. (3.305)-(3.308) by

$$\xi = \sqrt{2}(\xi'\xi'' - v'v''), \quad v = \sqrt{2}(\xi'v'' + v'\xi''), \quad \tau = \sqrt{2}(\tau'\tau'' - \zeta'\zeta''),$$
$$\zeta = \sqrt{2}(\tau'\zeta'' + \zeta'\xi''), \tag{3.416}$$

which show that, upon multiplication, the components ξ, v and τ, ζ obey, up to a normalization constant, the same rules as the real and imaginary components of usual, two-dimensional complex numbers.

If the coefficients in Eq. (3.406) are

$$a_l = a_{l0} + \alpha a_{l1} + \beta a_{l2} + \gamma a_{l3}, \tag{3.417}$$

and

$$A_{l1} = a_{l0} + \frac{a_{l1} - a_{l3}}{\sqrt{2}}, \quad \tilde{A}_{l1} = a_{l2} + \frac{a_{l1} + a_{l3}}{\sqrt{2}}, \quad A_{l2} = a_{l0} - \frac{a_{l1} - a_{l3}}{\sqrt{2}},$$
$$\tilde{A}_{l2} = -a_{l2} + \frac{a_{l1} + a_{l3}}{\sqrt{2}}, \tag{3.418}$$

the series (3.406) can be written as

$$\sum_{l=0}^{\infty} 2^{l/2} \left[(e_1 A_{l1} + \tilde{e}_1 \tilde{A}_{l1})(e_1\xi + \tilde{e}_1 v)^l + (e_2 A_{l2} + \tilde{e}_2 \tilde{A}_{l2})(e_2\tau + \tilde{e}_2\zeta)^l \right]. \tag{3.419}$$

Thus, the series in Eqs. (3.406) and (3.419) are absolutely convergent for

$$\rho_+ < c_1, \quad \rho_- < c_2, \tag{3.420}$$

where

$$c_1 = \lim_{l \to \infty} \frac{\left[A_{l1}^2 + \tilde{A}_{l1}^2\right]^{1/2}}{\sqrt{2}\left[A_{l+1,1}^2 + \tilde{A}_{l+1,1}^2\right]^{1/2}}, \quad c_2 = \lim_{l \to \infty} \frac{\left[A_{l2}^2 + \tilde{A}_{l2}^2\right]^{1/2}}{\sqrt{2}\left[A_{l+1,2}^2 + \tilde{A}_{l+1,2}^2\right]^{1/2}}.$$

$$(3.421)$$

It can be shown that $c = (1/\sqrt{2})\min(c_1, c_2)$, where min designates the smallest of the numbers c_1, c_2. Using the expression of $|u|$ in Eq. (3.415), it can be seen that the spherical region of convergence defined in Eqs. (3.409), (3.410) is included in the cylindrical region of convergence defined in Eqs. (3.420) and (3.421).

3.3.7 Analytic functions of planar fourcomplex variables

The fourcomplex function $f(u)$ of the fourcomplex variable u has been expressed in Eq. (3.117) in terms of the real functions $P(x, y, z, t), Q(x, y, z, t)$, $R(x, y, z, t), S(x, y, z, t)$ of real variables x, y, z, t. The relations between the partial derivatives of the functions P, Q, R, S are obtained by setting succesively in Eq. (3.118) $\Delta x \to 0, \Delta y = \Delta z = \Delta t = 0$; then $\Delta y \to 0, \Delta x = \Delta z = \Delta t = 0$; then $\Delta z \to 0, \Delta x = \Delta y = \Delta t = 0$; and finally $\Delta t \to 0, \Delta x = \Delta y = \Delta z = 0$. The relations are

$$\frac{\partial P}{\partial x} = \frac{\partial Q}{\partial y} = \frac{\partial R}{\partial z} = \frac{\partial S}{\partial t}, \tag{3.422}$$

$$\frac{\partial Q}{\partial x} = \frac{\partial R}{\partial y} = \frac{\partial S}{\partial z} = -\frac{\partial P}{\partial t}, \tag{3.423}$$

$$\frac{\partial R}{\partial x} = \frac{\partial S}{\partial y} = -\frac{\partial P}{\partial z} = -\frac{\partial Q}{\partial t}, \tag{3.424}$$

$$\frac{\partial S}{\partial x} = -\frac{\partial P}{\partial y} = -\frac{\partial Q}{\partial z} = -\frac{\partial R}{\partial t}. \tag{3.425}$$

The relations (3.422)-(3.425) are analogous to the Riemann relations for the real and imaginary components of a complex function. It can be shown from Eqs. (3.422)-(3.425) that the component P is a solution of the equations

$$\frac{\partial^2 P}{\partial x^2} + \frac{\partial^2 P}{\partial z^2} = 0, \quad \frac{\partial^2 P}{\partial y^2} + \frac{\partial^2 P}{\partial t^2} = 0, \tag{3.426}$$

and the components Q, R, S are solutions of similar equations. As can be seen from Eqs. (3.426), the components P, Q, R, S of an analytic function of planar fourcomplex variable are harmonic with respect to the pairs of variables x, y and z, t. The component P is also a solution of the mixed-derivative equations

$$\frac{\partial^2 P}{\partial x^2} = -\frac{\partial^2 P}{\partial y \partial t}, \quad \frac{\partial^2 P}{\partial y^2} = \frac{\partial^2 P}{\partial x \partial z}, \quad \frac{\partial^2 P}{\partial z^2} = \frac{\partial^2 P}{\partial y \partial t}, \quad \frac{\partial^2 P}{\partial t^2} = -\frac{\partial^2 P}{\partial x \partial z}, \quad (3.427)$$

and the components Q, R, S are solutions of similar equations. The component P is also a solution of the mixed-derivative equations

$$\frac{\partial^2 P}{\partial x \partial y} = -\frac{\partial^2 P}{\partial z \partial t}, \quad \frac{\partial^2 P}{\partial x \partial t} = \frac{\partial^2 P}{\partial y \partial z}, \quad (3.428)$$

and the components Q, R, S are solutions of similar equations.

3.3.8 Integrals of functions of planar fourcomplex variables

The singularities of planar fourcomplex functions arise from terms of the form $1/(u - u_0)^m$, with $m > 0$. Functions containing such terms are singular not only at $u = u_0$, but also at all points of the two-dimensional hyperplanes passing through u_0 and which are parallel to the nodal hyperplanes.

The integral of a planar fourcomplex function between two points A, B along a path situated in a region free of singularities is independent of path, which means that

$$\oint_\Gamma f(u) du = 0, \quad (3.429)$$

where it is supposed that a surface Σ spanning the closed loop Γ is not intersected by any of the two-dimensional hyperplanes associated with the singularities of the function $f(u)$. Using the expression, Eq. (3.117) for $f(u)$ and the fact that $du = dx + \alpha dy + \beta dz + \gamma dt$, the explicit form of the integral in Eq. (3.429) is

$$\oint_\Gamma f(u) du = \oint_\Gamma [(P dx - S dy - R dz - Q dt)$$
$$+ \alpha(Q dx + P dy - S dz - R dt) + \beta(R dx + Q dy + P dz - S dt)$$
$$+ \gamma(S dx + R dy + Q dz + P dt)]. \quad (3.430)$$

If the functions P, Q, R, S are regular on a surface Σ spanning the loop Γ, the integral along the loop Γ can be transformed with the aid of the theorem of Stokes in an integral over the surface Σ of terms of the form $\partial P/\partial y + \partial S/\partial x$, $\partial P/\partial z + \partial R/\partial x$, $\partial P/\partial t + \partial Q/\partial x$, $\partial R/\partial y - \partial S/\partial z$, $\partial S/\partial t - \partial Q/\partial y$, $\partial R/\partial t - \partial Q/\partial z$ and of similar terms arising from the α, β and γ

components, which are equal to zero by Eqs. (3.422)-(3.425), and this proves Eq. (3.429).

The integral of the function $(u - u_0)^m$ on a closed loop Γ is equal to zero for m a positive or negative integer not equal to -1,

$$\oint_\Gamma (u - u_0)^m du = 0, \quad m \text{ integer}, \ m \neq -1. \tag{3.431}$$

This is due to the fact that $\int (u - u_0)^m du = (u - u_0)^{m+1}/(m + 1)$, and to the fact that the function $(u - u_0)^{m+1}$ is singlevalued for m an integer.

The integral $\oint du/(u - u_0)$ can be calculated using the exponential form (3.376),

$$u - u_0 = \rho \exp \left[\frac{1}{2\sqrt{2}} (\alpha - \gamma) \ln \tan \psi + \frac{1}{2} \left(\beta + \frac{\alpha + \gamma}{\sqrt{2}} \right) \phi \right. \\ \left. - \frac{1}{2} \left(\beta - \frac{\alpha + \gamma}{\sqrt{2}} \right) \chi \right], \tag{3.432}$$

so that

$$\frac{du}{u - u_0} = \frac{d\rho}{\rho} + \frac{1}{2\sqrt{2}} (\alpha - \gamma) d \ln \tan \psi + \frac{1}{2} \left(\beta + \frac{\alpha + \gamma}{\sqrt{2}} \right) d\phi \\ - \frac{1}{2} \left(\beta - \frac{\alpha + \gamma}{\sqrt{2}} \right) d\chi. \tag{3.433}$$

Since ρ and ψ are singlevalued variables, it follows that $\oint_\Gamma d\rho/\rho = 0$, $\oint_\Gamma d \ln \tan \psi = 0$. On the other hand, ϕ and χ are cyclic variables, so that they may give a contribution to the integral around the closed loop Γ. Thus, if C_+ is a circle of radius r parallel to the $\xi O v$ plane, whose projection of the center of this circle on the $\xi O v$ plane coincides with the projection of the point u_0 on this plane, the points of the circle C_+ are described according to Eqs. (3.289)-(3.293) by the equations

$$\xi = \xi_0 + r \sin \psi \cos \phi, \ v = v_0 + r \sin \psi \sin \phi, \ \tau = \tau_0 + r \cos \psi \cos \chi, \\ \zeta = \zeta_0 + r \cos \psi \sin \chi, \tag{3.434}$$

for constant values of χ and ψ, $\psi \neq 0, \pi/2$, where $u_0 = x_0 + \alpha y_0 + \beta z_0 + \gamma t_0$, and $\xi_0, v_0, \tau_0, \zeta_0$ are calculated from x_0, y_0, z_0, t_0 according to Eqs. (3.289). Then

$$\oint_{C_+} \frac{du}{u - u_0} = \pi \left(\beta + \frac{\alpha + \gamma}{\sqrt{2}} \right). \tag{3.435}$$

If C_- is a circle of radius r parallel to the $\tau O \zeta$ plane, whose projection of the center of this circle on the $\tau O \zeta$ plane coincides with the projection of the point u_0 on this plane, the points of the circle C_- are described by the same Eqs. (3.434) but for constant values of ϕ and ψ, $\psi \neq 0, \pi/2$. Then

$$\oint_{C_-} \frac{du}{u - u_0} = -\pi \left(\beta - \frac{\alpha + \gamma}{\sqrt{2}} \right). \tag{3.436}$$

The expression of $\oint_\Gamma du/(u - u_0)$ can be written as a single equation with the aid of the functional $\mathrm{int}(M, C)$ defined in Eq. (3.134) as

$$\oint_\Gamma \frac{du}{u - u_0} = \pi \left(\beta + \frac{\alpha + \gamma}{\sqrt{2}} \right) \mathrm{int}(u_{0\xi v}, \Gamma_{\xi v})$$
$$- \pi \left(\beta - \frac{\alpha + \gamma}{\sqrt{2}} \right) \mathrm{int}(u_{0\tau\zeta}, \Gamma_{\tau\zeta}), \tag{3.437}$$

where $u_{0\xi v}, u_{0\tau\zeta}$ and $\Gamma_{\xi v}, \Gamma_{\tau\zeta}$ are respectively the projections of the point u_0 and of the loop Γ on the planes ξv and $\tau\zeta$.

If $f(u)$ is an analytic planar fourcomplex function which can be expanded in a series as written in Eq. (1.89), and the expansion holds on the curve Γ and on a surface spanning Γ, then from Eqs. (3.431) and (3.437) it follows that

$$\oint_\Gamma \frac{f(u)du}{u - u_0} = \pi \left[\left(\beta + \frac{\alpha + \gamma}{\sqrt{2}} \right) \mathrm{int}(u_{0\xi v}, \Gamma_{\xi v}) \right.$$
$$\left. - \left(\beta - \frac{\alpha + \gamma}{\sqrt{2}} \right) \mathrm{int}(u_{0\tau\zeta}, \Gamma_{\tau\zeta}) \right] f(u_0), \tag{3.438}$$

where $\Gamma_{\xi v}, \Gamma_{\tau\zeta}$ are the projections of the curve Γ on the planes ξv and respectively $\tau\zeta$, as shown in Fig. 3.2. As remarked previously, the definition of the variables in this section is different from the former definition for the circular hypercomplex numbers.

Substituting in the right-hand side of Eq. (3.438) the expression of $f(u)$ in terms of the real components P, Q, R, S, Eq. (3.117), yields

$$\oint_\Gamma \frac{f(u)du}{u - u_0}$$
$$= \pi \left[\left(\beta + \frac{\alpha + \gamma}{\sqrt{2}} \right) P - \left(1 + \frac{\alpha - \gamma}{\sqrt{2}} \right) R \right.$$
$$\left. - \left(\gamma - \frac{1 - \beta}{\sqrt{2}} \right) Q - \left(\alpha - \frac{1 + \beta}{\sqrt{2}} \right) S \right] \mathrm{int}(u_{0\xi v}, \Gamma_{\xi v})$$
$$- \pi \left[\left(\beta - \frac{\alpha + \gamma}{\sqrt{2}} \right) P - \left(1 - \frac{\alpha - \gamma}{\sqrt{2}} \right) R \right.$$
$$\left. - \left(\gamma - \frac{1 + \beta}{\sqrt{2}} \right) Q - \left(\alpha - \frac{1 - \beta}{\sqrt{2}} \right) S \right] \mathrm{int}(u_{0\tau\zeta}, \Gamma_{\tau\zeta}), $$

$$\tag{3.439}$$

where P, Q, R, S are the values of the components of f at $u = u_0$.

If $f(u)$ can be expanded as written in Eq. (1.89) on Γ and on a surface spanning Γ, then from Eqs. (3.431) and (3.437) it also results that

$$\oint_\Gamma \frac{f(u)du}{(u - u_0)^{m+1}} = \frac{\pi}{m!} \left[\left(\beta + \frac{\alpha + \gamma}{\sqrt{2}} \right) \text{int}(u_{0\xi v}, \Gamma_{\xi v}) \right.$$

$$\left. - \left(\beta - \frac{\alpha + \gamma}{\sqrt{2}} \right) \text{int}(u_{0\tau\varsigma}, \Gamma_{\tau\varsigma}) \right] f^{(m)}(u_0), \qquad (3.440)$$

where it has been used the fact that the derivative $f^{(m)}(u_0)$ of order n of $f(u)$ at $u = u_0$ is related to the expansion coefficient in Eq. (1.89) according to Eq. (1.93).

If a function $f(u)$ is expanded in positive and negative powers of $u - u_j$, where u_j are planar fourcomplex constants, j being an index, the integral of f on a closed loop Γ is determined by the terms in the expansion of f which are of the form $a_j/(u - u_j)$,

$$f(u) = \cdots + \sum_j \frac{a_j}{u - u_j} + \cdots . \qquad (3.441)$$

Then the integral of f on a closed loop Γ is

$$\oint_\Gamma f(u)du = \pi \left(\beta + \frac{\alpha + \gamma}{\sqrt{2}} \right) \sum_j \text{int}(u_{j\xi v}, \Gamma_{\xi v}) a_j$$

$$- \pi \left(\beta - \frac{\alpha + \gamma}{\sqrt{2}} \right) \sum_j \text{int}(u_{j\tau\varsigma}, \Gamma_{\tau\varsigma}) a_j. \qquad (3.442)$$

3.3.9 Factorization of planar fourcomplex polynomials

A polynomial of degree m of the planar fourcomplex variable $u = x + \alpha y + \beta z + \gamma t$ has the form

$$P_m(u) = u^m + a_1 u^{m-1} + \cdots + a_{m-1} u + a_m, \qquad (3.443)$$

where the constants are in general planar fourcomplex numbers.

It can be shown that any planar fourcomplex polynomial has a planar fourcomplex root, whence it follows that a polynomial of degree m can be written as a product of m linear factors of the form $u - u_j$, where the planar fourcomplex numbers u_j are the roots of the polynomials, although the factorization may not be unique,

$$P_m(u) = \prod_{j=1}^m (u - u_j). \qquad (3.444)$$

The fact that any planar fourcomplex polynomial has a root can be shown by considering the transformation of a fourdimensional sphere with the center at the origin by the function u^m. The points of the hypersphere of radius d are of the form written in Eq. (3.383), with d constant and ϕ, χ, ψ arbitrary. The point u^m is

$$
u^m = d^m \left[\cos\left(\psi - \frac{\pi}{4}\right) + \frac{\alpha - \gamma}{\sqrt{2}} \sin\left(\psi - \frac{\pi}{4}\right) \right]^m
$$
$$
\exp\left[\frac{1}{2}\left(\beta + \frac{\alpha + \gamma}{\sqrt{2}}\right) m\phi - \frac{1}{2}\left(\beta - \frac{\alpha + \gamma}{\sqrt{2}}\right) m\chi \right]. \tag{3.445}
$$

It can be shown with the aid of Eq. (3.387) that

$$
\left| u \exp\left[\frac{1}{2}\left(\beta + \frac{\alpha + \gamma}{\sqrt{2}}\right)\phi - \frac{1}{2}\left(\beta - \frac{\alpha + \gamma}{\sqrt{2}}\right)\chi \right] \right| = |u|, \tag{3.446}
$$

so that

$$
\left| \left[\cos(\psi - \pi/4) + \frac{\alpha - \gamma}{\sin}(\psi - \pi/4) \right]^m \right.
$$
$$
\left. \exp\left[\frac{1}{2}\left(\beta + \frac{\alpha + \gamma}{\sqrt{2}}\right) m\phi - \frac{1}{2}\left(\beta - \frac{\alpha + \gamma}{\sqrt{2}}\right) m\chi \right] \right|
$$
$$
= \left| \left(\cos(\psi - \pi/4) + \frac{\alpha - \gamma}{\sqrt{2}} \sin(\psi - \pi/4) \right)^m \right|. \tag{3.447}
$$

The right-hand side of Eq. (3.447) is

$$
\left| \left(\cos\epsilon + \frac{\alpha - \gamma}{\sqrt{2}} \sin\epsilon \right)^m \right|^2 = \sum_{k=0}^{m} C_{2m}^{2k} \cos^{2m-2k}\epsilon \sin^{2k}\epsilon, \tag{3.448}
$$

where $\epsilon = \psi - \pi/4$, and since $C_{2m}^{2k} \geq C_m^k$, it can be concluded that

$$
\left| \left(\cos\epsilon + \frac{\alpha - \gamma}{\sqrt{2}} \sin\epsilon \right)^m \right|^2 \geq 1. \tag{3.449}
$$

Then

$$
d^m \leq |u^m| \leq 2^{(m-1)/2} d^m, \tag{3.450}
$$

which shows that the image of a four-dimensional sphere via the transformation operated by the function u^m is a finite hypersurface.

If $u' = u^m$, and

$$
u' = d' \left[\cos(\psi' - \pi/4) + \frac{\alpha - \gamma}{\sqrt{2}} \sin(\psi' - \pi/4) \right]
$$
$$
\exp\left[\frac{1}{2}\left(\beta + \frac{\alpha + \gamma}{\sqrt{2}}\right)\phi' - \frac{1}{2}\left(\beta - \frac{\alpha + \gamma}{\sqrt{2}}\right)\chi' \right], \tag{3.451}
$$

then

$$\phi' = m\phi, \ \chi' = m\chi, \ \tan \psi' = \tan^m \psi. \tag{3.452}$$

Since for any values of the angles ϕ', χ', ψ' there is a set of solutions ϕ, χ, ψ of Eqs. (3.452), and since the image of the hypersphere is a finite hypersurface, it follows that the image of the four-dimensional sphere via the function u^m is also a closed hypersurface. A continuous hypersurface is called closed when any ray issued from the origin intersects that surface at least once in the finite part of the space.

A transformation of the four-dimensional space by the polynomial $P_m(u)$ will be considered further. By this transformation, a hypersphere of radius d having the center at the origin is changed into a certain finite closed surface, as discussed previously. The transformation of the four-dimensional space by the polynomial $P_m(u)$ associates to the point $u = 0$ the point $f(0) = a_m$, and the image of a hypersphere of very large radius d can be represented with good approximation by the image of that hypersphere by the function u^m. The origin of the axes is an inner point of the latter image. If the radius of the hypersphere is now reduced continuously from the initial very large values to zero, the image hypersphere encloses initially the origin, but the image shrinks to a_m when the radius approaches the value zero. Thus, the origin is initially inside the image hypersurface, and it lies outside the image hypersurface when the radius of the hypersphere tends to zero. Then since the image hypersurface is closed, the image surface must intersect at some stage the origin of the axes, which means that there is a point u_1 such that $f(u_1) = 0$. The factorization in Eq. (3.444) can then be obtained by iterations.

The roots of the polynomial P_m can be obtained by the following method. If the constants in Eq. (3.443) are $a_l = a_{l0} + \alpha a_{l1} + \beta a_{l2} + \gamma a_{l3}$, and with the notations of Eq. (3.418), the polynomial $P_m(u)$ can be written as

$$P_m = \sum_{l=0}^{m} 2^{(m-l)/2} (e_1 A_{l1} + \tilde{e}_1 \tilde{A}_{l1})(e_1 \xi + \tilde{e}_1 v)^{m-l}$$
$$+ \sum_{l=0}^{m} 2^{(m-l)/2} (e_2 A_{l2} + \tilde{e}_2 \tilde{A}_{l2})(e_2 \tau + \tilde{e}_2 \zeta)^{m-l}, \tag{3.453}$$

where the constants $A_{lk}, \tilde{A}_{lk}, k = 1, 2$ are real numbers. Each of the polynomials of degree m in $e_1\xi + \tilde{e}_1 v, e_2\tau + \tilde{e}_2\zeta$ in Eq. (3.453) can always be written as a product of linear factors of the form $e_1(\xi - \xi_p) + \tilde{e}_1(v - v_p)$ and respectively $e_2(\tau - \tau_p) + \tilde{e}_2(\zeta - \zeta_p)$, where the constants $\xi_p, v_p, \tau_p, \zeta_p$

are real,

$$\sum_{l=0}^{m} 2^{(m-l)/2}(e_1 A_{l1} + \tilde{e}_1 \tilde{A}_{l1})(e_1 \xi + \tilde{e}_1 \upsilon)^{m-l}$$

$$= \prod_{p=1}^{m} 2^{m/2} \left\{ e_1(\xi - \xi_p) + \tilde{e}_1(\upsilon - \upsilon_p) \right\}, \tag{3.454}$$

$$\sum_{l=0}^{m} 2^{(m-l)/2}(e_2 A_{l2} + \tilde{e}_2 \tilde{A}_{l2})(e_2 \tau + \tilde{e}_2 \zeta)^{m-l}$$

$$= \prod_{p=1}^{m} 2^{m/2} \left\{ e_2(\tau - \tau_p) + \tilde{e}_2(\zeta - \zeta_p) \right\}. \tag{3.455}$$

Due to the relations (3.413), the polynomial $P_m(u)$ can be written as a product of factors of the form

$$P_m(u) = \prod_{p=1}^{m} 2^{m/2} \left\{ e_1(\xi - \xi_p) + \tilde{e}_1(\upsilon - \upsilon_p) + e_2(\tau - \tau_p) + \tilde{e}_2(\zeta - \zeta_p) \right\}.$$

$$\tag{3.456}$$

This relation can be written with the aid of Eq. (3.411) in the form (3.444), where

$$u_p = \sqrt{2}(e_1 \xi_p + \tilde{e}_1 \upsilon_p + e_2 \tau_p + \tilde{e}_2 \zeta_p). \tag{3.457}$$

The roots $e_1\xi_p + \tilde{e}_1\upsilon_p$ and $e_2\tau_p + \tilde{e}_2\zeta_p$ defined in Eqs. (3.454) and respectively (3.455) may be ordered arbitrarily. This means that Eq. (3.457) gives sets of m roots $u_1, ..., u_m$ of the polynomial $P_m(u)$, corresponding to the various ways in which the roots $e_1\xi_p + \tilde{e}_1\upsilon_p$ and $e_2\tau_p + \tilde{e}_2\zeta_p$ are ordered according to p for each polynomial. Thus, while the hypercomplex components in Eqs. (3.454), (3.455) taken separately have unique factorizations, the polynomial $P_m(u)$ can be written in many different ways as a product of linear factors. The result of the planar fourcomplex integration, Eq. (3.442), is however unique.

If, for example, $P(u) = u^2 + 1$, the possible factorizations are $P = (u - \tilde{e}_1 - \tilde{e}_2)(u + \tilde{e}_1 + \tilde{e}_2)$ and $P = (u - \tilde{e}_1 + \tilde{e}_2)(u + \tilde{e}_1 - \tilde{e}_2)$, which can be written as $u^2 + 1 = (u - \beta)(u + \beta)$ or as $u^2 + 1 = \left\{ u - (\alpha + \gamma)/\sqrt{2} \right\} \left\{ (u + (\alpha + \gamma)/\sqrt{2} \right\}$. The result of the planar fourcomplex integration, Eq. (3.442), is however unique. It can be checked that $(\pm \tilde{e}_1 \pm \tilde{e}_2)^2 = -e_1 - e_2 = -1$.

3.3.10 Representation of planar fourcomplex numbers by irreducible matrices

If T is the unitary matrix,

$$T = \begin{pmatrix} \frac{1}{\sqrt{2}} & \frac{1}{2} & 0 & -\frac{1}{2} \\ 0 & \frac{1}{2} & \frac{1}{\sqrt{2}} & \frac{1}{2} \\ \frac{1}{\sqrt{2}} & -\frac{1}{2} & 0 & \frac{1}{2} \\ 0 & \frac{1}{2} & -\frac{1}{\sqrt{2}} & \frac{1}{2} \end{pmatrix}, \tag{3.458}$$

it can be shown that the matrix TUT^{-1} has the form

$$TUT^{-1} = \begin{pmatrix} V_1 & 0 \\ 0 & V_2 \end{pmatrix}, \tag{3.459}$$

where U is the matrix in Eq. (3.312) used to represent the planar fourcomplex number u. In Eq. (3.459), V_1, V_2 are the matrices

$$V_1 = \begin{pmatrix} x + \frac{y-t}{\sqrt{2}} & z + \frac{y+t}{\sqrt{2}} \\ -z - \frac{y+t}{\sqrt{2}} & x + \frac{y-t}{\sqrt{2}} \end{pmatrix}, \quad V_2 = \begin{pmatrix} x - \frac{y-t}{\sqrt{2}} & -z + \frac{y+t}{\sqrt{2}} \\ z - \frac{y+t}{\sqrt{2}} & x - \frac{y-t}{\sqrt{2}} \end{pmatrix}. \tag{3.460}$$

In Eq. (3.459), the symbols 0 denote the matrix

$$\begin{pmatrix} 0 & 0 \\ 0 & 0 \end{pmatrix}. \tag{3.461}$$

The relations between the variables $x + (y - t)/\sqrt{2}, z + (y + t)/\sqrt{2}, x - (y - t)/\sqrt{2}, -z + (y + t)/\sqrt{2}$ for the multiplication of planar fourcomplex numbers have been written in Eqs. (3.305)-(3.308). The matrix TUT^{-1} provides an irreducible representation [7] of the planar fourcomplex number u in terms of matrices with real coefficients.

3.4 Polar complex numbers in four dimensions

3.4.1 Operations with polar fourcomplex numbers

A polar fourcomplex number is determined by its four components (x, y, z, t). The sum of the polar fourcomplex numbers (x, y, z, t) and (x', y', z', t') is the polar fourcomplex number $(x + x', y + y', z + z', t + t')$. The product of the polar fourcomplex numbers (x, y, z, t) and (x', y', z', t') is defined in this section to be the polar fourcomplex number $(xx' + yt' + zz' + ty', xy' + yx' + zt' + tz', xz' + yy' + zx' + tt', xt' + yz' + zy' + tx')$. Polar fourcomplex numbers and their operations can be represented by writing the polar fourcomplex number (x, y, z, t) as $u = x + \alpha y + \beta z + \gamma t$, where α, β and γ are bases for which the multiplication rules are

$$\alpha^2 = \beta, \ \beta^2 = 1, \ \gamma^2 = \beta, \alpha\beta = \beta\alpha = \gamma, \ \alpha\gamma = \gamma\alpha = -1, \ \beta\gamma = \gamma\beta = \alpha. \tag{3.462}$$

Two polar fourcomplex numbers $u = x + \alpha y + \beta z + \gamma t, u' = x' + \alpha y' + \beta z' + \gamma t'$ are equal, $u = u'$, if and only if $x = x', y = y', z = z', t = t'$. If $u = x + \alpha y + \beta z + \gamma t, u' = x' + \alpha y' + \beta z' + \gamma t'$ are polar fourcomplex numbers, the sum $u + u'$ and the product uu' defined above can be obtained by applying the usual algebraic rules to the sum $(x + \alpha y + \beta z + \gamma t) + (x' + \alpha y' + \beta z' + \gamma t')$ and to the product $(x + \alpha y + \beta z + \gamma t)(x' + \alpha y' + \beta z' + \gamma t')$, and grouping of the resulting terms,

$$u + u' = x + x' + \alpha(y + y') + \beta(z + z') + \gamma(t + t'), \tag{3.463}$$

$$uu' = xx' + yt' + zz' + ty' + \alpha(xy' + yx' + zt' + tz') \\ + \beta(xz' + yy' + zx' + tt') + \gamma(xt' + yz' + zy' + tx'). \tag{3.464}$$

If u, u', u'' are polar fourcomplex numbers, the multiplication is associative

$$(uu')u'' = u(u'u'') \tag{3.465}$$

and commutative

$$uu' = u'u, \tag{3.466}$$

as can be checked through direct calculation. The polar fourcomplex zero is $0 + \alpha \cdot 0 + \beta \cdot 0 + \gamma \cdot 0$, denoted simply 0, and the polar fourcomplex unity is $1 + \alpha \cdot 0 + \beta \cdot 0 + \gamma \cdot 0$, denoted simply 1.

The inverse of the polar fourcomplex number $u = x + \alpha y + \beta z + \gamma t$ is a polar fourcomplex number $u' = x' + \alpha y' + \beta z' + \gamma t'$ having the property that

$$uu' = 1. \tag{3.467}$$

Written on components, the condition, Eq. (3.467), is

$$\begin{aligned} xx' + ty' + zz' + yt' &= 1, \\ yx' + xy' + tz' + zt' &= 0, \\ zx' + yy' + xz' + tt' &= 0, \\ tx' + zy' + yz' + xt' &= 0. \end{aligned} \tag{3.468}$$

The system (3.468) has the solution

$$x' = \frac{x(x^2 - z^2) + z(y^2 + t^2) - 2xyt}{\nu}, \tag{3.469}$$

$$y' = \frac{-y(x^2 + z^2) + t(y^2 - t^2) + 2xzt}{\nu}, \tag{3.470}$$

$$z' = \frac{-z(x^2 - z^2) + x(y^2 + t^2) - 2yzt}{\nu}, \tag{3.471}$$

$$t' = \frac{-t(x^2 + z^2) - y(y^2 - t^2) + 2xyz}{\nu}, \tag{3.472}$$

provided that $\nu \neq 0$, where

$$\begin{aligned} \nu = x^4 + z^4 - y^4 - t^4 \\ -2x^2z^2 + 2y^2t^2 - 4x^2yt - 4yz^2t + 4xy^2z + 4xzt^2. \end{aligned} \tag{3.473}$$

The quantity ν can be written as

$$\nu = v_+ v_- \mu_+^2, \tag{3.474}$$

where

$$v_+ = x + y + z + t, \quad v_- = x - y + z - t, \tag{3.475}$$

and

$$\mu_+^2 = (x - z)^2 + (y - t)^2. \tag{3.476}$$

Then a polar fourcomplex number $q = x + \alpha y + \beta z + \gamma t$ has an inverse, unless

$$v_+ = 0, \text{ or } v_- = 0, \text{ or } \mu_+ = 0. \tag{3.477}$$

The condition $v_+ = 0$ represents the 3-dimensional hyperplane $x + y + z + t = 0$, the condition $v_- = 0$ represents the 3-dimensional hyperplane $x - y + z - t = 0$, and the condition $\mu_+ = 0$ represents the 2-dimensional hyperplane $x = z, y = t$. For arbitrary values of the variables x, y, z, t, the quantity ν can be positive or negative. If $\nu \geq 0$, the quantity $\rho = \nu^{1/4}$ will be called amplitude of the polar fourcomplex number $x + \alpha y + \beta z + \gamma t$. Because of conditions (3.477), these hyperplanes will be called nodal hyperplanes.

It can be shown that if $uu' = 0$ then either $u = 0$, or $u' = 0$, or the polar fourcomplex numbers u, u' belong to different members of the pairs of orthogonal hypersurfaces listed further,

$$x + y + z + t = 0 \text{ and } x' = y' = z' = t', \tag{3.478}$$

$$x - y + z - t = 0 \text{ and } x' = -y' = z' = -t'. \tag{3.479}$$

Divisors of zero also exist if the polar fourcomplex numbers u, u' belong to different members of the pair of two-dimensional hypersurfaces,

$$x - z = 0, \ y - t = 0 \text{ and } x' + z' = 0, \ y' + t' = 0. \tag{3.480}$$

3.4.2   Geometric representation of polar fourcomplex numbers

The polar fourcomplex number $x + \alpha y + \beta z + \gamma t$ can be represented by the point A of coordinates (x, y, z, t). If O is the origin of the four-dimensional space x, y, z, t, the distance from A to the origin O can be taken as

$$d^2 = x^2 + y^2 + z^2 + t^2. \tag{3.481}$$

The distance d will be called modulus of the polar fourcomplex number $x + \alpha y + \beta z + \gamma t$, $d = |u|$.

If $u = x + \alpha y + \beta z + \gamma t$, $u_1 = x_1 + \alpha y_1 + \beta z_1 + \gamma t_1$, $u_2 = x_2 + \alpha y_2 + \beta z_2 + \gamma t_2$, and $u = u_1 u_2$, and if

$$s_{j+} = x_j + y_j + z_j + t_j, \ s_{j-} = x_j - y_j + z_j - t_j, \tag{3.482}$$

for $j = 1, 2$, it can be shown that

$$v_+ = s_{1+} s_{2+}, \ v_- = s_{1-} s_{2-}. \tag{3.483}$$

The relations (3.483) are a consequence of the identities

$$\begin{aligned}
(x_1 x_2 &+ z_1 z_2 + t_1 y_2 + y_1 t_2) + (x_1 y_2 + y_1 x_2 + z_1 t_2 + t_1 z_2) \\
&+ (x_1 z_2 + z_1 x_2 + y_1 y_2 + t_1 t_2) + (x_1 t_2 + t_1 x_2 + z_1 y_2 + y_1 z_2) \\
&= (x_1 + y_1 + z_1 + t_1)(x_2 + y_2 + z_2 + t_2), \tag{3.484}
\end{aligned}$$

$$\begin{aligned}
(x_1 x_2 &+ z_1 z_2 + t_1 y_2 + y_1 t_2) - (x_1 y_2 + y_1 x_2 + z_1 t_2 + t_1 z_2) \\
&+ (x_1 z_2 + z_1 x_2 + y_1 y_2 + t_1 t_2) - (x_1 t_2 + t_1 x_2 + z_1 y_2 + y_1 z_2) \\
&= (x_1 + z_1 - y_1 - t_1)(x_2 + z_2 - y_2 - t_2). \tag{3.485}
\end{aligned}$$

The differences

$$v_1 = x - z, \ \tilde{v}_1 = y - t \tag{3.486}$$

can be written with the aid of the radius μ_+, Eq. (3.476), and of the azimuthal angle ϕ, where $0 \leq \phi < 2\pi$, as

$$v_1 = \mu_+ \cos \phi, \ \tilde{v}_1 = \mu_+ \sin \phi. \tag{3.487}$$

The variables $v_+, v_-, v_1, \tilde{v}_1$ will be called canonical polar fourcomplex variables.

The distance d, Eq. (3.481), can be written as

$$d^2 = \frac{1}{4}v_+^2 + \frac{1}{4}v_-^2 + \frac{1}{2}\mu_+^2. \tag{3.488}$$

It can be shown that if $u_1 = x_1 + \alpha y_1 + \beta z_1 + \gamma t_1, u_2 = x_2 + \alpha y_2 + \beta z_2 + \gamma t_2$ are polar fourcomplex numbers of polar radii and angles ρ_{1-}, ϕ_1 and respectively ρ_{2-}, ϕ_2, then the polar radius ρ and the angle ϕ of the product polar fourcomplex number $u_1 u_2$ are

$$\mu_+ = \rho_{1-}\rho_{2-}, \tag{3.489}$$

$$\phi = \phi_1 + \phi_2. \tag{3.490}$$

The relation (3.489) is a consequence of the identity

$$\begin{aligned}
&[(x_1 x_2 + z_1 z_2 + y_1 t_2 + t_1 y_2) - (x_1 z_2 + z_1 x_2 + y_1 y_2 + t_1 t_2)]^2 \\
&\quad + [(x_1 y_2 + y_1 x_2 + z_1 t_2 + t_1 z_2) - (x_1 t_2 + t_1 x_2 + z_1 y_2 + y_1 z_2)]^2 \\
&= \left[(x_1 - z_1)^2 + (y_1 - t_1)^2\right]\left[(x_2 - z_2)^2 + (y_2 - t_2)^2\right],
\end{aligned} \tag{3.491}$$

and the relation (3.490) is a consequence of the identities

$$\begin{aligned}
&(x_1 x_2 + z_1 z_2 + y_1 t_2 + t_1 y_2) - (x_1 z_2 + z_1 x_2 + y_1 y_2 + t_1 t_2) \\
&= (x_1 - z_1)(x_2 - z_2) - (y_1 - t_1)(y_2 - t_2),
\end{aligned} \tag{3.492}$$

$$\begin{aligned}
&(x_1 y_2 + y_1 x_2 + z_1 t_2 + t_1 z_2) - (x_1 t_2 + t_1 x_2 + z_1 y_2 + y_1 z_2) \\
&= (y_1 - t_1)(x_2 - z_2) + (x_1 - z_1)(y_2 - t_2).
\end{aligned} \tag{3.493}$$

A consequence of Eqs. (3.483) and (3.489) is that if $u = u_1 u_2$, and $\nu_j = s_j s_j'' \rho_{j-}$, where $j = 1, 2$, then

$$\nu = \nu_1 \nu_2. \tag{3.494}$$

The angles θ_+, θ_- between the line OA and the v_+ and respectively v_- axes are

$$\tan\theta_+ = \frac{\sqrt{2}\mu_+}{v_+}, \tan\theta_- = \frac{\sqrt{2}\mu_+}{v_-}, \tag{3.495}$$

where $0 \le \theta_+ \le \pi$, $0 \le \theta_- \le \pi$. The variable μ_+ can be expressed with the aid of Eq. (3.488) as

$$\mu_+^2 = 2d^2\left(1 + \frac{1}{\tan^2\theta_+} + \frac{1}{\tan^2\theta_-}\right)^{-1}. \tag{3.496}$$

The coordinates x, y, z, t can then be expressed in terms of the distance d, of the polar angles θ_+, θ_- and of the azimuthal angle ϕ as

$$x = \frac{\mu_+(\tan\theta_+ + \tan\theta_-)}{2\sqrt{2}\tan\theta_+\tan\theta_-} + \frac{1}{2}\mu_+\cos\phi, \tag{3.497}$$

$$y = \frac{\mu_+(-\tan\theta_+ + \tan\theta_-)}{2\sqrt{2}\tan\theta_+\tan\theta_-} + \frac{1}{2}\mu_+\sin\phi, \tag{3.498}$$

$$z = \frac{\mu_+(\tan\theta_+ + \tan\theta_-)}{2\sqrt{2}\tan\theta_+\tan\theta_-} - \frac{1}{2}\mu_+\cos\phi, \tag{3.499}$$

$$t = \frac{\mu_+(-\tan\theta_+ + \tan\theta_-)}{2\sqrt{2}\tan\theta_+\tan\theta_-} - \frac{1}{2}\mu_+\sin\phi. \tag{3.500}$$

If $u = u_1 u_2$, then Eqs. (3.483) and (3.489) imply that

$$\tan\theta_+ = \frac{1}{\sqrt{2}}\tan\theta_{1+}\tan\theta_{2+}, \quad \tan\theta_- = \frac{1}{\sqrt{2}}\tan\theta_{1-}\tan\theta_{2-}, \tag{3.501}$$

where

$$\tan\theta_{j+} = \frac{\sqrt{2}\rho_{j+}}{s_j}, \quad \tan\theta_{j-} = \frac{\sqrt{2}\rho_{j-}}{s_j''}. \tag{3.502}$$

An alternative choice of the angular variables is

$$\mu_+ = \sqrt{2}d\cos\theta, \quad v_+ = 2d\sin\theta\cos\lambda, \quad v_- = 2d\sin\theta\sin\lambda, \tag{3.503}$$

where $0 \le \theta \le \pi/2, 0 \le \lambda < 2\pi$. If $u = u_1 u_2$, then

$$\tan\lambda = \tan\lambda_1\tan\lambda_2, \quad d\cos\theta = \sqrt{2}d_1 d_2\cos\theta_1\cos\theta_2, \tag{3.504}$$

where

$$\rho_{j-} = \sqrt{2}d_j\cos\theta_j, \quad s_j = 2d_j\sin\theta_j\cos\lambda_j, \quad s_j'' = 2d_j\sin\theta_j\sin\lambda_j, \tag{3.505}$$

for $j = 1, 2$. The coordinates x, y, z, t can then be expressed in terms of the distance d, of the polar angles θ, λ and of the azimuthal angle ϕ as

$$x = \frac{d}{\sqrt{2}}\sin\theta\sin(\lambda + \pi/4) + \frac{d}{\sqrt{2}}\cos\theta\cos\phi, \tag{3.506}$$

$$y = \frac{d}{\sqrt{2}}\sin\theta\cos(\lambda + \pi/4) + \frac{d}{\sqrt{2}}\cos\theta\sin\phi, \tag{3.507}$$

$$z = \frac{d}{\sqrt{2}}\sin\theta\sin(\lambda + \pi/4) - \frac{d}{\sqrt{2}}\cos\theta\cos\phi, \tag{3.508}$$

$$t = \frac{d}{\sqrt{2}} \sin\theta \cos(\lambda + \pi/4) - \frac{d}{\sqrt{2}} \cos\theta \sin\phi. \qquad (3.509)$$

The polar fourcomplex numbers

$$e_+ = \frac{1 + \alpha + \beta + \gamma}{4}, \ e_- = \frac{1 - \alpha + \beta - \gamma}{4}, \qquad (3.510)$$

have the property that

$$e_+^2 = e_+, \ e_-^2 = e_-, \ e_+ e_- = 0. \qquad (3.511)$$

The polar fourcomplex numbers

$$e_1 = \frac{1 - \beta}{2}, \tilde{e}_1 = \frac{\alpha - \gamma}{2} \qquad (3.512)$$

have the property that

$$e_1^2 = e_1, \ \tilde{e}_1^2 = -e_1, \ e_1 \tilde{e}_1 = \tilde{e}_1. \qquad (3.513)$$

The polar fourcomplex numbers e_+, e_- are orthogonal to $e_1, \tilde{e}_1$,

$$e_+ e_1 = 0, \ e_+ \tilde{e}_1 = 0, \ e_- e_1 = 0, \ e_- \tilde{e}_1 = 0. \qquad (3.514)$$

The polar fourcomplex number $q = x + \alpha y + \beta z + \gamma t$ can then be written as

$$x + \alpha y + \beta z + \gamma t = v_+ e_+ + v_- e_- + v_1 e_1 + \tilde{v}_1 \tilde{e}_1. \qquad (3.515)$$

The ensemble $e_+, e_-, e_1, \tilde{e}_1$ will be called the canonical polar fourcomplex base, and Eq. (3.515) gives the canonical form of the polar fourcomplex number. Thus, the product of the polar fourcomplex numbers u, u' can be expressed as

$$uu' = v_+ v'_+ e_+ + v_- v'_- e_- + (v_1 v'_1 - \tilde{v}_1 \tilde{v}'_1) e_1 + (v_1 \tilde{v}'_1 + v'_1 \tilde{v}_1) \tilde{e}_1, \ (3.516)$$

where $v'_+ = x' + y' + z' + t', v'_- = x' - y' + z' - t', v'_1 = x' - y', \tilde{v}'_1 = z' - t'$. The moduli of the bases used in Eq. (3.515) are

$$|e_+| = \frac{1}{2}, \ |e_-| = \frac{1}{2}, \ |e_1| = \frac{1}{\sqrt{2}}, \ |\tilde{e}_1| = \frac{1}{\sqrt{2}}. \qquad (3.517)$$

The fact that the amplitude of the product is equal to the product of the amplitudes, as written in Eq. (3.494), can be demonstrated also by using a representation of the multiplication of the polar fourcomplex numbers by matrices, in which the polar fourcomplex number $u = x + \alpha y + \beta z + \gamma t$ is represented by the matrix

$$A = \begin{pmatrix} x & y & z & t \\ t & x & y & z \\ z & t & x & y \\ y & z & t & x \end{pmatrix}. \tag{3.518}$$

The product $u = x + \alpha y + \beta z + \gamma t$ of the polar fourcomplex numbers $u_1 = x_1 + \alpha y_1 + \beta z_1 + \gamma t_1, u_2 = x_2 + \alpha y_2 + \beta z_2 + \gamma t_2$, can be represented by the matrix multiplication

$$A = A_1 A_2. \tag{3.519}$$

It can be checked that the determinant $\det(A)$ of the matrix A is

$$\det A = \nu. \tag{3.520}$$

The identity (3.494) is then a consequence of the fact the determinant of the product of matrices is equal to the product of the determinants of the factor matrices.

3.4.3   The polar fourdimensional cosexponential functions

The exponential function of a hypercomplex variable u and the addition theorem for the exponential function have been written in Eqs. (1.35)-(1.36). If $u = x + \alpha y + \beta z + \gamma t$, then $\exp u$ can be calculated as $\exp u = \exp x \cdot \exp(\alpha y) \cdot \exp(\beta z) \cdot \exp(\gamma t)$. According to Eq. (3.462),

$$\begin{aligned} \alpha^{4m} &= 1, \alpha^{4m+1} = \alpha, \alpha^{4m+2} = \beta, \alpha^{4m+3} = \gamma, \\ \beta^{2m} &= 1, \beta^{2m+1} = \beta, \\ \gamma^{4m} &= 1, \gamma^{4m+1} = \gamma, \gamma^{4m+2} = \beta, \gamma^{4m+3} = \alpha, \end{aligned} \tag{3.521}$$

where m is a natural number, so that $\exp(\alpha y)$, $\exp(\beta z)$ and $\exp(\gamma z)$ can be written as

$$\exp(\beta z) = \cosh z + \beta \sinh z, \tag{3.522}$$

and

$$\exp(\alpha y) = g_{40}(y) + \alpha g_{41}(y) + \beta g_{42}(y) + \gamma g_{43}(y), \tag{3.523}$$

$$\exp(\gamma t) = g_{40}(t) + \gamma g_{41}(t) + \beta g_{42}(t) + \alpha g_{43}(t), \tag{3.524}$$

where the four-dimensional cosexponential functions $g_{40}, g_{41}, g_{42}, g_{43}$ are defined by the series

$$g_{40}(x) = 1 + x^4/4! + x^8/8! + \cdots, \tag{3.525}$$

$$g_{41}(x) = x + x^5/5! + x^9/9! + \cdots, \tag{3.526}$$

$$g_{42}(x) = x^2/2! + x^6/6! + x^{10}/10! + \cdots, \tag{3.527}$$

$$g_{43}(x) = x^3/3! + x^7/7! + x^{11}/11! + \cdots. \tag{3.528}$$

The functions g_{40}, g_{42} are even, the functions g_{41}, g_{43} are odd,

$$g_{40}(-u) = g_{40}(u), \; g_{42}(-u) = g_{42}(u), \; g_{41}(-u) = -g_{41}(u),$$
$$g_{43}(-u) = -g_{43}(u). \tag{3.529}$$

It can be seen from Eqs. (3.525)-(3.528) that

$$g_{40} + g_{41} + g_{42} + g_{43} = e^x, \; g_{40} - g_{41} + g_{42} - g_{43} = e^{-x}, \tag{3.530}$$

and

$$g_{40} - g_{42} = \cos x, \; g_{41} - g_{43} = \sin x, \tag{3.531}$$

so that

$$(g_{40} + g_{41} + g_{42} + g_{43})(g_{40} - g_{41} + g_{42} - g_{43})$$
$$\left[(g_{40} - g_{42})^2 + (g_{41} - g_{43})^2 \right] = 1, \tag{3.532}$$

which can be also written as

$$g_{40}^4 - g_{41}^4 + g_{42}^4 - g_{43}^4 - 2(g_{40}^2 g_{42}^2 - g_{41}^2 g_{43}^2)$$
$$-4(g_{40}^2 g_{41} g_{43} + g_{42}^2 g_{41} g_{43} - g_{41}^2 g_{40} g_{42} - g_{43}^2 g_{40} g_{42}) = 1. \tag{3.533}$$

The combination of terms in Eq. (3.533) is similar to that in Eq. (3.473).

Addition theorems for the four-dimensional cosexponential functions can be obtained from the relation $\exp \alpha(x + y) = \exp \alpha x \cdot \exp \alpha y$, by substituting the expression of the exponentials as given in Eq. (3.523),

$$g_{40}(x + y) = g_{40}(x)g_{40}(y) + g_{41}(x)g_{43}(y) + g_{42}(x)g_{42}(y)$$
$$+ g_{43}(x)g_{41}(y), \tag{3.534}$$

$$g_{41}(x + y) = g_{40}(x)g_{41}(y) + g_{41}(x)g_{40}(y) + g_{42}(x)g_{43}(y)$$
$$+ g_{43}(x)g_{42}(y), \tag{3.535}$$

$$g_{42}(x + y) = g_{40}(x)g_{42}(y) + g_{41}(x)g_{41}(y) + g_{42}(x)g_{40}(y)$$
$$+ g_{43}(x)g_{43}(y), \tag{3.536}$$

$$g_{43}(x+y) = g_{40}(x)g_{43}(y) + g_{41}(x)g_{42}(y) + g_{42}(x)g_{41}(y)$$
$$+ g_{43}(x)g_{40}(y). \tag{3.537}$$

For $x = y$ the relations (3.534)-(3.537) take the form

$$g_{40}(2x) = g_{40}^2(x) + g_{42}^2(x) + 2g_{41}(x)g_{43}(x), \tag{3.538}$$

$$g_{41}(2x) = 2g_{40}(x)g_{41}(x) + 2g_{42}(x)g_{43}(x), \tag{3.539}$$

$$g_{42}(2x) = g_{41}^2(x) + g_{43}^2(x) + 2g_{40}(x)g_{42}(x), \tag{3.540}$$

$$g_{43}(2x) = 2g_{40}(x)g_{43}(x) + 2g_{41}(x)g_{42}(x). \tag{3.541}$$

For $x = -y$ the relations (3.534)-(3.537) and (3.529) yield

$$g_{40}^2(x) + g_{42}^2(x) - 2g_{41}(x)g_{43}(x) = 1, \tag{3.542}$$

$$g_{41}^2(x) + g_{43}^2(x) - 2g_{40}(x)g_{42}(x) = 0. \tag{3.543}$$

From Eqs. (3.522)-(3.524) it can be shown that, for m integer,

$$(\cosh z + \beta \sinh z)^m = \cosh mz + \beta \sinh mz, \tag{3.544}$$

and

$$[g_{40}(y) + \alpha g_{41}(y) + \beta g_{42}(y) + \gamma g_{43}(y)]^m$$
$$= g_{40}(my) + \alpha g_{41}(my) + \beta g_{42}(my) + \gamma g_{43}(my), \tag{3.545}$$

$$[g_{40}(t) + \gamma g_{41}(t) + \beta g_{42}(t) + \alpha g_{43}(t)]^m$$
$$= g_{40}(mt) + \gamma g_{41}(mt) + \beta g_{42}(mt) + \alpha g_{43}(mt). \tag{3.546}$$

Since

$$(\alpha + \gamma)^{2m} = 2^{2m-1}(1 + \beta), \ (\alpha + \gamma)^{2m+1} = 2^{2m}(\alpha + \gamma), \tag{3.547}$$

it can be shown from the definition of the exponential function, Eq. (1.35) that

$$\exp(\alpha + \gamma)x = e_1 + \frac{1 + \beta}{2} \cosh 2x + \frac{\alpha + \gamma}{2} \sinh 2x. \tag{3.548}$$

Substituting in the relation $\exp(\alpha + \gamma)x = \exp \alpha x \exp \gamma x$ the expression of the exponentials from Eqs. (3.523), (3.524) and (3.548) yields

$$g_{40}^2 + g_{41}^2 + g_{42}^2 + g_{43}^2 = \frac{1 + \cosh 2x}{2}, \tag{3.549}$$

$$g_{40}g_{42} + g_{41}g_{43} = \frac{-1 + \cosh 2x}{4}, \tag{3.550}$$

$$g_{40}g_{41} + g_{40}g_{43} + g_{41}g_{42} + g_{42}g_{43} = \frac{1}{2}\sinh 2x, \tag{3.551}$$

where $g_{40}, g_{41}, g_{42}, g_{43}$ are functions of x.

Similarly, since

$$(\alpha - \gamma)^{2m} = (-1)^m 2^{2m-1}(1 - \beta),$$
$$(\alpha - \gamma)^{2m+1} = (-1)^m 2^{2m}(\alpha - \gamma), \tag{3.552}$$

it can be shown from the definition of the exponential function, Eq. (1.35) that

$$\exp(\alpha - \gamma)x = \frac{1 + \beta}{2} + e_1 \cos 2x + \tilde{e}_1 \sin 2x. \tag{3.553}$$

Substituting in the relation $\exp(\alpha - \gamma)x = \exp \alpha x \exp(-\gamma x)$ the expression of the exponentials from Eqs. (3.523), (3.524) and (3.553) yields

$$g_{40}^2 - g_{41}^2 + g_{42}^2 - g_{43}^2 = \frac{1 + \cos 2x}{2}, \tag{3.554}$$

$$g_{40}g_{42} - g_{41}g_{43} = \frac{1 - \cos 2x}{2}, \tag{3.555}$$

$$g_{40}g_{41} - g_{40}g_{43} - g_{41}g_{42} + g_{42}g_{43} = \frac{1}{2}\sin 2x, \tag{3.556}$$

where $g_{40}, g_{41}, g_{42}, g_{43}$ are functions of x.

The expressions of the four-dimensional cosexponential functions are

$$g_{40}(x) = \frac{1}{2}(\cosh x + \cos x), \tag{3.557}$$

$$g_{41}(x) = \frac{1}{2}(\sinh x + \sin x), \tag{3.558}$$

$$g_{42}(x) = \frac{1}{2}(\cosh x - \cos x), \tag{3.559}$$

$$g_{43}(x) = \frac{1}{2}(\sinh x - \sin x). \tag{3.560}$$

The graphs of these four-dimensional cosexponential functions are shown in Fig. 3.4.

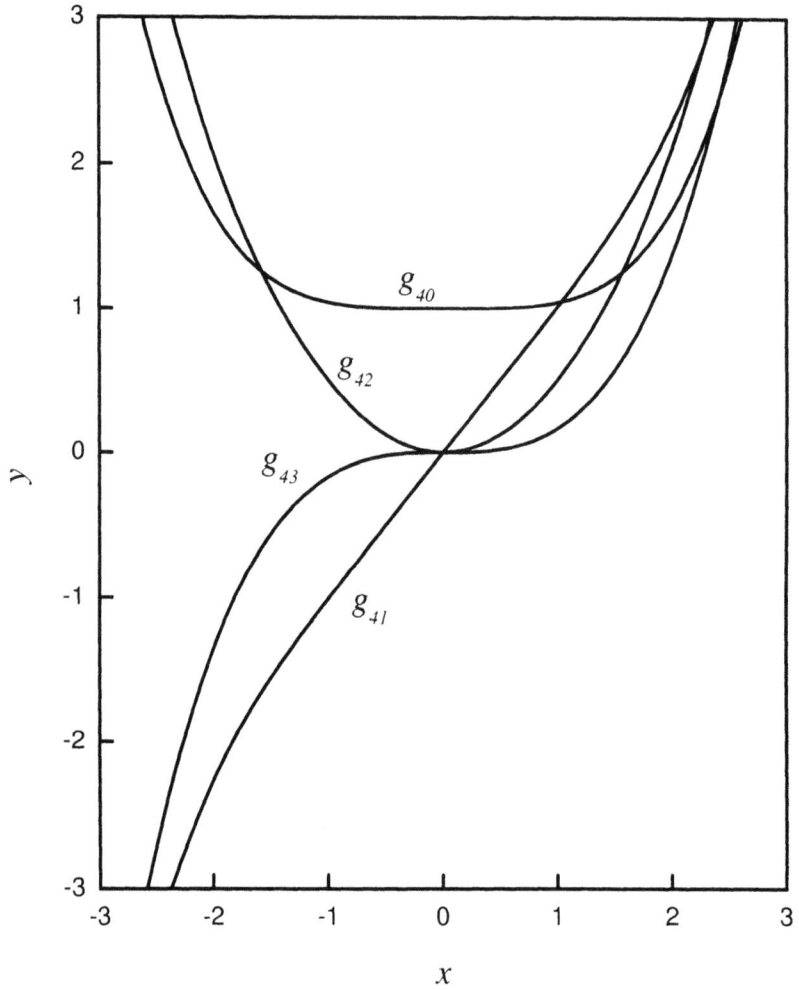

Figure 3.4:   The polar fourdimensional cosexponential functions $g_{40}, g_{41}, g_{42}, g_{43}$.

It can be checked that the cosexponential functions are solutions of the fourth-order differential equation

$$\frac{\mathrm{d}^4\zeta}{\mathrm{d}u^4} = \zeta, \tag{3.561}$$

whose solutions are of the form $\zeta(u) = Ag_{40}(u) + Bg_{41}(u) + Cg_{42(u)} + Dg_{43}(u)$. It can also be checked that the derivatives of the cosexponential functions are related by

$$\frac{dg_{40}}{dw} = g_{43}, \ \frac{dg_{41}}{dw} = g_{40}, \ \frac{dg_{42}}{dw} = g_{41}, \ \frac{dg_{43}}{dw} = g_{42}. \tag{3.562}$$

3.4.4 The exponential and trigonometric forms of a polar fourcomplex number

The polar fourcomplex numbers $u = x + \alpha y + \beta z + \gamma t$ for which $v_+ = x + y + z + t > 0$, $v_- = x - y + z - t > 0$ can be written in the form

$$x + \alpha y + \beta z + \gamma t = e^{x_1 + \alpha y_1 + \beta z_1 + \gamma t_1}. \tag{3.563}$$

The expressions of x_1, y_1, z_1, t_1 as functions of x, y, z, t can be obtained by developing $e^{\alpha y_1}, e^{\beta z_1}$ and $e^{\gamma t_1}$ with the aid of Eqs. (3.522)-(3.524), by multiplying these expressions and separating the hypercomplex components, and then substituting the expressions of the four-dimensional cosexponential functions, Eqs. (3.557)-(3.560),

$$x + y + z + t = e^{x_1 + y_1 + z_1 + t_1}, \tag{3.564}$$

$$x - z = e^{x_1 - z_1} \cos(y_1 - t_1), \tag{3.565}$$

$$x - y + z - t = e^{x_1 - y_1 + z_1 - t_1}, \tag{3.566}$$

$$y - t = e^{x_1 - z_1} \sin(y_1 - t_1). \tag{3.567}$$

It can be shown from Eqs. (3.564)-(3.567) that

$$x_1 = \frac{1}{2}\ln(\mu_+\mu_-), \ y_1 = \frac{1}{2}(\phi + \omega), \ z_1 = \frac{1}{2}\ln\frac{\mu_-}{\mu_+}, \ t_1 = \frac{1}{2}(\phi - \omega), \tag{3.568}$$

where

$$\mu_-^2 = (x + z)^2 - (y + t)^2 = v_+ v_-, \ v_+ > 0, \ v_- > 0. \tag{3.569}$$

The quantities ϕ and ω are determined by

$$\cos\phi = (x - z)/\mu_+, \ \sin\phi = (y - t)/\mu_+, \tag{3.570}$$

and

$$\cosh\omega = (x+z)/\mu_-, \quad \sinh\omega = (y+t)/\mu_-. \tag{3.571}$$

The explicit form of ω is

$$\omega = \frac{1}{2}\ln\frac{x+y+z+t}{x-y+z-t}. \tag{3.572}$$

If $u = u_1 u_2$, and $\mu_{j-}^2 = (x_j+z_j)^2 - (y_j+t_j)^2, j = 1,2$, it can be checked with the aid of Eqs. (3.483) that

$$\mu_- = \mu_{1-}\mu_{2-}. \tag{3.573}$$

Moreover, if $\cosh\omega_j = (x_j+z_j)/\mu_{j-}, \sinh\omega_j = (y_j+t_j)/\mu_{j-}, j = 1,2$, it can be checked that

$$\omega = \omega_1 + \omega_2. \tag{3.574}$$

The relation (3.574) is a consequence of the identities

$$(x_1 x_2 + z_1 z_2 + y_1 t_2 + t_1 y_2) + (x_1 z_2 + z_1 x_2 + y_1 y_2 + t_1 t_2)$$
$$= (x_1 + z_1)(x_2 + z_2) + (y_1 + t_1)(y_2 + t_2), \tag{3.575}$$

$$(x_1 y_2 + y_1 x_2 + z_1 t_2 + t_1 z_2) + (x_1 t_2 + t_1 x_2 + z_1 y_2 + y_1 z_2)$$
$$= (y_1 + t_1)(x_2 + z_2) + (x_1 + z_1)(y_2 + t_2). \tag{3.576}$$

According to Eq. (3.570), ϕ is a cyclic variable, $0 \le \phi < 2\pi$. As it has been assumed that $v_+ > 0, v_- > 0$, it follows that $x + z > 0$ and $x + z > |y + t|$. The range of the variable ω is $-\infty < \omega < \infty$. The exponential form of the polar fourcomplex number u is then

$$u = \rho\exp\left[\frac{1}{2}\beta\ln\frac{\mu_-}{\mu_+} + \frac{1}{2}\alpha(\omega + \phi) + \frac{1}{2}\gamma(\omega - \phi)\right], \tag{3.577}$$

where

$$\rho = (\mu_+\mu_-)^{1/2}. \tag{3.578}$$

If $u = u_1 u_2$, and $\rho_j = (\mu_{j+}\mu_{j-})^{1/2}, j = 1,2$, then from Eqs. (3.489) and (3.573) it results that

$$\rho = \rho_1\rho_2. \tag{3.579}$$

It can be checked with the aid of Eq. (3.513) that

$$\exp(\tilde{e}_1\phi) = \frac{1+\beta}{2} + e_1\cos\phi + \tilde{e}_1\sin\phi, \tag{3.580}$$

which shows that $e^{(\alpha-\gamma)\phi/2}$ is a periodic function of ϕ, with period 2π. The modulus has the property that

$$|u \exp(\tilde{e}_1 \phi)| = |u|. \tag{3.581}$$

By introducing in Eq. (3.577) the polar angles θ_+, θ_- defined in Eqs. (3.495), the exponential form of the fourcomplex number u becomes

$$u = \rho \exp\left[\frac{1}{4}(\alpha + \beta + \gamma)\ln\frac{\sqrt{2}}{\tan\theta_+}\right.$$
$$\left. -\frac{1}{4}(\alpha - \beta + \gamma)\ln\frac{\sqrt{2}}{\tan\theta_-} + \tilde{e}_1\phi\right], \tag{3.582}$$

where $0 < \theta_+ < \pi/2, 0 < \theta_- < \pi/2$. The relation between the amplitude ρ, Eq. (3.578), and the distance d, Eq. (3.481), is, according to Eqs. (3.495) and (3.496),

$$\rho = \frac{2^{3/4}d}{(\tan\theta_+\tan\theta_-)^{1/4}}\left(1 + \frac{1}{\tan^2\theta_+} + \frac{1}{\tan^2\theta_-}\right)^{-1/2}. \tag{3.583}$$

Using the properties of the vectors e_+, e_- written in Eq. (3.511), the first part of the exponential, Eq. (3.582) can be developed as

$$\exp\left[\frac{1}{4}(\alpha + \beta + \gamma)\ln\frac{\sqrt{2}}{\tan\theta_+} - \frac{1}{4}(\alpha - \beta + \gamma)\ln\frac{\sqrt{2}}{\tan\theta_-}\right]$$
$$= \left(\frac{1}{2}\tan\theta_+\tan\theta_-\right)^{1/4}\left(e_1 + e_+\frac{\sqrt{2}}{\tan\theta_+} + e_-\frac{\sqrt{2}}{\tan\theta_-}\right). \tag{3.584}$$

The fourcomplex number u, Eq. (3.582), can then be written as

$$u = d\sqrt{2}\left(1 + \frac{1}{\tan^2\theta_+} + \frac{1}{\tan^2\theta_-}\right)^{-1/2}\left(e_1 + e_+\frac{\sqrt{2}}{\tan\theta_+} + e_-\frac{\sqrt{2}}{\tan\theta_-}\right)$$
$$\exp(\tilde{e}_1\phi), \tag{3.585}$$

which is the trigonometric form of the fourcomplex number u.

The polar angles θ_+, θ_-, Eq. (3.495), can be expressed in terms of the variables θ, λ, Eq. (3.503), as

$$\tan\theta_+ = \frac{1}{\tan\theta\cos\lambda}, \quad \tan\theta_- = \frac{1}{\tan\theta\sin\lambda}, \tag{3.586}$$

so that

$$1 + \frac{1}{\tan^2\theta_+} + \frac{1}{\tan^2\theta_-} = \frac{1}{\cos^2\theta}. \tag{3.587}$$

The exponential form of the fourcomplex number u, written in terms of the amplitude ρ and of the angles θ, λ, ϕ is

$$u = \rho \exp \left[\frac{1}{4}(\alpha + \beta + \gamma) \ln(\sqrt{2} \tan \theta \cos \lambda) \right.$$
$$\left. - \frac{1}{4}(\alpha - \beta + \gamma) \ln(\sqrt{2} \tan \theta \sin \lambda) + \tilde{e}_1 \phi \right],$$

(3.588)

where $0 \leq \lambda < \pi/2$. The trigonometric form of the fourcomplex number u, written in terms of the amplitude ρ and of the angles θ, λ, ϕ is

$$u = d\sqrt{2} \left(e_1 \cos \theta + e_+ \sqrt{2} \sin \theta \cos \lambda + e_- \sqrt{2} \sin \theta \sin \lambda \right) \exp \left(\tilde{e}_1 \phi \right).$$

(3.589)

If $u = u_1 u_2$, it can be shown with the aid of the trigonometric form, Eq. (3.585), that the modulus of the product as a function of the polar angles is

$$d^2 = 4 d_1^2 d_2^2 \left(\frac{1}{2} + \frac{1}{\tan^2 \theta_{1+} \tan^2 \theta_{2+}} + \frac{1}{\tan^2 \theta_{1-} \tan^2 \theta_{2-}} \right)$$
$$\left(1 + \frac{1}{\tan^2 \theta_{1+}} + \frac{1}{\tan^2 \theta_{1-}} \right)^{-1} \left(1 + \frac{1}{\tan^2 \theta_{2+}} + \frac{1}{\tan^2 \theta_{2-}} \right)^{-1}.$$

(3.590)

The modulus d of the product $u_1 u_2$ can be expressed alternatively in terms of the angles θ, λ, ϕ with the aid of the trigonometric form, Eq. (3.589), as

$$d^2 = 4 d_1^2 d_2^2 \left(\frac{1}{2} \cos^2 \theta_1 \cos^2 \theta_2 + \sin^2 \theta_1 \sin^2 \theta_2 \cos^2 \lambda_1 \cos^2 \lambda_2 \right.$$
$$\left. + \sin^2 \theta_1 \sin^2 \theta_2 \sin^2 \lambda_1 \sin^2 \lambda_2 \right).$$

(3.591)

3.4.5   Elementary functions of polar fourcomplex variables

The logarithm u_1 of the polar fourcomplex number u, $u_1 = \ln u$, can be defined as the solution of the equation

$$u = e^{u_1},$$

(3.592)

written explicitly previously in Eq. (3.563), for u_1 as a function of u. From Eq. (3.577) it results that

$$\ln u = \ln \rho + \frac{1}{2} \beta \ln \frac{\mu_-}{\mu_+} + \frac{1}{2}\alpha(\omega + \phi) + \frac{1}{2}\gamma(\omega - \phi).$$

(3.593)

If the fourcomplex number u is written in terms of the amplitude ρ and of the angles θ_+, θ_-, ϕ, the logarithm is

$$\ln u = \ln \rho + \frac{1}{4}(\alpha + \beta + \gamma) \ln \frac{\sqrt{2}}{\tan \theta_+} - \frac{1}{4}(\alpha - \beta + \gamma) \ln \frac{\sqrt{2}}{\tan \theta_-} + \tilde{e}_1 \phi,$$

(3.594)

where $0 < \theta_+ < \pi/2, 0 < \theta_+ < \pi/2$. If the fourcomplex number u is written in terms of the amplitude ρ and of the angles θ, λ, ϕ, the logarithm is

$$\ln u = \ln \rho + \frac{1}{4}(\alpha + \beta + \gamma) \ln(\sqrt{2} \tan \theta \cos \lambda)$$

$$- \frac{1}{4}(\alpha - \beta + \gamma) \ln(\sqrt{2} \tan \theta \sin \lambda) + \tilde{e}_1 \phi,$$

(3.595)

where $0 < \theta < \pi/2, 0 \leq \lambda < \pi/2$. The logarithm is multivalued because of the term proportional to ϕ. It can be inferred from Eq. (3.594) that

$$\ln(u_1 u_2) = \ln u_1 + \ln u_2,$$

(3.596)

up to multiples of $\pi(\alpha - \gamma)$. If the expressions of ρ, μ_+, μ_- and ω in terms of x, y, z, t are introduced in Eq. (3.593), the logarithm of the polar fourcomplex number becomes

$$\ln u = \frac{1 + \alpha + \beta + \gamma}{4} \ln(x + y + z + t)$$

$$+ \frac{1 - \alpha + \beta - \gamma}{4} \ln(x - y + z - t) + e_1 \ln \mu_+ + \tilde{e}_1 \phi.$$

(3.597)

The power function u^m can be defined for $v_+ > 0, v_- > 0$ and real values of m as

$$u^m = e^{m \ln u}.$$

(3.598)

The power function is multivalued unless m is an integer. For integer m, it can be inferred from Eqs. (3.593) and (3.598) that

$$(u_1 u_2)^m = u_1^m u_2^m.$$

(3.599)

Using the expression (3.597) for $\ln u$ and the relations (3.511)-(3.514) it can be shown that

$$u^m = e_+ v_+^m + e_- v_-^m + \mu_+^m (e_1 \cos m\phi + \tilde{e}_1 \sin m\phi).$$

(3.600)

For integer m, the relation (3.600) is valid for any x, y, z, t. For natural m this relation can be written as

$$u^m = e_+ v_+^m + e_- v_-^m + [e_1(x - z) + \tilde{e}_1(y - t)]^m,$$

(3.601)

as can be shown with the aid of the relation

$$e_1 \cos m\phi + \tilde{e}_1 \sin m\phi = (e_1 \cos \phi + \tilde{e}_1 \sin \phi)^m, \qquad (3.602)$$

valid for natural m. For $m = -1$ the relation (3.600) becomes

$$\frac{1}{x + \alpha y + \beta z + \gamma t} = \frac{1}{4}\left(\frac{1 + \alpha + \beta + \gamma}{x + y + z + t} + \frac{1 - \alpha + \beta - \gamma}{x - y + z - t}\right)$$
$$+ \frac{1}{2}\frac{(1 - \beta)(x - z) - (\alpha - \gamma)(y - t)}{(x - z)^2 + (y - t)^2}. \qquad (3.603)$$

If $m = 2$, it can be checked that the right-hand side of Eq. (3.601) is equal to $(x + \alpha y + \beta z + \gamma t)^2 = x^2 + z^2 + 2yt + 2\alpha(xy + zt) + \beta(y^2 + t^2 + 2xz) + 2\gamma(xt + yz)$.

The trigonometric functions of the hypercomplex variable u and the addition theorems for these functions have been written in Eqs. (1.57)-(1.60). The cosine and sine functions of the hypercomplex variables $\alpha y, \beta z$ and γt can be expressed as

$$\cos \alpha y = g_{40} - \beta g_{42}, \sin \alpha y = \alpha g_{41} - \gamma g_{43}, \qquad (3.604)$$

$$\cos \beta y = \cos y, \sin \beta y = \beta \sin y, \qquad (3.605)$$

$$\cos \gamma y = g_{40} - \beta g_{42}, \sin \gamma y = \gamma g_{41} - \alpha g_{43}. \qquad (3.606)$$

The cosine and sine functions of a polar fourcomplex number $x + \alpha y + \beta z + \gamma t$ can then be expressed in terms of elementary functions with the aid of the addition theorems Eqs. (1.59), (1.60) and of the expressions in Eqs. (3.604)-(3.606).

The hyperbolic functions of the hypercomplex variable u and the addition theorems for these functions have been written in Eqs. (1.62)-(1.65). The hyperbolic cosine and sine functions of the hypercomplex variables $\alpha y, \beta z$ and γt can be expressed as

$$\cosh \alpha y = g_{40} + \beta g_{42}, \sinh \alpha y = \alpha g_{41} + \gamma g_{43}, \qquad (3.607)$$

$$\cosh \beta y = \cosh y, \sinh \beta y = \beta \sinh y, \qquad (3.608)$$

$$\cosh \gamma y = g_{40} + \beta g_{42}, \sinh \gamma y = \gamma g_{41} + \alpha g_{43}. \qquad (3.609)$$

The hyperbolic cosine and sine functions of a polar fourcomplex number $x + \alpha y + \beta z + \gamma t$ can then be expressed in terms of elementary functions with the aid of the addition theorems Eqs. (1.64), (1.65) and of the expressions in Eqs. (3.607)-(3.609).

3.4.6 Power series of polar fourcomplex variables

A polar fourcomplex series is an infinite sum of the form

$$a_0 + a_1 + a_2 + \cdots + a_l + \cdots, \tag{3.610}$$

where the coefficients a_l are polar fourcomplex numbers. The convergence of the series (3.610) can be defined in terms of the convergence of its 4 real components. The convergence of a polar fourcomplex series can however be studied using polar fourcomplex variables. The main criterion for absolute convergence remains the comparison theorem, but this requires a number of inequalities which will be discussed further.

The modulus of a polar fourcomplex number $u = x + \alpha y + \beta z + \gamma t$ can be defined as

$$|u| = (x^2 + y^2 + z^2 + t^2)^{1/2}, \tag{3.611}$$

so that according to Eq. (3.481) $d = |u|$. Since $|x| \leq |u|, |y| \leq |u|, |z| \leq |u|, |t| \leq |u|$, a property of absolute convergence established via a comparison theorem based on the modulus of the series (3.610) will ensure the absolute convergence of each real component of that series.

The modulus of the sum $u_1 + u_2$ of the polar fourcomplex numbers u_1, u_2 fulfils the inequality

$$||u_1| - |u_2|| \leq |u_1 + u_2| \leq |u_1| + |u_2|. \tag{3.612}$$

For the product the relation is

$$|u_1 u_2| \leq 2|u_1||u_2|, \tag{3.613}$$

which replaces the relation of equality extant for regular complex numbers. The equality in Eq. (3.613) takes place for $x_1 = y_1 = z_1 = t_1, x_2 = y_2 = z_2 = t_2$, or $x_1 = -y_1 = z_1 = -t_1, x_2 = -y_2 = z_2 = -t_2$. In particular

$$|u^2| \leq 2(x^2 + y^2 + z^2 + t^2). \tag{3.614}$$

The inequality in Eq. (3.613) implies that

$$|u^l| \leq 2^{l-1}|u|^l. \tag{3.615}$$

From Eqs. (3.613) and (3.615) it results that

$$|au^l| \leq 2^l|a||u|^l. \tag{3.616}$$

A power series of the polar fourcomplex variable u is a series of the form

$$a_0 + a_1 u + a_2 u^2 + \cdots + a_l u^l + \cdots. \tag{3.617}$$

Since

$$\left| \sum_{l=0}^{\infty} a_l u^l \right| \le \sum_{l=0}^{\infty} 2^l |a_l| |u|^l, \tag{3.618}$$

a sufficient condition for the absolute convergence of this series is that

$$\lim_{l\to\infty} \frac{2|a_{l+1}||u|}{|a_l|} < 1. \tag{3.619}$$

Thus the series is absolutely convergent for

$$|u| < c_0, \tag{3.620}$$

where

$$c_0 = \lim_{l\to\infty} \frac{|a_l|}{2|a_{l+1}|}. \tag{3.621}$$

The convergence of the series (3.617) can be also studied with the aid of the formula (3.601) which, for integer values of l, is valid for any x, y, z, t. If $a_l = a_{lx} + \alpha a_{ly} + \beta a_{lz} + \gamma a_{lt}$, and

$$\begin{aligned}
A_{l+} &= a_{lx} + a_{ly} + a_{lz} + a_{lt}, \\
A_{l-} &= a_{lx} - a_{ly} + a_{lz} - a_{lt}, \\
A_{l1} &= a_{lx} - a_{lz}, \\
\tilde{A}_{l1} &= a_{ly} - a_{lt},
\end{aligned} \tag{3.622}$$

it can be shown with the aid of relations (3.511)-(3.514) and (3.601) that the expression of the series (3.617) is

$$\sum_{l=0}^{\infty} \left[A_{l+} v_+^l e_+ + A_{l-} v_- e_- + \left(e_1 A_{l1} + \tilde{e}_1 \tilde{A}_{l1} \right) \left(e_1 v_1 + \tilde{e}_1 \tilde{v}_1 \right)^l \right], \tag{3.623}$$

where the quantities v_+, v_- have been defined in Eq. (3.475), and the quantities $v_1, \tilde{v}_1$ have been defined in Eq. (3.486).

The sufficient conditions for the absolute convergence of the series in Eq. (3.623) are that

$$\lim_{l\to\infty} \frac{|A_{l+1,+}||v_+|}{|A_{l+}|} < 1, \ \lim_{l\to\infty} \frac{|A_{l+1,-}||v_-|}{|A_{l-}|} < 1, \ \lim_{l\to\infty} \frac{A_{l+1}\mu_+}{A_l} < 1, \tag{3.624}$$

where the real and positive quantity $A_{l-} > 0$ is given by

$$A_l^2 = A_{l1}^2 + \tilde{A}_{l1}^2. \tag{3.625}$$

Thus the series in Eq. (3.623) is absolutely convergent for

$$|x + y + z + t| < c_+, \ |x - y + z - t| < c_-, \ \mu_+ < c_1, \tag{3.626}$$

where

$$c_+ = \lim_{l \to \infty} \frac{|A_{l+}|}{|A_{l+1,+}|}, \quad c_- = \lim_{l \to \infty} \frac{|A_{l-}|}{|A_{l+1,-}|}, \quad c_1 = \lim_{l \to \infty} \frac{A_{l-}}{A_{l+1,-}}. \qquad (3.627)$$

The relations (3.626) show that the region of convergence of the series (3.623) is a four-dimensional cylinder. It can be shown that $c_0 = (1/2) \min(c_+, c_-, c_1)$, where min designates the smallest of the numbers in the argument of this function. Using the expression of $|u|$ in Eq. (3.488), it can be seen that the spherical region of convergence defined in Eqs. (3.620), (3.621) is included in the cylindrical region of convergence defined in Eqs. (3.627).

3.4.7 Analytic functions of polar fourcomplex variables

The fourcomplex function $f(u)$ of the fourcomplex variable u has been expressed in Eq. (3.117) in terms of the real functions $P(x,y,z,t), Q(x,y,z,t), R(x,y,z,t), S(x,y,z,t)$ of real variables x,y,z,t. The relations between the partial derivatives of the functions P, Q, R, S are obtained by setting succesively in Eq. (3.118) $\Delta x \to 0, \Delta y = \Delta z = \Delta t = 0$; then $\Delta y \to 0, \Delta x = \Delta z = \Delta t = 0$; then $\Delta z \to 0, \Delta x = \Delta y = \Delta t = 0$; and finally $\Delta t \to 0, \Delta x = \Delta y = \Delta z = 0$. The relations are

$$\frac{\partial P}{\partial x} = \frac{\partial Q}{\partial y} = \frac{\partial R}{\partial z} = \frac{\partial S}{\partial t}, \qquad (3.628)$$

$$\frac{\partial Q}{\partial x} = \frac{\partial R}{\partial y} = \frac{\partial S}{\partial z} = \frac{\partial P}{\partial t}, \qquad (3.629)$$

$$\frac{\partial R}{\partial x} = \frac{\partial S}{\partial y} = \frac{\partial P}{\partial z} = \frac{\partial Q}{\partial t}, \qquad (3.630)$$

$$\frac{\partial S}{\partial x} = \frac{\partial P}{\partial y} = \frac{\partial Q}{\partial z} = \frac{\partial R}{\partial t}. \qquad (3.631)$$

The relations (3.628)-(3.631) are analogous to the Riemann relations for the real and imaginary components of a complex function. It can be shown from Eqs. (3.628)-(3.631) that the component P is a solution of the equations

$$\frac{\partial^2 P}{\partial x^2} - \frac{\partial^2 P}{\partial z^2} = 0, \quad \frac{\partial^2 P}{\partial y^2} - \frac{\partial^2 P}{\partial t^2} = 0, \qquad (3.632)$$

and the components Q, R, S are solutions of similar equations. As can be seen from Eqs. (3.632)-(3.632), the components P, Q, R, S of an analytic function of polar fourcomplex variable are harmonic with respect to the pairs of variables x, y and z, t. The component P is also a solution of the mixed-derivative equations

$$\frac{\partial^2 P}{\partial x^2} = \frac{\partial^2 P}{\partial y \partial t}, \quad \frac{\partial^2 P}{\partial y^2} = \frac{\partial^2 P}{\partial x \partial z}, \quad \frac{\partial^2 P}{\partial z^2} = \frac{\partial^2 P}{\partial y \partial t}, \quad \frac{\partial^2 P}{\partial t^2} = \frac{\partial^2 P}{\partial x \partial z}, \quad (3.633)$$

and the components Q, R, S are solutions of similar equations. The component P is also a solution of the mixed-derivative equations

$$\frac{\partial^2 P}{\partial x \partial y} = \frac{\partial^2 P}{\partial z \partial t}, \quad \frac{\partial^2 P}{\partial x \partial t} = \frac{\partial^2 P}{\partial y \partial z}, \quad (3.634)$$

and the components Q, R, S are solutions of similar equations.

3.4.8 Integrals of functions of polar fourcomplex variables

The singularities of polar fourcomplex functions arise from terms of the form $1/(u-u_0)^m$, with $m > 0$. Functions containing such terms are singular not only at $u = u_0$, but also at all points of the two-dimensional hyperplanes passing through u_0 and which are parallel to the nodal hyperplanes.

The integral of a polar fourcomplex function between two points A, B along a path situated in a region free of singularities is independent of path, which means that the integral of an analytic function along a loop situated in a region free from singularities is zero,

$$\oint_\Gamma f(u) du = 0, \quad (3.635)$$

where it is supposed that a surface Σ spanning the closed loop Γ is not intersected by any of the hyperplanes associated with the singularities of the function $f(u)$. Using the expression, Eq. (3.117) for $f(u)$ and the fact that $du = dx + \alpha dy + \beta dz + \gamma dt$, the explicit form of the integral in Eq. (3.635) is

$$\oint_\Gamma f(u) du = \oint_\Gamma [(P dx + S dy + R dz + Q dt)$$
$$+ \alpha(Q dx + P dy + S dz + R dt) + \beta(R dx + Q dy + P dz + S dt)$$
$$+ \gamma(S dx + R dy + Q dz + P dt)]. \quad (3.636)$$

If the functions P, Q, R, S are regular on a surface Σ spanning the loop Γ, the integral along the loop Γ can be transformed with the aid of the theorem of Stokes in an integral over the surface Σ of terms of the form $\partial P/\partial y - \partial S/\partial x$, $\partial P/\partial z - \partial R/\partial x$, $\partial P/\partial t - \partial Q/\partial x$, $\partial R/\partial y - \partial S/\partial z$, $\partial S/\partial t - \partial Q/\partial y$, $\partial R/\partial t - \partial Q/\partial z$ and of similar terms arising from the α, β and γ

components, which are equal to zero by Eqs. (3.628)-(3.631), and this proves Eq. (3.635).

The integral of the function $(u - u_0)^m$ on a closed loop Γ is equal to zero for m a positive or negative integer not equal to -1,

$$\oint_\Gamma (u - u_0)^m du = 0, \quad m \text{ integer}, \ m \neq -1. \tag{3.637}$$

This is due to the fact that $\int (u - u_0)^m du = (u - u_0)^{m+1}/(m + 1)$, and to the fact that the function $(u - u_0)^{m+1}$ is singlevalued for m an integer.

The integral $\oint_\Gamma du/(u - u_0)$ can be calculated using the exponential form (3.582),

$$u - u_0 = \rho \exp \left[\frac{1}{4}(\alpha + \beta + \gamma) \ln \frac{\sqrt{2}}{\tan \theta_+} \right.$$
$$\left. - \frac{1}{4}(\alpha - \beta + \gamma) \ln \frac{\sqrt{2}}{\tan \theta_-} + \tilde{e}_1 \phi \right], \tag{3.638}$$

so that

$$\frac{du}{u - u_0} = \frac{d\rho}{\rho} + \frac{1}{4}(\alpha + \beta + \gamma) d \ln \frac{\sqrt{2}}{\tan \theta_+} - \frac{1}{4}(\alpha - \beta + \gamma) d \ln \frac{\sqrt{2}}{\tan \theta_-}$$
$$+ \tilde{e}_1 d\phi. \tag{3.639}$$

Since ρ, $\tan \theta_+$ and $\tan \theta_-$ are singlevalued variables, it follows that $\oint_\Gamma d\rho/\rho = 0$, $\oint_\Gamma d \ln \sqrt{2}/\tan \theta_+ = 0$, and $\oint_\Gamma d \ln \sqrt{2}/\tan \theta_+ = 0$. On the other hand, ϕ is a cyclic variables, so that it may give a contribution to the integral around the closed loop Γ. The result of the integrations will be given in the rotated system of coordinates

$$\xi = \frac{1}{\sqrt{2}}(x - z), \ v = \frac{1}{\sqrt{2}}(y - t), \ \tau = \frac{1}{2}(x + y + z + t),$$
$$v = \frac{1}{2}(x - y + z - t). \tag{3.640}$$

Thus, if $C_\parallel$ is a circle of radius r parallel to the $\xi O v$ plane, and the projection of the center of this circle on the $\xi O v$ plane coincides with the projection of the point u_0 on this plane, the points of the circle $C_\parallel$ are described according to Eqs. (3.486)-(3.487) by the equations

$$\xi = \xi_0 + r \cos \phi, \ v = v_0 + r \sin \phi, \ \tau = \tau_0, \ \zeta = \zeta_0, \tag{3.641}$$

where $u_0 = x_0 + \alpha y_0 + \beta z_0 + \gamma t_0$, and $\xi_0, v_0, \tau_0, \zeta_0$ are calculated from x_0, y_0, z_0, t_0 according to Eqs. (3.640).

Then

$$\oint_{C_{\parallel}} \frac{du}{u - u_0} = 2\pi \tilde{e}_1. \qquad (3.642)$$

The expression of $\oint_{\Gamma} du/(u - u_0)$ can be written with the aid of the functional $\mathrm{int}(M, C)$ defined in Eq. (3.134) as

$$\oint_{\Gamma} \frac{du}{u - u_0} = 2\pi \tilde{e}_1 \, \mathrm{int}(u_{0\xi v}, \Gamma_{\xi v}), \qquad (3.643)$$

where $u_{0\xi v}$ and $\Gamma_{\xi v}$ are respectively the projections of the point u_0 and of the loop Γ on the plane ξv.

If $f(u)$ is an analytic polar fourcomplex function which can be expanded in a series as written in Eq. (1.89), and the expansion holds on the curve Γ and on a surface spanning Γ, then from Eqs. (3.637) and (3.643) it follows that

$$\oint_{\Gamma} \frac{f(u)du}{u - u_0} = 2\pi \tilde{e}_1 \, \mathrm{int}(u_{0\xi v}, \Gamma_{\xi v}) f(u_0), \qquad (3.644)$$

where $\Gamma_{\xi v}$ is the projection of the curve Γ on the plane ξv, as shown in Fig. 3.5. Substituting in the right-hand side of Eq. (3.644) the expression of $f(u)$ in terms of the real components P, Q, R, S, Eq. (3.117), yields

$$\oint_{\Gamma} \frac{f(u)du}{u - u_0} = \pi \left[(\beta - 1)(Q - S) + (\alpha - \gamma)(P - R) \right] \mathrm{int}(u_{0\xi v}, \Gamma_{\xi v}), \qquad (3.645)$$

where P, Q, R, S are the values of the components of f at $u = u_0$. If the integral is written as

$$\oint_{\Gamma} \frac{f(u)du}{u - u_0} = I + \alpha I_{\alpha} + \beta I_{\beta} + \gamma I_{\gamma}, \qquad (3.646)$$

it results from Eq. (3.645) that

$$I + I_{\alpha} + I_{\beta} + I_{\gamma} = 0. \qquad (3.647)$$

If $f(u)$ can be expanded as written in Eq. (1.89) on Γ and on a surface spanning Γ, then from Eqs. (3.637) and (3.643) it also results that

$$\oint_{\Gamma} \frac{f(u)du}{(u - u_0)^{m+1}} = \frac{2\pi}{m!} \tilde{e}_1 \, \mathrm{int}(u_{0\xi v}, \Gamma_{\xi v}) \, f^{(m)}(u_0), \qquad (3.648)$$

where the fact has been used that the derivative $f^{(m)}(u_0)$ of order m of $f(u)$ at $u = u_0$ is related to the expansion coefficient in Eq. (1.89) according to Eq. (1.93).

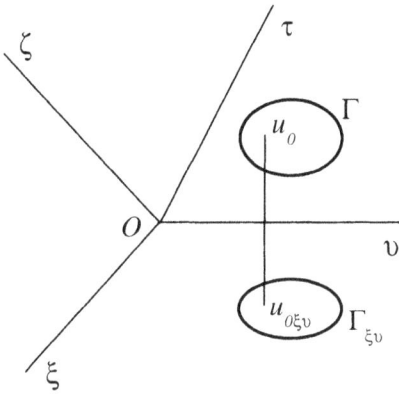

Figure 3.5: Integration path Γ and the pole u_0, and their projections $\Gamma_{\xi\upsilon}$ and $u_{0\xi\upsilon}$ on the plane $\xi\upsilon$.

If a function $f(u)$ is expanded in positive and negative powers of $u - u_j$, where u_j are polar fourcomplex constants, j being an index, the integral of f on a closed loop Γ is determined by the terms in the expansion of f which are of the form $a_j/(u - u_j)$,

$$f(u) = \cdots + \sum_j \frac{a_j}{u - u_j} + \cdots . \tag{3.649}$$

Then the integral of f on a closed loop Γ is

$$\oint_\Gamma f(u)du = 2\pi\tilde{e}_1 \sum_j \mathrm{int}(u_{j\xi\upsilon}, \Gamma_{\xi\upsilon})a_j. \tag{3.650}$$

3.4.9 Factorization of polar fourcomplex polynomials

A polynomial of degree m of the polar fourcomplex variable $u = x + \alpha y + \beta z + \gamma t$ has the form

$$P_m(u) = u^m + a_1 u^{m-1} + \cdots + a_{m-1}u + a_m, \tag{3.651}$$

where the constants are in general polar fourcomplex numbers.

If $a_m = a_{mx} + \alpha a_{my} + \beta a_{mz} + \gamma a_{mt}$, and with the notations of Eqs. (3.475) and (3.622) applied for $0, 1, \cdots, m$, the polynomial $P_m(u)$ can be written as

$$
\begin{aligned}
P_m = {} & \left[v_+^m + A_1 v_+^{m-1} + \cdots + A_{m-1} v_+ + A_m \right] e_+ \\
& + \left[v_-^m + A_1'' v_-^{m-1} + \cdots + A_{m-1}'' v_- + A_m'' \right] e_- \\
& + \left[(e_1 v_1 + \tilde{e}_1 \tilde{v}_1)^m + \sum_{l=1}^{m} \left(e_1 A_{l1} + \tilde{e}_1 \tilde{A}_{l1} \right) (e_1 v_1 + \tilde{e}_1 \tilde{v}_1)^{m-l} \right],
\end{aligned}
$$

$$(3.652)$$

where the constants $A_{l+}, A_{l-}, A_{l1}, \tilde{A}_{l1}$ are real numbers. The polynomial of degree m in $(e_1 v_1 + \tilde{e}_1 \tilde{v}_1)$ can always be written as a product of linear factors of the form $[e_1(v_1 - v_{1l}) + \tilde{e}_1(\tilde{v}_1 - \tilde{v}_{1l})]$, where the constants $v_{1l}, \tilde{v}_{1l}$ are real. The two polynomials of degree m with real coefficients in Eq. (3.652) which are multiplied by e_+ and e_- can be written as a product of linear or quadratic factors with real coefficients, or as a product of linear factors which, if imaginary, appear always in complex conjugate pairs. Using the latter form for the simplicity of notations, the polynomial P_m can be written as

$$
\begin{aligned}
P_m = {} & \prod_{l=1}^{m} (v_+ - s_{l+}) e_+ + \prod_{l=1}^{m} (v_- - s_{l-}) e_- \\
& + \prod_{l=1}^{m} \left[e_1(v_1 - v_{1l}) + \tilde{e}_1(\tilde{v}_1 - \tilde{v}_{1l}) \right],
\end{aligned}
$$

$$(3.653)$$

where the quantities s_{l+} appear always in complex conjugate pairs, and the same is true for the quantities s_{l-}. Due to the properties in Eqs. (3.511)-(3.514), the polynomial $P_m(u)$ can be written as a product of factors of the form

$$
\begin{aligned}
P_m(u) = {} & \prod_{l=1}^{m} \left[(v_+ - s_{l+}) e_+ + (v_- - s_{l-}) e_- \right. \\
& \left. + (e_1(v_1 - v_{1l}) + \tilde{e}_1(\tilde{v}_1 - \tilde{v}_{1l})) \right].
\end{aligned}
$$

$$(3.654)$$

These relations can be written with the aid of Eq. (3.515) as

$$
P_m(u) = \prod_{p=1}^{m} (u - u_p),
$$

$$(3.655)$$

where

$$
u_p = s_{p+} e_+ + s_{p-} e_- + e_1 v_{1p} + \tilde{e}_1 \tilde{v}_{1p}, \quad p = 1, ..., m.
$$

$$(3.656)$$

The roots $s_{p+}, s_{p-}, v_{1p}e_1 + \tilde{v}_{1p}\tilde{e}_1$ of the corresponding polynomials in Eq. (3.653) may be ordered arbitrarily. This means that Eq. (3.656) gives sets of m roots $u_1, ..., u_m$ of the polynomial $P_m(u)$, corresponding to the various ways in which the roots $s_{p+}, s_{p-}, v_{1p}e_1 + \tilde{v}_{1p}\tilde{e}_1$ are ordered according to p in each group. Thus, while the hypercomplex components in Eq. (3.652) taken separately have unique factorizations, the polynomial $P_m(u)$ can be written in many different ways as a product of linear factors. The result of the polar fourcomplex integration, Eq. (3.650), is however unique.

If $P(u) = u^2 - 1$, the factorization in Eq. (3.655) is $u^2 - 1 = (u - u_1)(u - u_2)$, where $u_1 = \pm e_+ \pm e_- \pm e_1, u_2 = -u_1$, so that there are 4 distinct factorizations of $u^2 - 1$,

$$
\begin{aligned}
u^2 - 1 &= (u - 1)(u + 1), \\
u^2 - 1 &= (u - \beta)(u + \beta), \\
u^2 - 1 &= \left(u - \tfrac{1+\alpha-\beta+\gamma}{2}\right)\left(u + \tfrac{1+\alpha-\beta+\gamma}{2}\right), \\
u^2 - 1 &= \left(u - \tfrac{-1+\alpha+\beta+\gamma}{2}\right)\left(u + \tfrac{-1+\alpha+\beta+\gamma}{2}\right).
\end{aligned}
\tag{3.657}
$$

It can be checked that $\{\pm e_+ \pm e_- \pm e_1\}^2 = e_+ + e_- + e_1 = 1$.

3.4.10 Representation of polar fourcomplex numbers by irreducible matrices

If T is the unitary matrix,

$$
T = \begin{pmatrix}
\frac{1}{2} & \frac{1}{2} & \frac{1}{2} & \frac{1}{2} \\
\frac{1}{2} & -\frac{1}{2} & \frac{1}{2} & -\frac{1}{2} \\
\frac{1}{\sqrt{2}} & 0 & -\frac{1}{\sqrt{2}} & 0 \\
0 & \frac{1}{\sqrt{2}} & 0 & -\frac{1}{\sqrt{2}}
\end{pmatrix},
\tag{3.658}
$$

it can be shown that the matrix TUT^{-1} has the form

$$
TUT^{-1} = \begin{pmatrix}
x + y + z + t & 0 & 0 \\
0 & x - y + z - t & 0 \\
0 & 0 & V_1
\end{pmatrix},
\tag{3.659}
$$

where U is the matrix in Eq. (3.513) used to represent the polar fourcomplex number u. In Eq. (3.659), V_1 is the matrix

$$
V_1 = \begin{pmatrix}
x - z & y - t \\
-y + t & x - z
\end{pmatrix}.
\tag{3.660}
$$

The relations between the variables $x + y + z + t, x - y + z - t, x - z, y - t$ for the multiplication of polar fourcomplex numbers have been written in Eqs. (3.484),(3.485), (3.492), (3.493). The matrix TUT^{-1} provides an irreducible representation [7] of the polar fourcomplex number u in terms of matrices with real coefficients.

Chapter 4

Complex Numbers in 5 Dimensions

A system of complex numbers in 5 dimensions is described in this chapter, for which the multiplication is associative and commutative, which have exponential and trigonometric forms, and for which the concepts of analytic 5-complex function, contour integration and residue can be defined. The 5-complex numbers introduced in this chapter have the form $u = x_0 + h_1 x_1 + h_2 x_2 + h_3 x_3 + h_4 x_4$, the variables x_0, x_1, x_2, x_3, x_4 being real numbers. If the 5-complex number u is represented by the point A of coordinates x_0, x_1, x_2, x_3, x_4, the position of the point A can be described by the modulus $d = (x_0^2 + x_1^2 + x_2^2 + x_3^2 + x_4^2)^{1/2}$, by 2 azimuthal angles ϕ_1, ϕ_2, by 1 planar angle ψ_1, and by 1 polar angle θ_+.

The exponential function of a 5-complex number can be expanded in terms of the polar 5-dimensional cosexponential functions $g_{5k}(y)$, $k = 0, 1, 2, 3, 4$. The expressions of these functions are obtained from the properties of the exponential function of a 5-complex variable. Addition theorems and other relations are obtained for the polar 5-dimensional cosexponential functions. Exponential and trigonometric forms are given for the 5-complex numbers. Expressions are obtained for the elementary functions of 5-complex variable. The functions $f(u)$ of 5-complex variable which are defined by power series have derivatives independent of the direction of approach to the point under consideration. If the 5-complex function $f(u)$ of the 5-complex variable u is written in terms of the real functions $P_k(x_0, x_1, x_2, x_3, x_4), k = 0, 1, 2, 3, 4$, then relations of equality exist between partial derivatives of the functions P_k. The integral $\int_A^B f(u) du$ of a 5-complex function between two points A, B is independent of the path connecting A, B, in regions where f is regular. The fact that the exponential form of the 5-complex numbers depends on the cyclic variables ϕ_1, ϕ_2 leads to the concept of pole and residue for integrals on closed paths, and if $f(u)$ is an analytic 5-complex function, then $\oint_\Gamma f(u)du/(u-u_0)$ is expressed in this chapter in terms of the 5-complex residue $f(u_0)$. The polynomials of 5-complex variables can be written as products of linear or quadratic factors.

The 5-complex numbers described in this chapter are a particular case for $n = 5$ of the polar complex numbers in n dimensions discussed in Sec. 6.1.

4.1 Operations with polar complex numbers in 5 dimensions

A polar hypercomplex number u in 5 dimensions is represented as

$$u = x_0 + h_1 x_1 + h_2 x_2 + h_3 x_3 + h_4 x_4. \tag{4.1}$$

The multiplication rules for the bases h_1, h_2, h_3, h_4 are

$$h_1^2 = h_2, \ h_2^2 = h_4, \ h_3^2 = h_1, \ h_4^2 = h_3,$$
$$h_1 h_2 = h_3, \ h_1 h_3 = h_4, \ h_1 h_4 = 1,$$
$$h_2 h_3 = 1, \ h_2 h_4 = h_1, \ h_3 h_4 = h_2. \tag{4.2}$$

The significance of the composition laws in Eq. (4.2) can be understood by representing the bases h_j, h_k by points on a circle at the angles $\alpha_j = 2\pi j/5, \alpha_k = 2\pi k/5$, as shown in Fig. 4.1, and the product $h_j h_k$ by the point of the circle at the angle $2\pi(j + k)/5$. If $2\pi \leq 2\pi(j + k)/5 < 4\pi$, the point represents the basis h_l of angle $\alpha_l = 2\pi(j + k)/5 - 2\pi$.

The sum of the 5-complex numbers u and u' is

$$u+u' = x_0+x_0'+h_1(x_1+x_1')+h_2(x_2+x_2')+h_3(x_3+x_3')+h_4(x_4+x_4'). \tag{4.3}$$

The product of the numbers u, u' is then

$$\begin{aligned} uu' = &\ x_0 x_0' + x_1 x_4' + x_2 x_3' + x_3 x_2' + x_4 x_1' \\ &+ h_1(x_0 x_1' + x_1 x_0' + x_2 x_4' + x_3 x_3' + x_4 x_2') \\ &+ h_2(x_0 x_2' + x_1 x_1' + x_2 x_0' + x_3 x_4' + x_4 x_3') \\ &+ h_3(x_0 x_3' + x_1 x_2' + x_2 x_1' + x_3 x_0' + x_4 x_4') \\ &+ h_4(x_0 x_4' + x_1 x_3' + x_2 x_2' + x_3 x_1' + x_4 x_0'). \end{aligned} \tag{4.4}$$

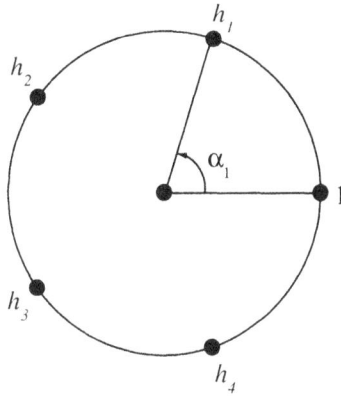

Figure 4.1: Representation of the polar hypercomplex bases $1, h_1, h_2, h_3, h_4$ by points on a circle at the angles $\alpha_k = 2\pi k/5$. The product $h_j h_k$ will be represented by the point of the circle at the angle $2\pi(j+k)/5$, $i, k = 0, 1, ..., 4$, where $h_0 = 1$. If $2\pi \leq 2\pi(j+k)/5 \leq 4\pi$, the point represents the basis h_l of angle $\alpha_l = 2\pi(j+k)/5 - 2\pi$.

The relation between the variables $v_+, v_1, \tilde{v}_1, v_2, \tilde{v}_2$ and x_0, x_1, x_2, x_3, x_4 can be written with the aid of the parameters $p = (\sqrt{5} - 1)/4, q = \sqrt{(5 + \sqrt{5})/8}$ as

$$\begin{pmatrix} v_+ \\ v_1 \\ \tilde{v}_1 \\ v_2 \\ \tilde{v}_2 \end{pmatrix} = \begin{pmatrix} 1 & 1 & 1 & 1 & 1 \\ 1 & p & 2p^2 - 1 & 2p^2 - 1 & p \\ 0 & q & 2pq & -2pq & -q \\ 1 & 2p^2 - 1 & p & p & 2p^2 - 1 \\ 0 & 2pq & -q & q & -2pq \end{pmatrix} \begin{pmatrix} x_0 \\ x_1 \\ x_2 \\ x_3 \\ x_4 \end{pmatrix}.$$

(4.5)

The other variables are $v_3 = v_2, \tilde{v}_3 = -\tilde{v}_2, v_4 = v_1, \tilde{v}_4 = -\tilde{v}_1$. The variables $v_+, v_1, \tilde{v}_1, v_2, \tilde{v}_2$ will be called canonical 5-complex variables.

4.2 Geometric representation of polar complex numbers in 5 dimensions

The 5-complex number $x_0 + h_1 x_1 + h_2 x_2 + h_3 x_3 + h_4 x_4$ can be represented by the point A of coordinates $(x_0, x_1, x_2, x_3, x_4)$. If O is the origin of the

5-dimensional space, the distance from the origin O to the point A of co-ordinates $(x_0, x_1, x_2, x_3, x_4)$ has the expression

$$d^2 = x_0^2 + x_1^2 + x_2^2 + x_3^2 + x_4^2. \tag{4.6}$$

The quantity d will be called modulus of the 5-complex number u. The modulus of a 5-complex number u will be designated by $d = |u|$. The modulus has the property that

$$|u'u''| \leq \sqrt{5}|u'||u''|. \tag{4.7}$$

The exponential and trigonometric forms of the 5-complex number u can be obtained conveniently in a rotated system of axes defined by the transformation

$$\begin{pmatrix} \xi_+ \\ \xi_1 \\ \eta_1 \\ \xi_2 \\ \eta_2 \end{pmatrix} = \sqrt{\frac{2}{5}} \begin{pmatrix} \frac{1}{\sqrt{2}} & \frac{1}{\sqrt{2}} & \frac{1}{\sqrt{2}} & \frac{1}{\sqrt{2}} & \frac{1}{\sqrt{2}} \\ 1 & p & 2p^2-1 & 2p^2-1 & p \\ 0 & q & 2pq & -2pq & -q \\ 1 & 2p^2-1 & p & p & 2p^2-1 \\ 0 & 2pq & -q & q & -2pq \end{pmatrix} \begin{pmatrix} x_0 \\ x_1 \\ x_2 \\ x_3 \\ x_4 \end{pmatrix}. \tag{4.8}$$

The lines of the matrices in Eq. (4.8) gives the components of the 5 basis vectors of the new system of axes. These vectors have unit length and are orthogonal to each other. The relations between the two sets of variables are

$$v_+ = \sqrt{5}\xi_+, v_k = \sqrt{\frac{5}{2}}\xi_k, \tilde{v}_k = \sqrt{\frac{5}{2}}\eta_k, k = 1, 2. \tag{4.9}$$

The radius ρ_k and the azimuthal angle ϕ_k in the plane of the axes $v_k, \tilde{v}_k$ are

$$\rho_k^2 = v_k^2 + \tilde{v}_k^2, \cos\phi_k = v_k/\rho_k, \sin\phi_k = \tilde{v}_k/\rho_k, \tag{4.10}$$

$0 \leq \phi_k < 2\pi$, $k = 1, 2$, so that there are 2 azimuthal angles. The planar angle ψ_1 is

$$\tan\psi_1 = \rho_1/\rho_2, \tag{4.11}$$

where $0 \leq \psi_1 \leq \pi/2$. There is a polar angle θ_+,

$$\tan\theta_+ = \frac{\sqrt{2}\rho_1}{v_+}, \tag{4.12}$$

where $0 \leq \theta_+ \leq \pi$. It can be checked that

$$\frac{1}{5}v_+^2 + \frac{2}{5}(\rho_1^2 + \rho_2^2) = d^2. \tag{4.13}$$

The amplitude of a 5-complex number u is

$$\rho = \left(v_+\rho_1^2\rho_2^2\right)^{1/5}. \tag{4.14}$$

If $u = u'u''$, the parameters of the hypercomplex numbers are related by

$$v_+ = v'_+ v''_+, \tag{4.15}$$

$$\rho_k = \rho'_k \rho''_k, \tag{4.16}$$

$$\tan\theta_+ = \frac{1}{\sqrt{2}}\tan\theta'_+ \tan\theta''_+, \tag{4.17}$$

$$\tan\psi_1 = \tan\psi'_1 \tan\psi''_1, \tag{4.18}$$

$$\phi_k = \phi'_k + \phi''_k, \tag{4.19}$$

$$v_k = v'_k v''_k - \tilde{v}'_k \tilde{v}''_k, \quad \tilde{v}_k = v'_k \tilde{v}''_k + \tilde{v}'_k v''_k, \tag{4.20}$$

$$\rho = \rho'\rho'', \tag{4.21}$$

where $k = 1, 2$.

The 5-complex number $u = x_0 + h_1x_1 + h_2x_2 + h_3x_3 + h_4x_4$ can be represented by the matrix

$$U = \begin{pmatrix} x_0 & x_1 & x_2 & x_3 & x_4 \\ x_4 & x_0 & x_1 & x_2 & x_3 \\ x_3 & x_4 & x_0 & x_1 & x_2 \\ x_2 & x_3 & x_4 & x_0 & x_1 \\ x_1 & x_2 & x_3 & x_4 & x_0 \end{pmatrix}. \tag{4.22}$$

The product $u = u'u''$ is represented by the matrix multiplication $U = U'U''$.

4.3   The polar 5-dimensional cosexponential functions

The polar cosexponential functions in 5 dimensions are

$$g_{5k}(y) = \sum_{p=0}^{\infty} y^{k+5p}/(k+5p)!, \tag{4.23}$$

for $k = 0, ..., 4$. The polar cosexponential functions g_{5k} do not have a definite parity. It can be checked that

$$\sum_{k=0}^{4} g_{5k}(y) = e^{y}. \tag{4.24}$$

The exponential of the quantity $h_k y, k = 1, ..., 4$ can be written as

$$\begin{aligned}
e^{h_1 y} &= g_{50}(y) + h_1 g_{51}(y) + h_2 g_{52}(y) + h_3 g_{53}(y) + h_4 g_{54}(y), \\
e^{h_2 y} &= g_{50}(y) + h_1 g_{53}(y) + h_2 g_{51}(y) + h_3 g_{54}(y) + h_4 g_{52}(y), \\
e^{h_3 y} &= g_{50}(y) + h_1 g_{52}(y) + h_2 g_{54}(y) + h_3 g_{51}(y) + h_4 g_{53}(y), \\
e^{h_4 y} &= g_{50}(y) + h_1 g_{54}(y) + h_2 g_{53}(y) + h_3 g_{52}(y) + h_4 g_{51}(y).
\end{aligned} \tag{4.25}$$

The polar cosexponential functions in 5 dimensions can be obtained by calculating $e^{(h_1+h_4)y}$ and $e^{(h_1-h_4)y}$ and then by multiplying the resulting expression. The series expansions for $e^{(h_1+h_4)y}$ and $e^{(h_1-h_4)y}$ are

$$e^{(h_1+h_4)y} = \sum_{m=0}^{\infty} \frac{1}{m!}(h_1+h_4)^m y^m, \tag{4.26}$$

$$e^{(h_1-h_4)y} = \sum_{m=0}^{\infty} \frac{1}{m!}(h_1-h_4)^m y^m. \tag{4.27}$$

The powers of $h_1 + h_4$ have the form

$$(h_1+h_4)^m = A_m(h_1+h_4) + B_m(h_2+h_3) + C_m. \tag{4.28}$$

The recurrence relations for A_m, B_m, C_m are

$$A_{m+1} = B_m + C_m, B_{m+1} = A_m + B_m, C_{m+1} = 2A_m, \tag{4.29}$$

and $A_1 = 1, B_1 = 0, C_1 = 0, A_2 = 0, B_2 = 1, C_2 = 2, A_3 = 3, B_3 = 1, C_3 = 0$. The expressions of the coefficients are

$$A_m = \frac{2^m}{5} + \frac{2-3a}{5}a^{m-3} + (-1)^{m-3}\frac{5+3a}{5}(1+a)^{m-3}, m \geq 3, \tag{4.30}$$

$$B_m = \frac{2^m}{5} + \frac{a-1}{5}a^{m-3} - (-1)^{m-3}\frac{a+2}{5}(1+a)^{m-3}, m \geq 3, \quad (4.31)$$

$$C_m = \frac{2^m}{5} + \frac{4-6a}{5}a^{m-4} + (-1)^{m-4}\frac{10+6a}{5}(1+a)^{m-4}, m \geq 4, (4.32)$$

where a is a solution of the equation $a^2 + a - 1 = 0$. Substituting the expressions of A_m, B_m, C_m from Eqs. (4.30)-(4.32) in Eq. (4.26) and grouping the terms yields

$$\begin{aligned}
e^{(h_1+h_4)y} =& \frac{1}{5}e^{2y} + \frac{2}{5}e^{ay} + \frac{2}{5}e^{-(1+a)y} \\
&+ (h_1 + h_4)\left[\frac{1}{5}e^{2y} + \frac{a}{5}e^{ay} - \frac{a+1}{5}e^{-(1+a)y}\right] \\
&+ (h_2 + h_3)\left[\frac{1}{5}e^{2y} - \frac{a+1}{5}e^{ay} + \frac{a}{5}e^{-(1+a)y}\right].
\end{aligned} \quad (4.33)$$

The odd powers of $h_1 - h_4$ have the form

$$(h_1 - h_4)^{2m+1} = D_m(h_1 - h_4) + E_m(h_2 - h_3). \quad (4.34)$$

The recurrence relations for D_m, E_m are

$$D_{m+1} = -3D_m - E_m, E_{m+1} = -D_m - 2E_m, \quad (4.35)$$

and $D_1 = -3, E_1 = -1, D_2 = 10, E_2 = 5$. The expressions of the coefficients are

$$D_m = (b+1)b^{m-1} + (-1)^{m-2}(b+4)(5+b)^{m-1}, m \geq 1, \quad (4.36)$$

$$E_m = -\frac{b+1}{b+2}b^{m-1} + \frac{(-1)^{m-2}}{b+2}(5+b)^{m-1}, m \geq 1, \quad (4.37)$$

where b is a solution of the equation $b^2 + 5b + 5 = 0$. The even powers of $h_1 - h_4$ have the form

$$(h_1 - h_4)^{2m} = F_m(h_1 + h_4) + G_m(h_2 + h_3) + H_m. \quad (4.38)$$

The recurrence relations for F_m, G_m, H_m are

$$F_{m+1} = -F_m + G_m, G_{m+1} = F_m - 2G_m + H_m, H_{m+1} = 2(G_m - H_m), \quad (4.39)$$

and $F_1 = 0, G_1 = 1, H_1 = -2, F_2 = 1, G_2 = -4, H_2 = 6$. The expressions of the coefficients are

$$F_m = -\frac{1}{5(b+2)}b^m + (-1)^{m-1}\frac{b+1}{5(b+2)}(5+b)^m, m \geq 1, \quad (4.40)$$

$$G_m = \frac{4b+5}{5(b+2)}b^{m-1} + (-1)^{m-1}\frac{1}{5(b+2)}(5+b)^m, m \geq 1, \qquad (4.41)$$

$$H_m = -\frac{6b+10}{5(b+2)}b^{m-1} + (-1)^m\frac{2}{5}(5+b)^m, m \geq 1, \qquad (4.42)$$

where b is a solution of the equation $b^2 + 5b + 5 = 0$.

Substituting the expressions of D_m, E_m, F_m, G_m, H_m from Eqs. (4.36)-(4.37) and (4.40)-(4.42) in Eq. (4.27) and grouping the terms yields

$$\begin{aligned}
e^{(h_1-h_4)y} = {} & \frac{1}{5} + \frac{2}{5}\cos(\sqrt{-b}y) + \frac{2}{5}\cos(\sqrt{5+b}y) \\
& + (h_1 + h_4)\left[\frac{1}{5} - \frac{b+3}{5}\cos(\sqrt{-b}y) + \frac{b+2}{5}\cos(\sqrt{5+b}y)\right] \\
& + (h_2 + h_3)\left[\frac{1}{5} + \frac{b+2}{5}\cos(\sqrt{-b}y) - \frac{b+3}{5}\cos(\sqrt{5+b}y)\right] \\
& + (h_1 - h_4)\left[\frac{\sqrt{-b}}{5}\sin(\sqrt{-b}y) + \frac{1}{\sqrt{-5b}}\sin(\sqrt{5+b}y)\right] \\
& + (h_2 - h_3)\left[-\frac{2b+5}{5\sqrt{-b}}\sin(\sqrt{-b}y) + \frac{b+2}{\sqrt{-5b}}\sin(\sqrt{5+b}y)\right]. \quad (4.43)
\end{aligned}$$

On the other hand, e^{2h_1y} can be written with the aid of the 5-dimensional polar cosexponential functions as

$$e^{2h_1y} = g_{50}(2y) + h_1g_{51}(2y) + h_2g_{52}(2y) + h_3g_{53}(2y) + h_4g_{54}(2y). (4.44)$$

The multiplication of the expressions of $e^{(h_1+h_4)y}$ and $e^{(h_1-h_4)y}$ in Eqs. (4.33) and (4.43) and the separation of the real components yields the expressions of the 5-dimensional cosexponential functions, for $a = (\sqrt{5}-1)/2, b = -(5+\sqrt{5})/2$, as

$$g_{50}(2y) = \frac{1}{5}e^{2y} + \frac{2}{5}e^{ay}\cos(\sqrt{-b}y) + \frac{2}{5}e^{-(1+a)y}\cos(\sqrt{5+b}y), \quad (4.45)$$

$$\begin{aligned}
g_{51}(2y) = {} & \frac{1}{5}e^{2y} + \frac{1}{5}e^{ay}\left[\frac{-1+\sqrt{5}}{2}\cos(\sqrt{-b}y) + \frac{5+\sqrt{5}}{2\sqrt{-b}}\sin(\sqrt{-b}y)\right] \\
& + \frac{1}{5}e^{-(1+a)y}\left[-\frac{1+\sqrt{5}}{2}\cos(\sqrt{5+b}y) + \sqrt{\frac{5}{-b}}\sin(\sqrt{5+b}y)\right],
\end{aligned}$$

$$(4.46)$$

$$g_{52}(2y) = \frac{1}{5}e^{2y} + \frac{1}{5}e^{ay}\left[-\frac{1+\sqrt{5}}{2}\cos(\sqrt{-b}y) + \sqrt{\frac{5}{-b}}\sin(\sqrt{-b}y)\right]$$
$$+\frac{1}{5}e^{-(1+a)y}\left[\frac{-1+\sqrt{5}}{2}\cos(\sqrt{5+b}y) - \frac{5+\sqrt{5}}{2\sqrt{-b}}\sin(\sqrt{5+b}y)\right],$$

$$(4.47)$$

$$g_{53}(2y) = \frac{1}{5}e^{2y} + \frac{1}{5}e^{ay}\left[-\frac{1+\sqrt{5}}{2}\cos(\sqrt{-b}y) - \sqrt{\frac{5}{-b}}\sin(\sqrt{-b}y)\right]$$
$$+\frac{1}{5}e^{-(1+a)y}\left[\frac{-1+\sqrt{5}}{2}\cos(\sqrt{5+b}y) + \frac{5+\sqrt{5}}{2\sqrt{-b}}\sin(\sqrt{5+b}y)\right],$$

$$(4.48)$$

$$g_{54}(2y) = \frac{1}{5}e^{2y} + \frac{1}{5}e^{ay}\left[\frac{-1+\sqrt{5}}{2}\cos(\sqrt{-b}y) - \frac{5+\sqrt{5}}{2\sqrt{-b}}\sin(\sqrt{-b}y)\right]$$
$$+\frac{1}{5}e^{-(1+a)y}\left[-\frac{1+\sqrt{5}}{2}\cos(\sqrt{5+b}y) - \sqrt{\frac{5}{-b}}\sin(\sqrt{5+b}y)\right].$$

$$(4.49)$$

The polar 5-dimensional cosexponential functions can be written as

$$g_{5k}(y) = \frac{1}{5}\sum_{l=0}^{4}\exp\left[y\cos\left(\frac{2\pi l}{5}\right)\right]\cos\left[y\sin\left(\frac{2\pi l}{5}\right) - \frac{2\pi kl}{5}\right],$$
$$k = 0, ..., 4.$$

$$(4.50)$$

The graphs of the polar 5-dimensional cosexponential functions are shown in Fig. 4.2.

It can be checked that

$$\sum_{k=0}^{4}g_k^2(y) = \frac{1}{5}e^{2y} + \frac{2}{5}e^{(\sqrt{5}-1)y/2} + \frac{2}{5}e^{-(\sqrt{5}+1)y/2}.$$

$$(4.51)$$

The addition theorems for the polar 5-dimensional cosexponential functions are

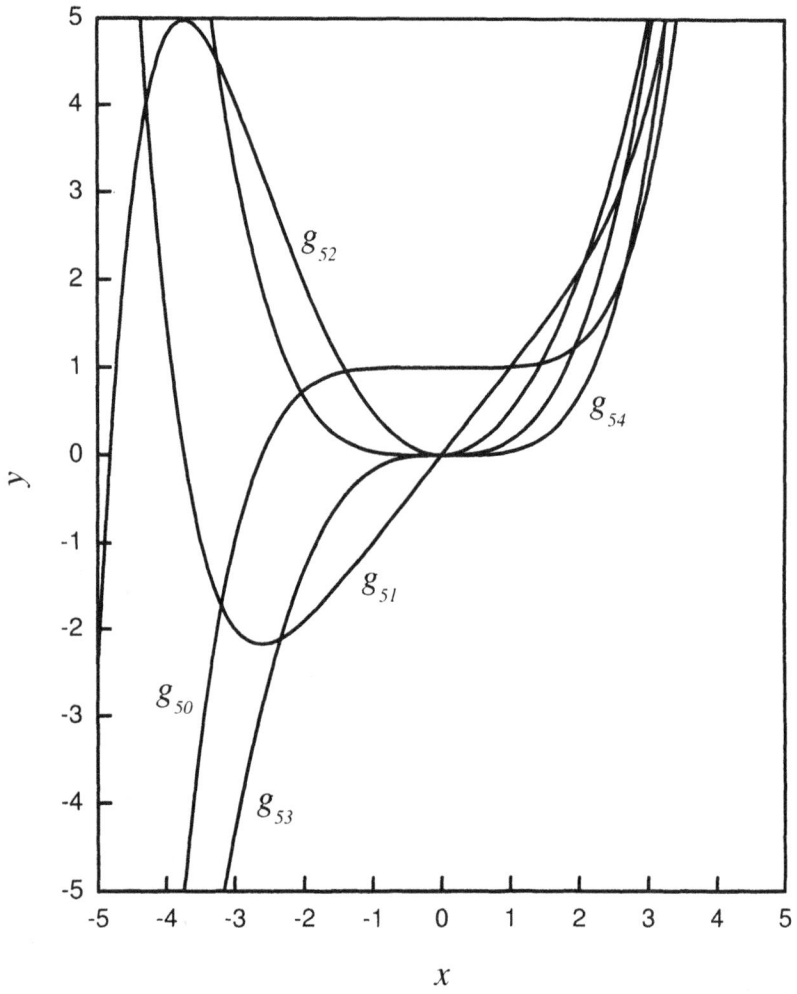

Figure 4.2: Polar cosexponential functions $g_{50}, g_{51}, g_{52}, g_{53}, g_{54}$.

$$g_{50}(y+z) = g_{50}(y)g_{50}(z) + g_{51}(y)g_{54}(z) + g_{52}(y)g_{53}(z)$$
$$+ g_{53}(y)g_{52}(z) + g_{54}(y)g_{51}(z),$$
$$g_{51}(y+z) = g_{50}(y)g_{51}(z) + g_{51}(y)g_{50}(z) + g_{52}(y)g_{54}(z)$$
$$+ g_{53}(y)g_{53}(z) + g_{54}(y)g_{52}(z),$$
$$g_{52}(y+z) = g_{50}(y)g_{52}(z) + g_{51}(y)g_{51}(z) + g_{52}(y)g_{50}(z)$$
$$+ g_{53}(y)g_{54}(z) + g_{54}(y)g_{53}(z),$$
$$g_{53}(y+z) = g_{50}(y)g_{53}(z) + g_{51}(y)g_{52}(z) + g_{52}(y)g_{51}(z)$$
$$+ g_{53}(y)g_{50}(z) + g_{54}(y)g_{54}(z),$$
$$g_{54}(y+z) = g_{50}(y)g_{54}(z) + g_{51}(y)g_{53}(z) + g_{52}(y)g_{52}(z)$$
$$+ g_{53}(y)g_{51}(z) + g_{54}(y)g_{50}(z).$$

$$(4.52)$$

It can be shown that

$$\{g_{50}(y) + h_1 g_{51}(y) + h_2 g_{52}(y) + h_3 g_{53}(y) + h_4 g_{54}(y)\}^l$$
$$= g_{50}(ly) + h_1 g_{51}(ly) + h_2 g_{52}(ly) + h_3 g_{53}(ly) + h_4 g_{54}(ly),$$
$$\{g_{50}(y) + h_1 g_{53}(y) + h_2 g_{51}(y) + h_3 g_{54}(y) + h_4 g_{52}(y)\}^l$$
$$= g_{50}(ly) + h_1 g_{53}(ly) + h_2 g_{51}(ly) + h_3 g_{54}(ly) + h_4 g_{52}(ly),$$
$$\{g_{50}(y) + h_1 g_{52}(y) + h_2 g_{54}(y) + h_3 g_{51}(y) + h_4 g_{53}(y)\}^l$$
$$= g_{50}(ly) + h_1 g_{52}(ly) + h_2 g_{54}(ly) + h_3 g_{51}(ly) + h_4 g_{53}(ly),$$
$$\{g_{50}(y) + h_1 g_{54}(y) + h_2 g_{53}(y) + h_3 g_{52}(y) + h_4 g_{51}(y)\}^l$$
$$= g_{50}(ly) + h_1 g_{54}(ly) + h_2 g_{53}(ly) + h_3 g_{52}(ly) + h_4 g_{51}(ly).$$

$$(4.53)$$

The derivatives of the polar cosexponential functions are related by

$$\frac{dg_{50}}{du} = g_{54}, \ \frac{dg_{51}}{du} = g_{50}, \ \frac{dg_{52}}{du} = g_{51}, \ \frac{dg_{53}}{du} = g_{52}, \ \frac{dg_{54}}{du} = g_{53}. \quad (4.54)$$

4.4 Exponential and trigonometric forms of polar 5-complex numbers

The exponential and trigonometric forms of 5-complex numbers can be expressed with the aid of the hypercomplex bases

$$\begin{pmatrix} e_+ \\ e_1 \\ \tilde{e}_1 \\ e_2 \\ \tilde{e}_2 \end{pmatrix} = \frac{2}{5} \begin{pmatrix} \frac{1}{2} & \frac{1}{2} & \frac{1}{2} & \frac{1}{2} & \frac{1}{2} \\ 1 & p & 2p^2-1 & 2p^2-1 & p \\ 0 & q & 2pq & -2pq & -q \\ 1 & 2p^2-1 & p & p & 2p^2-1 \\ 0 & 2pq & -q & q & -2pq \end{pmatrix} \begin{pmatrix} 1 \\ h_1 \\ h_2 \\ h_3 \\ h_4 \end{pmatrix}.$$

$$(4.55)$$

The multiplication relations for these bases are

$$
\begin{aligned}
&e_+^2 = e_+, \ e_+ e_k = 0, \ e_+ \tilde{e}_k = 0, \\
&e_k^2 = e_k, \ \tilde{e}_k^2 = -e_k, \ e_k \tilde{e}_k = \tilde{e}_k, \ e_k e_l = 0, \ e_k \tilde{e}_l = 0, \ \tilde{e}_k \tilde{e}_l = 0, \\
&k, l = 1, 2, \ k \neq l.
\end{aligned}
\tag{4.56}
$$

The bases have the property that

$$
e_+ + e_1 + e_2 = 1.
\tag{4.57}
$$

The moduli of the new bases are

$$
|e_+| = \frac{1}{\sqrt{5}}, \ |e_k| = \sqrt{\frac{2}{5}}, \ |\tilde{e}_k| = \sqrt{\frac{2}{5}},
\tag{4.58}
$$

for $k = 1, 2$.

It can be checked that

$$
x_0 + h_1 x_1 + h_2 x_2 + h_3 x_3 + h_4 x_4 = e_+ v_+ + e_1 v_1 + \tilde{e}_1 \tilde{v}_1 + e_2 v_2 + \tilde{e}_2 \tilde{v}_2.
\tag{4.59}
$$

The ensemble $e_+, e_1, \tilde{e}_1, e_2, \tilde{e}_2$ will be called the canonical 5-complex base, and Eq. (4.59) gives the canonical form of the 5-complex number.

The exponential form of the 5-complex number u is

$$
\begin{aligned}
u = \rho \exp \Bigg\{ &\frac{1}{5}(h_1 + h_2 + h_3 + h_4) \ln \frac{\sqrt{2}}{\tan \theta_+} \\
&+ \left[\frac{\sqrt{5}+1}{10}(h_1 + h_4) - \frac{\sqrt{5}-1}{10}(h_2 + h_3) \right] \ln \tan \psi_1 \\
&+ \tilde{e}_1 \phi_1 + \tilde{e}_2 \phi_2 \Bigg\},
\end{aligned}
\tag{4.60}
$$

for $0 < \theta_+ < \pi/2$.

The trigonometric form of the 5-complex number u is

$$
\begin{aligned}
u = d \left(\frac{5}{2} \right)^{1/2} &\left(\frac{1}{\tan^2 \theta_+} + 1 + \frac{1}{\tan^2 \psi_1} \right)^{-1/2} \left(\frac{e_+ \sqrt{2}}{\tan \theta_+} + e_1 + \frac{e_2}{\tan \psi_1} \right) \\
&\exp \left(\tilde{e}_1 \phi_1 + \tilde{e}_2 \phi_2 \right).
\end{aligned}
\tag{4.61}
$$

The modulus d and the amplitude ρ are related by

$$
d = \rho \frac{2^{2/5}}{\sqrt{5}} \left(\tan \theta_+ \tan^2 \psi_1 \right)^{1/5} \left(\frac{1}{\tan^2 \theta_+} + 1 + \frac{1}{\tan^2 \psi_1} \right)^{1/2}.
\tag{4.62}
$$

4.5 Elementary functions of a polar 5-complex variable

The logarithm and power function exist for $v_+ > 0$, which means that $0 < \theta_+ < \pi/2$, and are given by

$$\ln u = \ln \rho + \frac{1}{5}(h_1 + h_2 + h_3 + h_4) \ln \frac{\sqrt{2}}{\tan \theta_+}$$
$$+ \left[\frac{\sqrt{5}+1}{10}(h_1 + h_4) - \frac{\sqrt{5}-1}{10}(h_2 + h_3) \right] \ln \tan \psi_1$$
$$+ \tilde{e}_1 \phi_1 + \tilde{e}_2 \phi_2, \tag{4.63}$$

$$u^m = e_+ v_+^m + \rho_1^m (e_1 \cos m\phi_1 + \tilde{e}_1 \sin m\phi_1)$$
$$+ \rho_2^m (e_2 \cos m\phi_2 + \tilde{e}_2 \sin m\phi_2). \tag{4.64}$$

The exponential of the 5-complex variable u is

$$e^u = e_+ e^{v_+} + e^{v_1} (e_1 \cos \tilde{v}_1 + \tilde{e}_1 \sin \tilde{v}_1) + e^{v_2} (e_2 \cos \tilde{v}_2 + \tilde{e}_2 \sin \tilde{v}_2). \tag{4.65}$$

The trigonometric functions of the 5-complex variable u are

$$\cos u = e_+ \cos v_+ + \sum_{k=1}^{2} (e_k \cos v_k \cosh \tilde{v}_k - \tilde{e}_k \sin v_k \sinh \tilde{v}_k), \tag{4.66}$$

$$\sin u = e_+ \sin v_+ + \sum_{k=1}^{2} (e_k \sin v_k \cosh \tilde{v}_k + \tilde{e}_k \cos v_k \sinh \tilde{v}_k). \tag{4.67}$$

The hyperbolic functions of the 5-complex variable u are

$$\cosh u = e_+ \cosh v_+ + \sum_{k=1}^{2} (e_k \cosh v_k \cos \tilde{v}_k + \tilde{e}_k \sinh v_k \sin \tilde{v}_k), \tag{4.68}$$

$$\sinh u = e_+ \sinh v_+ + \sum_{k=1}^{2} (e_k \sinh v_k \cos \tilde{v}_k + \tilde{e}_k \cosh v_k \sin \tilde{v}_k). \tag{4.69}$$

4.6 Power series of 5-complex numbers

A power series of the 5-complex variable u is a series of the form

$$a_0 + a_1 u + a_2 u^2 + \cdots + a_l u^l + \cdots. \tag{4.70}$$

Since

$$|au^l| \leq 5^{l/2}|a||u|^l, \tag{4.71}$$

the series is absolutely convergent for

$$|u| < c, \tag{4.72}$$

where

$$c = \lim_{l\to\infty} \frac{|a_l|}{\sqrt{5}|a_{l+1}|}. \tag{4.73}$$

If $a_l = \sum_{p=0}^{4} h_p a_{lp}$, where $h_0 = 1$, and

$$A_{l+} = \sum_{p=0}^{4} a_{lp}, \tag{4.74}$$

$$A_{lk} = \sum_{p=0}^{4} a_{lp} \cos\left(\frac{2\pi kp}{5}\right), \tag{4.75}$$

$$\tilde{A}_{lk} = \sum_{p=0}^{4} a_{lp} \sin\left(\frac{2\pi kp}{5}\right), \tag{4.76}$$

for $k = 1, 2$, the series (4.70) can be written as

$$\sum_{l=0}^{\infty}\left[e_+ A_{l+} v_+^l + \sum_{k=1}^{2}(e_k A_{lk} + \tilde{e}_k \tilde{A}_{lk})(e_k v_k + \tilde{e}_k \tilde{v}_k)^l\right]. \tag{4.77}$$

The series in Eq. (4.70) is absolutely convergent for

$$|v_+| < c_+, \; \rho_k < c_k, k = 1, 2, \tag{4.78}$$

where

$$c_+ = \lim_{l\to\infty} \frac{|A_{l+}|}{|A_{l+1,+}|}, \; c_k = \lim_{l\to\infty} \frac{\left(A_{lk}^2 + \tilde{A}_{lk}^2\right)^{1/2}}{\left(A_{l+1,k}^2 + \tilde{A}_{l+1,k}^2\right)^{1/2}}. \tag{4.79}$$

4.7 Analytic functions of a polar 5-complex variable

If $f(u) = \sum_{k=0}^{4} h_k P_k(x_0, x_1, x_2, x_3, x_4)$, then

$$\frac{\partial P_0}{\partial x_0} = \frac{\partial P_1}{\partial x_1} = \frac{\partial P_2}{\partial x_2} = \frac{\partial P_3}{\partial x_3} = \frac{\partial P_4}{\partial x_4}, \tag{4.80}$$

$$\frac{\partial P_1}{\partial x_0} = \frac{\partial P_2}{\partial x_1} = \frac{\partial P_3}{\partial x_2} = \frac{\partial P_4}{\partial x_3} = \frac{\partial P_0}{\partial x_4}, \tag{4.81}$$

$$\frac{\partial P_2}{\partial x_0} = \frac{\partial P_3}{\partial x_1} = \frac{\partial P_4}{\partial x_2} = \frac{\partial P_0}{\partial x_3} = \frac{\partial P_1}{\partial x_4}, \tag{4.82}$$

$$\frac{\partial P_3}{\partial x_0} = \frac{\partial P_4}{\partial x_1} = \frac{\partial P_0}{\partial x_2} = \frac{\partial P_1}{\partial x_3} = \frac{\partial P_2}{\partial x_4}, \tag{4.83}$$

$$\frac{\partial P_4}{\partial x_0} = \frac{\partial P_0}{\partial x_1} = \frac{\partial P_1}{\partial x_2} = \frac{\partial P_2}{\partial x_3} = \frac{\partial P_3}{\partial x_4}, \tag{4.84}$$

and

$$\frac{\partial^2 P_k}{\partial x_0 \partial x_l} = \frac{\partial^2 P_k}{\partial x_1 \partial x_{l-1}} = \cdots = \frac{\partial^2 P_k}{\partial x_{[l/2]} \partial x_{l-[l/2]}}$$

$$= \frac{\partial^2 P_k}{\partial x_{l+1} \partial x_4} = \frac{\partial^2 P_k}{\partial x_{l+2} \partial x_3} = \cdots = \frac{\partial^2 P_k}{\partial x_{l+1+[(3-l)/2]} \partial x_{4-[(3-l)/2]}}, \tag{4.85}$$

for $k, l = 0, ..., 4$. In Eq. (4.85), $[a]$ denotes the integer part of a, defined as $[a] \leq a < [a] + 1$. In this chapter, brackets larger than the regular brackets $[\,]$ do not have the meaning of integer part.

4.8 Integrals of polar 5-complex functions

If $f(u)$ is an analytic 5-complex function, then

$$\oint_\Gamma \frac{f(u)du}{u - u_0} = 2\pi f(u_0) \left\{ \tilde{e}_1 \, \text{int}(u_{0\xi_1\eta_1}, \Gamma_{\xi_1\eta_1}) + \tilde{e}_2 \, \text{int}(u_{0\xi_2\eta_2}, \Gamma_{\xi_2\eta_2}) \right\}, \tag{4.86}$$

where

$$\text{int}(M, C) = \begin{cases} 1 & \text{if } M \text{ is an interior point of } C, \\ 0 & \text{if } M \text{ is exterior to } C, \end{cases} \tag{4.87}$$

and $u_{0\xi_k\eta_k}$, $\Gamma_{\xi_k\eta_k}$ are respectively the projections of the pole u_0 and of the loop Γ on the plane defined by the axes ξ_k and η_k, $k = 1, 2$.

4.9   Factorization of polar 5-complex polynomials

A polynomial of degree m of the 5-complex variable u has the form

$$P_m(u) = u^m + a_1 u^{m-1} + \cdots + a_{m-1} u + a_m, \tag{4.88}$$

where a_l, for $l = 1, ..., m$, are 5-complex constants. If $a_l = \sum_{p=0}^{4} h_p a_{lp}$, and with the notations of Eqs. (4.74)-(4.76) applied for $l = 1, \cdots, m$, the polynomial $P_m(u)$ can be written as

$$P_m = e_+ \left(v_+^m + \sum_{l=1}^{m} A_{l+} v_+^{m-l} \right)$$
$$+ \sum_{k=1}^{2} \left[(e_k v_k + \tilde{e}_k \tilde{v}_k)^m + \sum_{l=1}^{m} (e_k A_{lk} + \tilde{e}_k \tilde{A}_{lk})(e_k v_k + \tilde{e}_k \tilde{v}_k)^{m-l} \right]. \tag{4.89}$$

The polynomial $P_m(u)$ can be written, as

$$P_m(u) = \prod_{p=1}^{m} (u - u_p), \tag{4.90}$$

where

$$u_p = e_+ v_{p+} + (e_1 v_{1p} + \tilde{e}_1 \tilde{v}_{1p}) + (e_2 v_{2p} + \tilde{e}_2 \tilde{v}_{2p}), p = 1, ..., m. \tag{4.91}$$

The quantities v_{p+}, $e_k v_{kp} + \tilde{e}_k \tilde{v}_{kp}$, $p = 1, ..., m, k = 1, 2$, are the roots of the corresponding polynomial in Eq. (4.89). The roots v_{p+} appear in complex-conjugate pairs, and $v_{kp}, \tilde{v}_{kp}$ are real numbers. Since all these roots may be ordered arbitrarily, the polynomial $P_m(u)$ can be written in many different ways as a product of linear factors.

If $P(u) = u^2 - 1$, the degree is $m = 2$, the coefficients of the polynomial are $a_1 = 0, a_2 = -1$, the coefficients defined in Eqs. (4.74)-(4.76) are $A_{2+} = -1, A_{21} = -1, \tilde{A}_{21} = 0, A_{22} = -1, \tilde{A}_{22} = 0$. The expression of $P(u)$, Eq. (4.89), is $v_+^2 - e_+ + (e_1 v_1 + \tilde{e}_1 \tilde{v}_1)^2 - e_1 + (e_2 v_2 + \tilde{e}_2 \tilde{v}_2)^2 - e_2$. The factorization of $P(u)$, Eq. (4.90), is $P(u) = (u - u_1)(u - u_2)$, where the roots are $u_1 = \pm e_+ \pm e_1 \pm e_2, u_2 = -u_1$. If e_+, e_1, e_2 are expressed with the aid of Eq. (4.55) in terms of h_1, h_2, h_3, h_4, the factorizations of $P(u)$ are obtained as

$$
\begin{aligned}
u^2 - 1 &= (u+1)(u-1), \\
u^2 - 1 &= \left[u + \tfrac{1}{5} + \tfrac{\sqrt{5}+1}{5}(h_1 + h_4) - \tfrac{\sqrt{5}-1}{5}(h_2 + h_3)\right] \\
&\quad \left[u - \tfrac{1}{5} - \tfrac{\sqrt{5}+1}{5}(h_1 + h_4) + \tfrac{\sqrt{5}-1}{5}(h_2 + h_3)\right], \\
u^2 - 1 &= \left[u + \tfrac{1}{5} - \tfrac{\sqrt{5}-1}{5}(h_1 + h_4) + \tfrac{\sqrt{5}+1}{5}(h_2 + h_3)\right] \\
&\quad \left[u - \tfrac{1}{5} + \tfrac{\sqrt{5}-1}{5}(h_1 + h_4) - \tfrac{\sqrt{5}+1}{5}(h_2 + h_3)\right], \\
u^2 - 1 &= \left[u + \tfrac{3}{5} - \tfrac{2}{5}(h_1 + h_2 + h_3 + h_4)\right] \\
&\quad \left[u - \tfrac{3}{5} + \tfrac{2}{5}(h_1 + h_2 + h_3 + h_4)\right].
\end{aligned}
$$

$$(4.92)$$

It can be checked that $(\pm e_+ \pm e_1 \pm e_2)^2 = e_+ + e_1 + e_2 = 1$.

4.10   Representation of polar 5-complex numbers by irreducible matrices

If the unitary matrix which can be obtained from the expression, Eq. (4.8), of the variables $\xi_+, \xi_1, \eta_1, \xi_k, \eta_k$ in terms of x_0, x_1, x_2, x_3, x_4 is called T, the irreducible representation [7] of the hypercomplex number u is

$$
TUT^{-1} = \begin{pmatrix} v_+ & 0 & 0 \\ 0 & V_1 & 0 \\ 0 & 0 & V_2 \end{pmatrix},
$$

$$(4.93)$$

where U is the matrix in Eq. (4.22), and V_k are the matrices

$$
V_k = \begin{pmatrix} v_k & \tilde{v}_k \\ -\tilde{v}_k & v_k \end{pmatrix}, \quad k = 1, 2.
$$

$$(4.94)$$

Chapter 5

Complex Numbers in 6 Dimensions

Two distinct systems of commutative complex numbers in 6 dimensions having the form $u = x_0 + h_1 x_1 + h_2 x_2 + h_3 x_3 + h_4 x_4 + h_5 x_5$ are described in this chapter, for which the multiplication is associative and commutative, where the variables $x_0, x_1, x_2, x_3, x_4, x_5$ are real numbers. The first type of 6-complex numbers described in this article is characterized by the presence of two polar axes, so that these numbers will be called polar 6-complex numbers. The other type of 6-complex numbers described in this paper will be called planar n-complex numbers.

The polar 6-complex numbers introduced in this chapter can be specified by the modulus d, the amplitude ρ, and the polar angles θ_+, θ_-, the planar angle ψ_1, and the azimuthal angles ϕ_1, ϕ_2. The planar 6-complex numbers introduced in this paper can be specified by the modulus d, the amplitude ρ, the planar angles ψ_1, ψ_2, and the azimuthal angles ϕ_1, ϕ_2, ϕ_3. Exponential and trigonometric forms are given for the 6-complex numbers. The 6-complex functions defined by series of powers are analytic, and the partial derivatives of the components of the 6-complex functions are closely related. The integrals of polar 6-complex functions are independent of path in regions where the functions are regular. The fact that the exponential form of ther 6-complex numbers depends on cyclic variables leads to the concept of pole and residue for integrals on closed paths. The polynomials of polar 6-complex variables can be written as products of linear or quadratic factors, the polynomials of planar 6-complex variables can always be written as products of linear factors, although the factorization is not unique.

The polar 6-complex numbers described in this paper are a particular case for $n = 6$ of the polar hypercomplex numbers in n dimensions discussed in Sec. 6.1, and the planar 6-complex numbers described in this section are a particular case for $n = 6$ of the planar hypercomplex numbers in n dimensions discussed in Sec. 6.2.

5.1 Polar complex numbers in 6 dimensions

5.1.1 Operations with polar complex numbers in 6 dimensions

The polar hypercomplex number u in 6 dimensions is represented as

$$u = x_0 + h_1 x_1 + h_2 x_2 + h_3 x_3 + h_4 x_4 + h_5 x_5. \tag{5.1}$$

The multiplication rules for the bases h_1, h_2, h_3, h_4, h_5 are

$$h_1^2 = h_2, \; h_2^2 = h_4, \; h_3^2 = 1, \; h_4^2 = h_2, \; h_5^2 = h_4, \; h_1 h_2 = h_3, \; h_1 h_3 = h_4,$$
$$h_1 h_4 = h_5, \; h_1 h_5 = 1, \; h_2 h_3 = h_5, \; h_2 h_4 = 1, \; h_2 h_5 = h_1,$$
$$h_3 h_4 = h_1, \; h_3 h_5 = h_2, \; h_4 h_5 = h_3. \tag{5.2}$$

The significance of the composition laws in Eq. (5.2) can be understood by representing the bases h_j, h_k by points on a circle at the angles $\alpha_j = \pi j/3, \alpha_k = \pi k/3$, as shown in Fig. 5.1, and the product $h_j h_k$ by the point of the circle at the angle $\pi(j + k)/3$. If $2\pi \leq \pi(j + k)/3 < 4\pi$, the point represents the basis h_l of angle $\alpha_l = \pi(j + k)/3 - 2\pi$.

The sum of the 6-complex numbers u and u' is

$$u + u' = x_0 + x_0' + h_1(x_1 + x_1') + h_1(x_2 + x_2') + h_3(x_3 + x_3')$$
$$+ h_4(x_4 + x_4') + h_5(x_5 + x_5'). \tag{5.3}$$

The product of the numbers u, u' is

$$\begin{aligned}
uu' = &x_0 x_0' + x_1 x_5' + x_2 x_4' + x_3 x_3' + x_4 x_2' + x_5 x_1' \\
&+ h_1(x_0 x_1' + x_1 x_0' + x_2 x_5' + x_3 x_4' + x_4 x_3' + x_5 x_2') \\
&+ h_2(x_0 x_2' + x_1 x_1' + x_2 x_0' + x_3 x_5' + x_4 x_4' + x_5 x_3') \\
&+ h_3(x_0 x_3' + x_1 x_2' + x_2 x_1' + x_3 x_0' + x_4 x_5' + x_5 x_4') \\
&+ h_4(x_0 x_4' + x_1 x_3' + x_2 x_2' + x_3 x_1' + x_4 x_0' + x_5 x_5') \\
&+ h_5(x_0 x_5' + x_1 x_4' + x_2 x_3' + x_3 x_2' + x_4 x_1' + x_5 x_0').
\end{aligned} \tag{5.4}$$

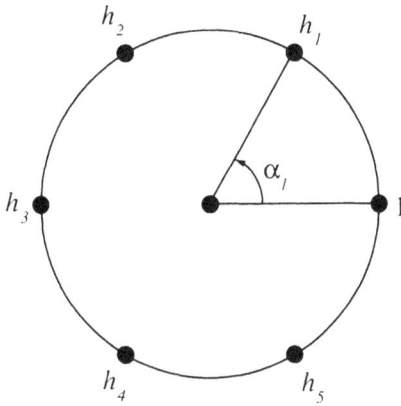

Figure 5.1: Representation of the polar hypercomplex bases $1, h_1, h_2, h_3, h_4, h_5$ by points on a circle at the angles $\alpha_k = 2\pi k/6$. The product $h_j h_k$ will be represented by the point of the circle at the angle $2\pi(j+k)/6$, $i, k = 0, 1, ..., 5$, where $h_0 = 1$. If $2\pi \leq 2\pi(j+k)/6 \leq 4\pi$, the point represents the basis h_l of angle $\alpha_l = 2\pi(j+k)/6 - 2\pi$.

The relation between the variables $v_+, v_-, v_1, \tilde{v}_1, v_2, \tilde{v}_2$ and $x_0, x_1, x_2, x_3,$ x_4, x_5 are

$$
\begin{pmatrix} v_+ \\ v_- \\ v_1 \\ \tilde{v}_1 \\ v_2 \\ \tilde{v}_2 \end{pmatrix} = \begin{pmatrix} 1 & 1 & 1 & 1 & 1 & 1 \\ 1 & -1 & 1 & -1 & 1 & -1 \\ 1 & \frac{1}{2} & -\frac{1}{2} & -1 & -\frac{1}{2} & \frac{1}{2} \\ 0 & \frac{\sqrt{3}}{2} & \frac{\sqrt{3}}{2} & 0 & -\frac{\sqrt{3}}{2} & -\frac{\sqrt{3}}{2} \\ 1 & -\frac{1}{2} & -\frac{1}{2} & 1 & -\frac{1}{2} & -\frac{1}{2} \\ 0 & \frac{\sqrt{3}}{2} & -\frac{\sqrt{3}}{2} & 0 & \frac{\sqrt{3}}{2} & -\frac{\sqrt{3}}{2} \end{pmatrix} \begin{pmatrix} x_0 \\ x_1 \\ x_2 \\ x_3 \\ x_4 \\ x_5 \end{pmatrix}. \tag{5.5}
$$

The other variables are $v_4 = v_2, \tilde{v}_4 = -\tilde{v}_2, v_5 = v_1, \tilde{v}_5 = -\tilde{v}_1$. The variables $v_+, v_-, v_1, \tilde{v}_1, v_2, \tilde{v}_2$ will be called canonical polar 6-complex variables.

5.1.2 Geometric representation of polar complex numbers in 6 dimensions

The 6-complex number $u = x_0 + h_1 x_1 + h_2 x_2 + h_3 x_3 + h_4 x_4 + h_5 x_5$ is represented by the point A of coordinates $(x_0, x_1, x_2, x_3, x_4, x_5)$. The distance from the origin O of the 6-dimensional space to the point A has the expression

$$d^2 = x_0^2 + x_1^2 + x_2^2 + x_3^2 + x_4^2 + x_5^2. \tag{5.6}$$

The distance d is called modulus of the 6-complex number u, and is designated by $d = |u|$. The modulus has the property that

$$|u'u''| \leq \sqrt{6}|u'||u''|. \tag{5.7}$$

The exponential and trigonometric forms of the 6-complex number u can be obtained conveniently in a rotated system of axes defined by a transformation which has the form

$$\begin{pmatrix} \xi_+ \\ \xi_- \\ \xi_1 \\ \tilde{\xi}_1 \\ \xi_2 \\ \tilde{\xi}_2 \end{pmatrix} = \begin{pmatrix} \frac{1}{\sqrt{6}} & \frac{1}{\sqrt{6}} & \frac{1}{\sqrt{6}} & \frac{1}{\sqrt{6}} & \frac{1}{\sqrt{6}} & \frac{1}{\sqrt{6}} \\ \frac{1}{\sqrt{6}} & -\frac{1}{\sqrt{6}} & \frac{1}{\sqrt{6}} & -\frac{1}{\sqrt{6}} & \frac{1}{\sqrt{6}} & -\frac{1}{\sqrt{6}} \\ \frac{\sqrt{3}}{3} & \frac{\sqrt{3}}{6} & -\frac{\sqrt{3}}{6} & -\frac{\sqrt{3}}{3} & -\frac{\sqrt{3}}{6} & \frac{\sqrt{3}}{6} \\ 0 & \frac{1}{2} & \frac{1}{2} & 0 & -\frac{1}{2} & -\frac{1}{2} \\ \frac{\sqrt{3}}{3} & -\frac{\sqrt{3}}{6} & -\frac{\sqrt{3}}{6} & \frac{\sqrt{3}}{3} & -\frac{\sqrt{3}}{6} & -\frac{\sqrt{3}}{6} \\ 0 & \frac{1}{2} & -\frac{1}{2} & 0 & \frac{1}{2} & -\frac{1}{2} \end{pmatrix} \begin{pmatrix} x_0 \\ x_1 \\ x_2 \\ x_3 \\ x_4 \\ x_5 \end{pmatrix}. \tag{5.8}$$

The lines of the matrices in Eq. (5.8) gives the components of the 6 basis vectors of the new system of axes. These vectors have unit length and are orthogonal to each other. The relations between the two sets of variables are

$$v_+ = \sqrt{6}\xi_+, v_- = \sqrt{6}\xi_-, v_k = \sqrt{3}\xi_k, \tilde{v}_k = \sqrt{3}\eta_k, k = 1, 2. \tag{5.9}$$

The radius ρ_k and the azimuthal angle ϕ_k in the plane of the axes $v_k, \tilde{v}_k$ are

$$\rho_k^2 = v_k^2 + \tilde{v}_k^2, \cos\phi_k = v_k/\rho_k, \sin\phi_k = \tilde{v}_k/\rho_k, 0 \leq \phi_k < 2\pi, \ k = 1, 2, \tag{5.10}$$

so that there are 2 azimuthal angles. The planar angle ψ_1 is

$$\tan\psi_1 = \rho_1/\rho_2, 0 \leq \psi_1 \leq \pi/2. \tag{5.11}$$

There is a polar angle θ_+,

$$\tan\theta_+ = \frac{\sqrt{2}\rho_1}{v_+}, 0 \leq \theta_+ \leq \pi, \tag{5.12}$$

and there is also a polar angle θ_-,

$$\tan\theta_- = \frac{\sqrt{2}\rho_1}{v_-}, 0 \leq \theta_- \leq \pi. \tag{5.13}$$

The amplitude of a 6-complex number u is

$$\rho = \left(v_+ v_- \rho_1^2 \rho_2^2\right)^{1/6}.$$

(5.14)

It can be checked that

$$d^2 = \frac{1}{6}v_+^2 + \frac{1}{6}v_-^2 + \frac{1}{3}(\rho_1^2 + \rho_2^2).$$

(5.15)

If $u = u'u''$, the parameters of the hypercomplex numbers are related by

$$v_+ = v'_+ v''_+,$$

(5.16)

$$\tan \theta_+ = \frac{1}{\sqrt{2}} \tan \theta'_+ \tan \theta''_+,$$

(5.17)

$$v_- = v'_- v''_-,$$

(5.18)

$$\tan \theta_- = \frac{1}{\sqrt{2}} \tan \theta'_- \tan \theta''_-,$$

(5.19)

$$\tan \psi_1 = \tan \psi'_1 \tan \psi''_1,$$

(5.20)

$$\rho_k = \rho'_k \rho''_k,$$

(5.21)

$$\phi_k = \phi'_k + \phi''_k,$$

(5.22)

$$v_k = v'_k v''_k - \tilde{v}'_k \tilde{v}''_k, \quad \tilde{v}_k = v'_k \tilde{v}''_k + \tilde{v}'_k v''_k,$$

(5.23)

$$\rho = \rho' \rho'',$$

(5.24)

where $k = 1, 2$.

The 6-complex number $u = x_0 + h_1 x_1 + h_2 x_2 + h_3 x_3 + h_4 x_4 + h_5 x_5$ can be represented by the matrix

$$U = \begin{pmatrix} x_0 & x_1 & x_2 & x_3 & x_4 & x_5 \\ x_5 & x_0 & x_1 & x_2 & x_3 & x_4 \\ x_4 & x_5 & x_0 & x_1 & x_2 & x_3 \\ x_3 & x_4 & x_5 & x_0 & x_1 & x_2 \\ x_2 & x_3 & x_4 & x_5 & x_0 & x_1 \\ x_1 & x_2 & x_3 & x_4 & x_5 & x_0 \end{pmatrix}.$$

(5.25)

The product $u = u'u''$ is represented by the matrix multiplication $U = U'U''$.

5.1.3   The polar 6-dimensional cosexponential functions

The polar cosexponential functions in 6 dimensions are

$$g_{6k}(y) = \sum_{p=0}^{\infty} y^{k+6p}/(k+6p)!, \tag{5.26}$$

for $k = 0, ..., 5$. The polar cosexponential functions g_{6k} of even index k are
even functions, $g_{6,2p}(-y) = g_{6,2p}(y)$, and the polar cosexponential functions
of odd index k are odd functions, $g_{6,2p+1}(-y) = -g_{6,2p+1}(y)$, $p = 0, 1, 2$.
It can be checked that

$$\sum_{k=0}^{5} g_{6k}(y) = e^y, \tag{5.27}$$

$$\sum_{k=0}^{5} (-1)^k g_{6k}(y) = e^{-y}. \tag{5.28}$$

The exponential function of the quantity $h_k y$ is

$$
\begin{aligned}
e^{h_1 y} &= g_{60}(y) + h_1 g_{61}(y) + h_2 g_{62}(y) + h_3 g_{63}(y) + h_4 g_{64}(y) + h_5 g_{65}(y), \\
e^{h_2 y} &= g_{60}(y) + g_{63}(y) + h_2\{g_{61}(y) + g_{64}(y)\} + h_4\{g_{62}(y) + g_{65}(y)\}, \\
e^{h_3 y} &= g_{60}(y) + g_{62}(y) + g_{64}(y) + h_3\{g_{61}(y) + g_{63}(y) + g_{65}(y)\}, \\
e^{h_4 y} &= g_{60}(y) + g_{63}(y) + h_2\{g_{62}(y) + g_{65}(y)\} + h_4\{g_{61}(y) + g_{64}(y)\}, \\
e^{h_5 y} &= g_{60}(y) + h_1 g_{65}(y) + h_2 g_{64}(y) + h_3 g_{63}(y) + h_4 g_{62}(y) + h_5 g_{61}(y).
\end{aligned}
\tag{5.29}
$$

The relations for h_2 and h_4 can be written equivalently as $e^{h_2 y} = g_{30} + h_2 g_{31} + h_4 g_{32}$, $e^{h_4 y} = g_{30} + h_2 g_{32} + h_4 g_{31}$, and the relation for h_3 can be written as $e^{h_3 y} = g_{20} + h_3 g_{21}$, which is the same as $e^{h_3 y} = \cosh y + h_3 \sinh y$.
The expressions of the polar 6-dimensional cosexponential functions are

$$
\begin{aligned}
g_{60}(y) &= \tfrac{1}{3}\cosh y + \tfrac{2}{3}\cosh \tfrac{y}{2}\cos \tfrac{\sqrt{3}}{2}y. \\
g_{61}(y) &= \tfrac{1}{3}\sinh y + \tfrac{1}{3}\sinh \tfrac{y}{2}\cos \tfrac{\sqrt{3}}{2}y + \tfrac{\sqrt{3}}{3}\cosh \tfrac{y}{2}\sin \tfrac{\sqrt{3}}{2}y, \\
g_{62}(y) &= \tfrac{1}{3}\cosh y - \tfrac{1}{3}\cosh \tfrac{y}{2}\cos \tfrac{\sqrt{3}}{2}y + \tfrac{\sqrt{3}}{3}\sinh \tfrac{y}{2}\sin \tfrac{\sqrt{3}}{2}y, \\
g_{63}(y) &= \tfrac{1}{3}\sinh y - \tfrac{2}{3}\sinh \tfrac{y}{2}\cos \tfrac{\sqrt{3}}{2}y, \\
g_{64}(y) &= \tfrac{1}{3}\cosh y - \tfrac{1}{3}\cosh \tfrac{y}{2}\cos \tfrac{\sqrt{3}}{2}y - \tfrac{\sqrt{3}}{3}\sinh \tfrac{y}{2}\sin \tfrac{\sqrt{3}}{2}y, \\
g_{65}(y) &= \tfrac{1}{3}\sinh y + \tfrac{1}{3}\sinh \tfrac{y}{2}\cos \tfrac{\sqrt{3}}{2}y - \tfrac{\sqrt{3}}{3}\cosh \tfrac{y}{2}\sin \tfrac{\sqrt{3}}{2}y.
\end{aligned}
\tag{5.30}
$$

The cosexponential functions (5.30) can be written as

$$g_{6k}(y) = \frac{1}{6}\sum_{l=0}^{5}\exp\left[y\cos\left(\frac{2\pi l}{6}\right)\right]\cos\left[y\sin\left(\frac{2\pi l}{6}\right) - \frac{2\pi kl}{6}\right], \tag{5.31}$$

for $k = 0, ..., 5$. The graphs of the polar 6-dimensional cosexponential functions are shown in Fig. 5.2.

It can be checked that

$$\sum_{k=0}^{5} g_{6k}^2(y) = \frac{1}{3}\cosh 2y + \frac{2}{3}\cosh y. \tag{5.32}$$

The addition theorems for the polar 6-dimensional cosexponential functions are

$$\begin{aligned}
g_{60}(y+z) &= g_{60}(y)g_{60}(z) + g_{61}(y)g_{65}(z) + g_{62}(y)g_{64}(z) + g_{63}(y)g_{63}(z) \\
&\quad + g_{64}(y)g_{62}(z) + g_{65}(y)g_{61}(z), \\
g_{61}(y+z) &= g_{60}(y)g_{61}(z) + g_{61}(y)g_{60}(z) + g_{62}(y)g_{65}(z) + g_{63}(y)g_{64}(z) \\
&\quad + g_{64}(y)g_{63}(z) + g_{65}(y)g_{62}(z), \\
g_{62}(y+z) &= g_{60}(y)g_{62}(z) + g_{61}(y)g_{61}(z) + g_{62}(y)g_{60}(z) + g_{63}(y)g_{65}(z) \\
&\quad + g_{64}(y)g_{64}(z) + g_{65}(y)g_{63}(z), \\
g_{63}(y+z) &= g_{60}(y)g_{63}(z) + g_{61}(y)g_{62}(z) + g_{62}(y)g_{61}(z) + g_{63}(y)g_{60}(z) \\
&\quad + g_{64}(y)g_{65}(z) + g_{65}(y)g_{64}(z), \\
g_{64}(y+z) &= g_{60}(y)g_{64}(z) + g_{61}(y)g_{63}(z) + g_{62}(y)g_{62}(z) + g_{63}(y)g_{61}(z) \\
&\quad + g_{64}(y)g_{60}(z) + g_{65}(y)g_{65}(z), \\
g_{65}(y+z) &= g_{60}(y)g_{65}(z) + g_{61}(y)g_{64}(z) + g_{62}(y)g_{63}(z) + g_{63}(y)g_{62}(z) \\
&\quad + g_{64}(y)g_{61}(z) + g_{65}(y)g_{60}(z).
\end{aligned}$$

$$\tag{5.33}$$

It can be shown that

$$\begin{aligned}
&\{g_{60}(y) + h_1 g_{61}(y) + h_2 g_{62}(y) + h_3 g_{63}(y) + h_4 g_{64}(y) + h_5 g_{65}(y)\}^l \\
&= g_{60}(ly) + h_1 g_{61}(ly) + h_2 g_{62}(ly) + h_3 g_{63}(ly) + h_4 g_{64}(ly) \\
&\quad + h_5 g_{65}(ly), \\
&\{g_{60}(y) + g_{63}(y) + h_2\{g_{61}(y) + g_{64}(y)\} + h_4\{g_{62}(y) + g_{65}(y)\}\}^l \\
&= g_{60}(ly) + g_{63}(ly) + h_2\{g_{61}(ly) + g_{64}(ly)\} + h_4\{g_{62}(ly) \\
&\quad + g_{65}(ly)\}, \\
&\{g_{60}(y) + g_{62}(y) + g_{64}(y) + h_3\{g_{61}(y) + g_{63}(y) + g_{65}(y)\}\}^l \\
&= g_{60}(ly) + g_{62}(ly) + g_{64}(ly) + h_3\{g_{61}(ly) + g_{63}(ly) \\
&\quad + g_{65}(ly)\}, \\
&\{g_{60}(y) + g_{63}(y) + h_2\{g_{62}(y) + g_{65}(y)\} + h_4\{g_{61}(y) + g_{64}(y)\}\}^l \\
&= g_{60}(ly) + g_{63}(ly) + h_2\{g_{62}(ly) + g_{65}(ly)\} + h_4\{g_{61}(ly) \\
&\quad + g_{64}(ly)\}, \\
&\{g_{60}(y) + h_1 g_{65}(y) + h_2 g_{64}(y) + h_3 g_{63}(y) + h_4 g_{62}(y) + h_5 g_{61}(y)\}^l \\
&= g_{60}(ly) + h_1 g_{65}(ly) + h_2 g_{64}(ly) + h_3 g_{63}(ly) + h_4 g_{62}(ly) \\
&\quad + h_5 g_{61}(ly).
\end{aligned}$$

$$\tag{5.34}$$

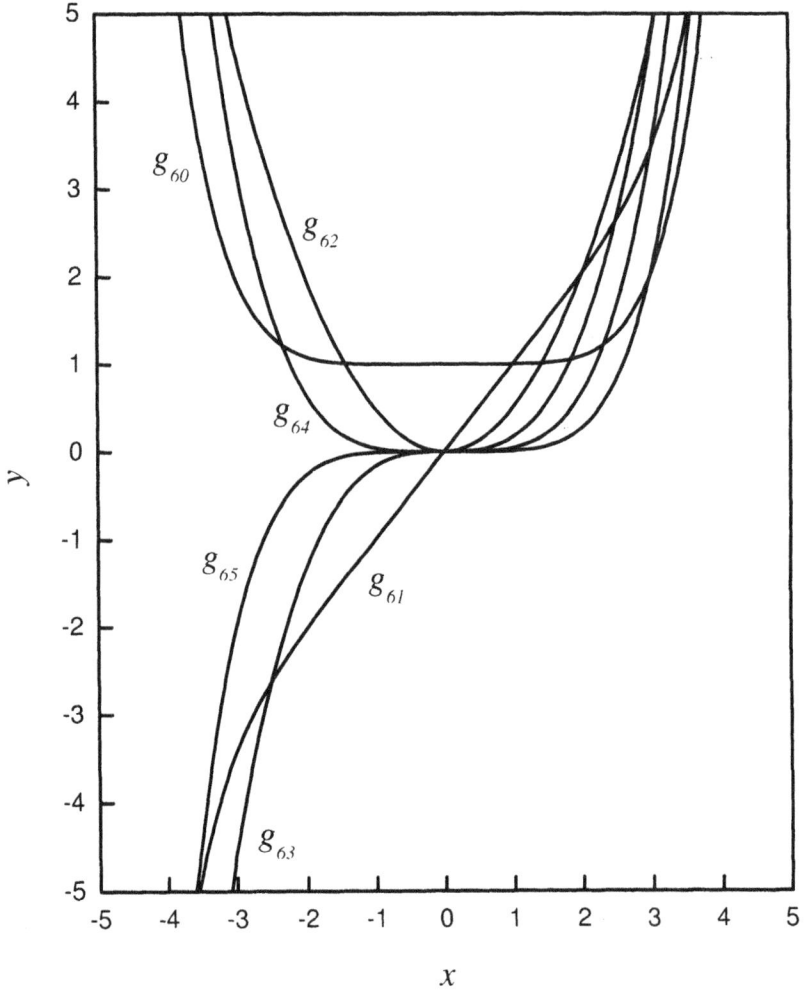

Figure 5.2: Polar cosexponential functions $g_{60}, g_{61}, g_{62}, g_{63}, g_{64}, g_{65}$.

The derivatives of the polar cosexponential functions are related by

$$\frac{dg_{60}}{du} = g_{65}, \ \frac{dg_{61}}{du} = g_{60}, \ \frac{dg_{62}}{du} = g_{61}, \ \frac{dg_{63}}{du} = g_{62}, \ \frac{dg_{64}}{du} = g_{63},$$

$$\frac{dg_{65}}{du} = g_{64}. \tag{5.35}$$

5.1.4 Exponential and trigonometric forms of polar 6-complex numbers

The exponential and trigonometric forms of polar 6-complex numbers can be expressed with the aid of the hypercomplex bases

$$\begin{pmatrix} e_+ \\ e_- \\ e_1 \\ \tilde{e}_1 \\ e_2 \\ \tilde{e}_2 \end{pmatrix} = \begin{pmatrix} \frac{1}{6} & \frac{1}{6} & \frac{1}{6} & \frac{1}{6} & \frac{1}{6} & \frac{1}{6} \\ \frac{1}{6} & -\frac{1}{6} & \frac{1}{6} & -\frac{1}{6} & \frac{1}{6} & -\frac{1}{6} \\ \frac{1}{3} & \frac{1}{6} & -\frac{1}{6} & -\frac{1}{3} & -\frac{1}{6} & \frac{1}{6} \\ 0 & \frac{\sqrt{3}}{6} & \frac{\sqrt{3}}{6} & 0 & -\frac{\sqrt{3}}{6} & -\frac{\sqrt{3}}{6} \\ \frac{1}{3} & -\frac{1}{6} & -\frac{1}{6} & \frac{1}{3} & -\frac{1}{6} & -\frac{1}{6} \\ 0 & \frac{\sqrt{3}}{6} & -\frac{\sqrt{3}}{6} & 0 & \frac{\sqrt{3}}{6} & -\frac{\sqrt{3}}{6} \end{pmatrix} \begin{pmatrix} 1 \\ h_1 \\ h_2 \\ h_3 \\ h_4 \\ h_5 \end{pmatrix}. \tag{5.36}$$

The multiplication relations for these bases are

$$e_+^2 = e_+, \ e_-^2 = e_-, \ e_+e_- = 0, \ e_+e_k = 0, \ e_+\tilde{e}_k = 0, \ e_-e_k = 0,$$

$$e_-\tilde{e}_k = 0, \ e_k^2 = e_k, \ \tilde{e}_k^2 = -e_k, \ e_k\tilde{e}_k = \tilde{e}_k, \ e_ke_l = 0, \ e_k\tilde{e}_l = 0,$$

$$\tilde{e}_k\tilde{e}_l = 0, \ k,l = 1,2, k \neq l. \tag{5.37}$$

The bases have the property that

$$e_+ + e_- + e_1 + e_2 = 1. \tag{5.38}$$

The moduli of the new bases are

$$|e_+| = \frac{1}{\sqrt{6}}, \ |e_-| = \frac{1}{\sqrt{6}}, \ |e_k| = \frac{1}{\sqrt{3}}, \ |\tilde{e}_k| = \frac{1}{\sqrt{3}}, k = 1,2. \tag{5.39}$$

It can be shown that

$$x_0 + h_1x_1 + h_2x_2 + h_3x_3 + h_4x_4 + h_5x_5$$

$$= e_+v_+ + e_-v_- + e_1v_1 + \tilde{e}_1\tilde{v}_1 + e_2v_2 + \tilde{e}_2\tilde{v}_2. \tag{5.40}$$

The ensemble $e_+, e_-, e_1, \tilde{e}_1, e_2, \tilde{e}_2$ will be called the canonical polar 6-complex base, and Eq. (5.40) gives the canonical form of the polar 6-complex number.

The exponential form of the 6-complex number u is

$$u = \rho \exp \left\{ \frac{1}{6}(h_1 + h_2 + h_3 + h_4 + h_5) \ln \frac{\sqrt{2}}{\tan \theta_+} \right.$$

$$-\frac{1}{6}(h_1 - h_2 + h_3 - h_4 + h_5) \ln \frac{\sqrt{2}}{\tan \theta_-}$$

$$\left. +\frac{1}{6}(h_1 + h_2 - 2h_3 + h_4 + h_5) \ln \tan \psi_1 + \tilde{e}_1 \phi_1 + \tilde{e}_2 \phi_2 \right\}, \quad (5.41)$$

for $0 < \theta_+ < \pi/2, 0 < \theta_- < \pi/2$.

The trigonometric form of the 6-complex number u is

$$u = d\sqrt{3} \left(\frac{1}{\tan^2 \theta_+} + \frac{1}{\tan^2 \theta_-} + 1 + \frac{1}{\tan^2 \psi_1} \right)^{-1/2}$$

$$\left(\frac{e_+ \sqrt{2}}{\tan \theta_+} + \frac{e_- \sqrt{2}}{\tan \theta_-} + e_1 + \frac{e_2}{\tan \psi_1} \right) \exp \left(\tilde{e}_1 \phi_1 + \tilde{e}_2 \phi_2 \right). \quad (5.42)$$

The modulus d and the amplitude ρ are related by

$$d = \rho \frac{2^{1/3}}{\sqrt{6}} \left(\tan \theta_+ \tan \theta_- \tan^2 \psi_1 \right)^{1/6}$$

$$\left(\frac{1}{\tan^2 \theta_+} + \frac{1}{\tan^2 \theta_-} + 1 + \frac{1}{\tan^2 \psi_1} \right)^{1/2}. \quad (5.43)$$

5.1.5   Elementary functions of a polar 6-complex variable

The logarithm and power functions of the 6-complex number u exist for $v_+ > 0, v_- > 0$, which means that $0 < \theta_+ < \pi/2, 0 < \theta_- < \pi/2$, and are given by

$$\ln u = \ln \rho + \frac{1}{6}(h_1 + h_2 + h_3 + h_4 + h_5) \ln \frac{\sqrt{2}}{\tan \theta_+}$$

$$-\frac{1}{6}(h_1 - h_2 + h_3 - h_4 + h_5) \ln \frac{\sqrt{2}}{\tan \theta_-}$$

$$+\frac{1}{6}(h_1 + h_2 - 2h_3 + h_4 + h_5) \ln \tan \psi_1 + \tilde{e}_1 \phi_1 + \tilde{e}_2 \phi_2, \quad (5.44)$$

$$u^m = e_+ v_+^m + e_- v_-^m + \rho_1^m (e_1 \cos m\phi_1 + \tilde{e}_1 \sin m\phi_1)$$

$$+\rho_2^m (e_2 \cos m\phi_2 + \tilde{e}_2 \sin m\phi_2). \quad (5.45)$$

The exponential of the 6-complex variable u is

$$e^u = e_+ e^{v_+} + e_- e^{v_-} + e^{v_1} (e_1 \cos \tilde{v}_1 + \tilde{e}_1 \sin \tilde{v}_1)$$

$$+e^{v_2} (e_2 \cos \tilde{v}_2 + \tilde{e}_2 \sin \tilde{v}_2). \quad (5.46)$$

The trigonometric functions of the 6-complex variable u are

$$\cos u = e_+ \cos v_+ + e_- \cos v_-$$
$$+ \sum_{k=1}^{2} \left(e_k \cos v_k \cosh \tilde{v}_k - \tilde{e}_k \sin v_k \sinh \tilde{v}_k \right), \qquad (5.47)$$

$$\sin u = e_+ \sin v_+ + e_- \sin v_-$$
$$+ \sum_{k=1}^{2} \left(e_k \sin v_k \cosh \tilde{v}_k + \tilde{e}_k \cos v_k \sinh \tilde{v}_k \right). \qquad (5.48)$$

The hyperbolic functions of the 6-complex variable u are

$$\cosh u = e_+ \cosh v_+ + e_- \cosh v_-$$
$$+ \sum_{k=1}^{2} \left(e_k \cosh v_k \cos \tilde{v}_k + \tilde{e}_k \sinh v_k \sin \tilde{v}_k \right), \qquad (5.49)$$

$$\sinh u = e_+ \sinh v_+ + e_- \sinh v_-$$
$$+ \sum_{k=1}^{2} \left(e_k \sinh v_k \cos \tilde{v}_k + \tilde{e}_k \cosh v_k \sin \tilde{v}_k \right). \qquad (5.50)$$

5.1.6   Power series of polar 6-complex numbers

A power series of the 6-complex variable u is a series of the form

$$a_0 + a_1 u + a_2 u^2 + \cdots + a_l u^l + \cdots. \qquad (5.51)$$

Since

$$|au^l| \leq 6^{l/2} |a| |u|^l, \qquad (5.52)$$

the series is absolutely convergent for

$$|u| < c, \qquad (5.53)$$

where

$$c = \lim_{l \to \infty} \frac{|a_l|}{\sqrt{6} |a_{l+1}|}. \qquad (5.54)$$

If $a_l = \sum_{p=0}^{5} h_p a_{lp}$, where $h_0 = 1$, and

$$A_{l+} = \sum_{p=0}^{5} a_{lp}, \qquad (5.55)$$

$$A_{l-} = \sum_{p=0}^{5}(-1)^p a_{lp},\qquad(5.56)$$

$$A_{lk} = \sum_{p=0}^{5} a_{lp}\cos\frac{\pi kp}{3},\qquad(5.57)$$

$$\tilde{A}_{lk} = \sum_{p=0}^{5} a_{lp}\sin\frac{\pi kp}{3},\qquad(5.58)$$

for $k=1,2$, the series (5.51) can be written as

$$\sum_{l=0}^{\infty}\left[e_+A_{l+}v_+^l + e_-A_{l-}v_-^l + \sum_{k=1}^{2}(e_k A_{lk}+\tilde{e}_k\tilde{A}_{lk})(e_k v_k+\tilde{e}_k\tilde{v}_k)^l\right].(5.59)$$

The series in Eq. (5.51) is absolutely convergent for

$$|v_+| < c_+,\ |v_-| < c_-,\ \rho_k < c_k, k=1,2,\qquad(5.60)$$

where

$$c_+ = \lim_{l\to\infty}\frac{|A_{l+}|}{|A_{l+1,+}|},\ c_- = \lim_{l\to\infty}\frac{|A_{l-}|}{|A_{l+1,-}|},$$

$$c_k = \lim_{l\to\infty}\frac{\left(A_{lk}^2+\tilde{A}_{lk}^2\right)^{1/2}}{\left(A_{l+1,k}^2+\tilde{A}_{l+1,k}^2\right)^{1/2}},\ k=1,2.\qquad(5.61)$$

5.1.7 Analytic functions of a polar 6-complex variable

If $f(u)=\sum_{k=0}^{5}h_k P_k(x_0,x_1,x_2,x_3,x_4,x_5)$, then

$$\frac{\partial P_0}{\partial x_0}=\frac{\partial P_1}{\partial x_1}=\frac{\partial P_2}{\partial x_2}=\frac{\partial P_3}{\partial x_3}=\frac{\partial P_4}{\partial x_4}=\frac{\partial P_5}{\partial x_5},\qquad(5.62)$$

$$\frac{\partial P_1}{\partial x_0}=\frac{\partial P_2}{\partial x_1}=\frac{\partial P_3}{\partial x_2}=\frac{\partial P_4}{\partial x_3}=\frac{\partial P_5}{\partial x_4}=\frac{\partial P_0}{\partial x_5},\qquad(5.63)$$

$$\frac{\partial P_2}{\partial x_0}=\frac{\partial P_3}{\partial x_1}=\frac{\partial P_4}{\partial x_2}=\frac{\partial P_5}{\partial x_3}=\frac{\partial P_0}{\partial x_4}=\frac{\partial P_1}{\partial x_5},\qquad(5.64)$$

$$\frac{\partial P_3}{\partial x_0}=\frac{\partial P_4}{\partial x_1}=\frac{\partial P_5}{\partial x_2}=\frac{\partial P_0}{\partial x_3}=\frac{\partial P_1}{\partial x_4}=\frac{\partial P_2}{\partial x_5},\qquad(5.65)$$

$$\frac{\partial P_4}{\partial x_0}=\frac{\partial P_5}{\partial x_1}=\frac{\partial P_0}{\partial x_2}=\frac{\partial P_1}{\partial x_3}=\frac{\partial P_2}{\partial x_4}=\frac{\partial P_3}{\partial x_5},\qquad(5.66)$$

$$\frac{\partial P_5}{\partial x_0} = \frac{\partial P_0}{\partial x_1} = \frac{\partial P_1}{\partial x_2} = \frac{\partial P_2}{\partial x_3} = \frac{\partial P_3}{\partial x_4} = \frac{\partial P_4}{\partial x_5}, \tag{5.67}$$

and

$$\frac{\partial^2 P_k}{\partial x_0 \partial x_l} = \frac{\partial^2 P_k}{\partial x_1 \partial x_{l-1}} = \cdots = \frac{\partial^2 P_k}{\partial x_{[l/2]} \partial x_{l-[l/2]}}$$

$$= \frac{\partial^2 P_k}{\partial x_{l+1} \partial x_5} = \frac{\partial^2 P_k}{\partial x_{l+2} \partial x_4} = \cdots = \frac{\partial^2 P_k}{\partial x_{l+1+[(4-l)/2]} \partial x_{5-[(4-l)/2]}}, \tag{5.68}$$

for $k, l = 0, ..., 5$. In Eq. (5.68), $[a]$ denotes the integer part of a, defined as $[a] \leq a < [a] + 1$. In this work, brackets larger than the regular brackets [ ] do not have the meaning of integer part.

5.1.8   Integrals of polar 6-complex functions

If $f(u)$ is an analytic 6-complex function, then

$$\oint_\Gamma \frac{f(u)du}{u - u_0} = 2\pi f(u_0) \left[\tilde{e}_1 \, \text{int}(u_{0\xi_1\eta_1}, \Gamma_{\xi_1\eta_1}) + \tilde{e}_2 \, \text{int}(u_{0\xi_2\eta_2}, \Gamma_{\xi_2\eta_2}) \right], \tag{5.69}$$

where

$$\text{int}(M, C) = \begin{cases} 1 \text{ if } M \text{ is an interior point of } C, \\ 0 \text{ if } M \text{ is exterior to } C, \end{cases} \tag{5.70}$$

and $u_{0\xi_k\eta_k}$ and $\Gamma_{\xi_k\eta_k}$ are respectively the projections of the pole u_0 and of the loop Γ on the plane defined by the axes ξ_k and η_k, $k = 1, 2$.

5.1.9   Factorization of polar 6-complex polynomials

A polynomial of degree m of the polar 6-complex variable u has the form

$$P_m(u) = u^m + a_1 u^{m-1} + \cdots + a_{m-1}u + a_m, \tag{5.71}$$

where a_l, for $l = 1, ..., m$, are 6-complex constants. If $a_l = \sum_{p=0}^5 h_p a_{lp}$, and with the notations of Eqs. (5.55)-(5.58) applied for $l = 1, \cdots, m$, the polynomial $P_m(u)$ can be written as

$$P_m = e_+ \left(v_+^m + \sum_{l=1}^m A_{l+} v_+^{m-l} \right) + e_- \left(v_-^m + \sum_{l=1}^m A_{l-} v_-^{m-l} \right)$$

$$+ \sum_{k=1}^2 \left[(e_k v_k + \tilde{e}_k \tilde{v}_k)^m + \sum_{l=1}^m (e_k A_{lk} + \tilde{e}_k \tilde{A}_{lk})(e_k v_k + \tilde{e}_k \tilde{v}_k)^{m-l} \right], \tag{5.72}$$

where the constants $A_{l+}, A_{l-}, A_{lk}, \tilde{A}_{lk}$ are real numbers.

The polynomial $P_m(u)$ can be written, as

$$P_m(u) = \prod_{p=1}^{m} (u - u_p), \tag{5.73}$$

where

$$u_p = e_+ v_{p+} + e_- v_{p-} + (e_1 v_{1p} + \tilde{e}_1 \tilde{v}_{1p}) + (e_2 v_{2p} + \tilde{e}_2 \tilde{v}_{2p}), p = 1, ..., m. \tag{5.74}$$

The quantities $v_{p+}, v_{p-}, e_k v_{kp} + \tilde{e}_k \tilde{v}_{kp}, p = 1, ..., m, k = 1, 2$, are the roots of the corresponding polynomial in Eq. (5.72). The roots v_{p+}, v_{p-} appear in complex-conjugate pairs, and $v_{kp}, \tilde{v}_{kp}$ are real numbers. Since all these roots may be ordered arbitrarily, the polynomial $P_m(u)$ can be written in many different ways as a product of linear factors.

If $P(u) = u^2 - 1$, the degree is $m = 2$, the coefficients of the polynomial are $a_1 = 0, a_2 = -1$, the coefficients defined in Eqs. (5.55)-(5.58) are $A_{2+} = -1, A_{2-} = -1, A_{21} = -1, \tilde{A}_{21} = 0, A_{22} = -1, \tilde{A}_{22} = 0$. The expression of $P(u)$, Eq. (5.72), is $v_+^2 - e_+ + v_-^2 - e_- + (e_1 v_1 + \tilde{e}_1 \tilde{v}_1)^2 - e_1 + (e_2 v_2 + \tilde{e}_2 \tilde{v}_2)^2 - e_2$. The factorization of $P(u)$, Eq. (5.73), is $P(u) = (u - u_1)(u - u_2)$, where the roots are $u_1 = \pm e_+ \pm e_- \pm e_1 \pm e_2, u_2 = -u_1$. If e_+, e_-, e_1, e_2 are expressed with the aid of Eq. (5.36) in terms of h_1, h_2, h_3, h_4, h_5, the factorizations of $P(u)$ are obtained as

$$u^2 - 1 = (u + 1)(u - 1),$$
$$u^2 - 1 = \left[u + \tfrac{1}{3}(1 + h_1 + h_2 - 2h_3 + h_4 + h_5)\right]$$
$$\left[u - \tfrac{1}{3}(1 + h_1 + h_2 - 2h_3 + h_4 + h_5)\right],$$
$$u^2 - 1 = \left[u + \tfrac{1}{3}(1 - h_1 + h_2 + 2h_3 + h_4 - h_5)\right]$$
$$\left[u - \tfrac{1}{3}(1 - h_1 + h_2 + 2h_3 + h_4 - h_5)\right],$$
$$u^2 - 1 = \left[u + \tfrac{1}{3}(2 + h_1 - h_2 + h_3 - h_4 + h_5)\right]$$
$$\left[u - \tfrac{1}{3}(2 + h_1 - h_2 + h_3 - h_4 + h_5)\right],$$
$$u^2 - 1 = \left[u + \tfrac{1}{3}(-1 + 2h_2 + 2h_4)\right]\left[u - \tfrac{1}{3}(-1 + 2h_2 + 2h_4)\right],$$
$$u^2 - 1 = \left[u + \tfrac{1}{3}(2h_1 - h_3 + 2h_5)\right]\left[u - \tfrac{1}{3}(2h_1 - h_3 + 2h_5)\right],$$
$$u^2 - 1 = (u + h_3)(u - h_3),$$
$$u^2 - 1 = \left[u + \tfrac{1}{3}(-2 + h_1 + h_2 + h_3 + h_4 + h_5)\right]$$
$$\left[u - \tfrac{1}{3}(-2 + h_1 + h_2 + h_3 + h_4 + h_5)\right]. \tag{5.75}$$

It can be checked that $(\pm e_+ \pm e_- \pm e_1 \pm e_2)^2 = e_+ + e_- + e_1 + e_2 = 1$.

5.1.10 Representation of polar 6-complex numbers by irreducible matrices

If the unitary matrix which appears in the expression, Eq. (5.8), of the variables $\xi_+, \xi_-, \ \xi_1, \eta_1, \ \xi_k, \eta_k$ in terms of $x_0, x_1, x_2, x_3, x_4, x_5$ is called T, the irreducible representation of the hypercomplex number u is

$$
TUT^{-1} = \begin{pmatrix} v_+ & 0 & 0 & 0 \\ 0 & v_- & 0 & 0 \\ 0 & 0 & V_1 & 0 \\ 0 & 0 & 0 & V_2 \end{pmatrix},
\tag{5.76}
$$

where U is the matrix in Eq. (5.25), and V_k are the matrices

$$
V_k = \begin{pmatrix} v_k & \tilde{v}_k \\ -\tilde{v}_k & v_k \end{pmatrix}, \quad k = 1, 2.
\tag{5.77}
$$

5.2 Planar complex numbers in 6 dimensions

5.2.1 Operations with planar complex numbers in 6 dimensions

The planar hypercomplex number u in 6 dimensions is represented as

$$
u = x_0 + h_1 x_1 + h_2 x_2 + h_3 x_3 + h_4 x_4 + h_5 x_5.
\tag{5.78}
$$

The multiplication rules for the bases h_1, h_2, h_3, h_4, h_5 are

$$
\begin{aligned}
& h_1^2 = h_2, \ h_2^2 = h_4, \ h_3^2 = 1, \ h_4^2 = -h_2, \ h_5^2 = -h_4, \ h_1 h_2 = h_3, \\
& \quad h_1 h_3 = h_4, \ h_1 h_4 = h_5, \ h_1 h_5 = -1, \ h_2 h_3 = h_5, \ h_2 h_4 = -1, \\
& \quad h_2 h_5 = -h_1, \ h_3 h_4 = -h_1, \ h_3 h_5 = -h_2, \ h_4 h_5 = -h_3.
\end{aligned}
\tag{5.79}
$$

The significance of the composition laws in Eq. (5.79) can be understood by representing the bases $1, h_1, h_2, h_3, h_4, h_5$ by points on a circle at the angles $\alpha_k = \pi k/6$, as shown in Fig. 5.3. The product $h_j h_k$ will be represented by the point of the circle at the angle $\pi(j + k)/12$, $j, k = 0, 1, ..., 5$. If $\pi \leq \pi(j + k)/12 \leq 2\pi$, the point is opposite to the basis h_l of angle $\alpha_l = \pi(j + k)/6 - \pi$.

The sum of the 6-complex numbers u and u' is

$$
\begin{aligned}
u + u' = x_0 + x_0' &+ h_1(x_1 + x_1') + h_1(x_2 + x_2') + h_3(x_3 + x_3') \\
&+ h_4(x_4 + x_4') + h_5(x_5 + x_5').
\end{aligned}
\tag{5.80}
$$

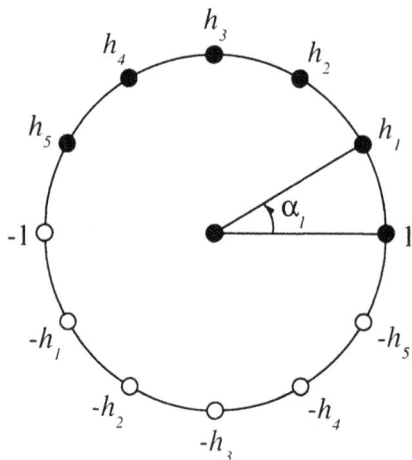

Figure 5.3: Representation of the planar hypercomplex bases $1, h_1, h_2, h_3, h_4, h_5$ by points on a circle at the angles $\alpha_k = \pi k/6$. The product $h_j h_k$ will be represented by the point of the circle at the angle $\pi(j+k)/12$, $i, k = 0, 1, ..., 5$. If $\pi \leq \pi(j+k)/12 \leq 2\pi$, the point is opposite to the basis h_l of angle $\alpha_l = \pi(j+k)/6 - \pi$.

The product of the numbers u, u' is

$$
\begin{aligned}
uu' = {} & x_0 x_0' - x_1 x_5' - x_2 x_4' - x_3 x_3' - x_4 x_2' - x_5 x_1' \\
& + h_1(x_0 x_1' + x_1 x_0' - x_2 x_5' - x_3 x_4' - x_4 x_3' - x_5 x_2') \\
& + h_2(x_0 x_2' + x_1 x_1' + x_2 x_0' - x_3 x_5' - x_4 x_4' - x_5 x_3') \\
& + h_3(x_0 x_3' + x_1 x_2' + x_2 x_1' + x_3 x_0' - x_4 x_5' - x_5 x_4') \\
& + h_4(x_0 x_4' + x_1 x_3' + x_2 x_2' + x_3 x_1' + x_4 x_0' - x_5 x_5') \\
& + h_5(x_0 x_5' + x_1 x_4' + x_2 x_3' + x_3 x_2' + x_4 x_1' + x_5 x_0').
\end{aligned}
\tag{5.81}
$$

The relation between the variables $v_1, \tilde{v}_1, v_2, \tilde{v}_2, v_3, \tilde{v}_3$ and $x_0, x_1, x_2, x_3, x_4, x_5$ are

$$
\begin{pmatrix} v_1 \\ \tilde{v}_1 \\ v_2 \\ \tilde{v}_2 \\ v_3 \\ \tilde{v}_3 \end{pmatrix}
=
\begin{pmatrix}
1 & \frac{\sqrt{3}}{2} & \frac{1}{2} & 0 & -\frac{1}{2} & -\frac{\sqrt{3}}{2} \\
0 & \frac{1}{2} & \frac{\sqrt{3}}{2} & 1 & \frac{\sqrt{3}}{2} & \frac{1}{2} \\
1 & 0 & -1 & 0 & 1 & 0 \\
0 & 1 & 0 & -1 & 0 & 1 \\
1 & -\frac{\sqrt{3}}{2} & \frac{1}{2} & 0 & -\frac{1}{2} & \frac{\sqrt{3}}{2} \\
0 & \frac{1}{2} & -\frac{\sqrt{3}}{2} & 1 & -\frac{\sqrt{3}}{2} & \frac{1}{2}
\end{pmatrix}
\begin{pmatrix} x_0 \\ x_1 \\ x_2 \\ x_3 \\ x_4 \\ x_5 \end{pmatrix}.
\tag{5.82}
$$

The other variables are $v_4 = v_3, \tilde{v}_4 = -\tilde{v}_3, v_5 = v_2, \tilde{v}_5 = -\tilde{v}_2, v_6 = v_1, \tilde{v}_6 = -\tilde{v}_1$. The variables $v_1, \tilde{v}_1, v_2, \tilde{v}_2, v_3, \tilde{v}_3$ will be called canonical planar 6-complex variables.

5.2.2 Geometric representation of planar complex numbers in 6 dimensions

The 6-complex number $u = x_0 + h_1 x_1 + h_2 x_2 + h_3 x_3 + h_4 x_4 + h_5 x_5$ is represented by the point A of coordinates $(x_0, x_1, x_2, x_3, x_4, x_5)$. The distance from the origin O of the 6-dimensional space to the point A has the expression

$$d^2 = x_0^2 + x_1^2 + x_2^2 + x_3^2 + x_4^2 + x_5^2, \tag{5.83}$$

is called modulus of the 6-complex number u, and is designated by $d = |u|$. The modulus has the property that

$$|u'u''| \leq \sqrt{3}|u'||u''|. \tag{5.84}$$

The exponential and trigonometric forms of the 6-complex number u can be obtained conveniently in a rotated system of axes defined by a transformation which has the form

$$\begin{pmatrix} \xi_1 \\ \tilde{\xi}_1 \\ \xi_2 \\ \tilde{\xi}_2 \\ \xi_3 \\ \tilde{\xi}_3 \end{pmatrix} = \begin{pmatrix} \frac{1}{\sqrt{3}} & \frac{1}{2} & \frac{1}{2\sqrt{3}} & 0 & -\frac{1}{2\sqrt{3}} & -\frac{1}{2} \\ 0 & \frac{1}{2\sqrt{3}} & \frac{1}{2} & \frac{1}{\sqrt{3}} & \frac{1}{2} & \frac{1}{2\sqrt{3}} \\ \frac{1}{\sqrt{3}} & 0 & -\frac{1}{\sqrt{3}} & 0 & \frac{1}{\sqrt{3}} & 0 \\ 0 & \frac{1}{\sqrt{3}} & 0 & -\frac{1}{\sqrt{3}} & 0 & \frac{1}{\sqrt{3}} \\ \frac{1}{\sqrt{3}} & -\frac{1}{2} & \frac{1}{2\sqrt{3}} & 0 & -\frac{1}{2\sqrt{3}} & \frac{1}{2} \\ 0 & \frac{1}{2\sqrt{3}} & -\frac{1}{2} & \frac{1}{\sqrt{3}} & -\frac{1}{2} & \frac{1}{2\sqrt{3}} \end{pmatrix} \begin{pmatrix} x_0 \\ x_1 \\ x_2 \\ x_3 \\ x_4 \\ x_5 \end{pmatrix}. \tag{5.85}$$

The lines of the matrices in Eq. (5.85) give the components of the 6 vectors of the new basis system of axes. These vectors have unit length and are orthogonal to each other. The relations between the two sets of variables are

$$v_k = \sqrt{3}\xi_k, \tilde{v}_k = \sqrt{3}\eta_k, \tag{5.86}$$

for $k = 1, 2, 3$.

The radius ρ_k and the azimuthal angle ϕ_k in the plane of the axes $v_k, \tilde{v}_k$ are

$$\rho_k^2 = v_k^2 + \tilde{v}_k^2, \cos \phi_k = v_k/\rho_k, \sin \phi_k = \tilde{v}_k/\rho_k, \tag{5.87}$$

where $0 \leq \phi_k < 2\pi$, $k = 1, 2, 3$, so that there are 3 azimuthal angles. The planar angles ψ_{k-1} are

$$\tan \psi_1 = \rho_1/\rho_2, \ \tan \psi_2 = \rho_1/\rho_3, \tag{5.88}$$

where $0 \leq \psi_1 \leq \pi/2$, $0 \leq \psi_2 \leq \pi/2$, so that there are 2 planar angles. The amplitude of an 6-complex number u is

$$\rho = (\rho_1 \rho_2 \rho_3)^{1/3}. \tag{5.89}$$

It can be checked that

$$d^2 = \frac{1}{3}(\rho_1^2 + \rho_2^2 + \rho_3^2). \tag{5.90}$$

If $u = u'u''$, the parameters of the hypercomplex numbers are related by

$$\rho_k = \rho_k' \rho_k'', \tag{5.91}$$

$$\tan \psi_k = \tan \psi_k' \tan \psi_k'', \tag{5.92}$$

$$\phi_k = \phi_k' + \phi_k'', \tag{5.93}$$

$$v_k = v_k' v_k'' - \tilde{v}_k' \tilde{v}_k'', \ \tilde{v}_k = v_k' \tilde{v}_k'' + \tilde{v}_k' v_k'', \tag{5.94}$$

$$\rho = \rho' \rho'', \tag{5.95}$$

where $k = 1, 2, 3$.

The 6-complex planar number $u = x_0 + h_1 x_1 + h_2 x_2 + h_3 x_3 + h_4 x_4 + h_5 x_5$ can be represented by the matrix

$$U = \begin{pmatrix} x_0 & x_1 & x_2 & x_3 & x_4 & x_5 \\ -x_5 & x_0 & x_1 & x_2 & x_3 & x_4 \\ -x_4 & -x_5 & x_0 & x_1 & x_2 & x_3 \\ -x_3 & -x_4 & -x_5 & x_0 & x_1 & x_2 \\ -x_2 & -x_3 & -x_4 & -x_5 & x_0 & x_1 \\ -x_1 & -x_2 & -x_3 & -x_4 & -x_5 & x_0 \end{pmatrix}. \tag{5.96}$$

The product $u = u'u''$ is represented by the matrix multiplication $U = U'U''$.

5.2.3 The planar 6-dimensional cosexponential functions

The planar cosexponential functions in 6 dimensions are

$$f_{6k}(y) = \sum_{p=0}^{\infty} (-1)^p \frac{y^{k+6p}}{(k+6p)!}, \tag{5.97}$$

for $k = 0, ..., 5$. The planar cosexponential functions of even index k are even functions, $f_{6,2l}(-y) = f_{6,2l}(y)$, and the planar cosexponential functions of odd index are odd functions, $f_{6,2l+1}(-y) = -f_{6,2l+1}(y)$, $l = 0, 1, 2$. The exponential function of the quantity $h_k y$ is

$$\begin{aligned}
e^{h_1 y} &= f_{60}(y) + h_1 f_{61}(y) + h_2 f_{62}(y) + h_3 f_{63}(y) + h_4 f_{64}(y) + h_5 f_{65}(y), \\
e^{h_2 y} &= g_{60}(y) - g_{63}(y) + h_2\{g_{61}(y) - g_{64}(y)\} + h_4\{g_{62}(y) - g_{65}(y)\}, \\
e^{h_3 y} &= f_{60}(y) - f_{62}(y) + f_{64}(y) + h_3\{f_{61}(y) - f_{63}(y) + f_{65}(y)\}, \\
e^{h_4 y} &= g_{60}(y) + g_{63}(y) - h_2\{g_{62}(y) + g_{65}(y)\} + h_4\{g_{61}(y) + g_{64}(y)\}, \\
e^{h_5 y} &= f_{60}(y) + h_1 f_{65}(y) - h_2 f_{64}(y) + h_3 f_{63}(y) - h_4 f_{62}(y) + h_5 f_{61}(y).
\end{aligned} \tag{5.98}$$

The relations for h_2 and h_4 can be written equivalently as $e^{h_2 y} = f_{30} + h_2 f_{31} + h_4 f_{32}, e^{h_4 y} = g_{30} - h_2 f_{32} + h_4 g_{31}$, and the relation for h_3 can be written as $e^{h_3 y} = f_{20} + h_3 f_{21}$, which is the same as $e^{h_3 y} = \cos y + h_3 \sin y$.

The planar 6-dimensional cosexponential functions $f_{6k}(y)$ are related to the polar 6-dimensional cosexponential function $g_{6k}(y)$ by the relations

$$f_{6k}(y) = e^{-i\pi k/6} g_{6k}\left(e^{i\pi/6} y\right), \tag{5.99}$$

for $k = 0, ..., 5$. The planar 6-dimensional cosexponential functions $f_{6k}(y)$ are related to the polar 6-dimensional cosexponential function $g_{6k}(y)$ also by the relations

$$f_{6k}(y) = e^{-i\pi k/2} g_{6k}(iy), \tag{5.100}$$

for $k = 0, ..., 5$. The expressions of the planar 6-dimensional cosexponential functions are

$$\begin{aligned}
f_{60}(y) &= \tfrac{1}{3}\cos y + \tfrac{2}{3}\cosh \tfrac{\sqrt{3}}{2}y \cos \tfrac{y}{2}, \\
f_{61}(y) &= \tfrac{1}{3}\sin y + \tfrac{\sqrt{3}}{3}\sinh \tfrac{\sqrt{3}}{2}y \cos \tfrac{y}{2} + \tfrac{1}{3}\cosh \tfrac{\sqrt{3}}{2}y \sin \tfrac{y}{2}, \\
f_{62}(y) &= -\tfrac{1}{3}\cos y + \tfrac{1}{3}\cosh \tfrac{\sqrt{3}}{2}y \cos \tfrac{y}{2} + \tfrac{\sqrt{3}}{3}\sinh \tfrac{\sqrt{3}}{2}y \sin \tfrac{y}{2}, \\
f_{63}(y) &= -\tfrac{1}{3}\sin y + \tfrac{2}{3}\cosh \tfrac{\sqrt{3}}{2}y \sin \tfrac{y}{2}, \\
f_{64}(y) &= \tfrac{1}{3}\cos y - \tfrac{1}{3}\cosh \tfrac{\sqrt{3}}{2}y \cos \tfrac{y}{2} + \tfrac{\sqrt{3}}{3}\sinh \tfrac{\sqrt{3}}{2}y \sin \tfrac{y}{2}, \\
f_{65}(y) &= \tfrac{1}{3}\sin y - \tfrac{\sqrt{3}}{3}\sinh \tfrac{\sqrt{3}}{2}y \cos \tfrac{y}{2} + \tfrac{1}{3}\cosh \tfrac{\sqrt{3}}{2}y \sin \tfrac{y}{2}.
\end{aligned} \tag{5.101}$$

The planar 6-dimensional cosexponential functions can be written as

$$f_{6k}(y) = \frac{1}{6} \sum_{l=1}^{6} \exp\left[y\cos\left(\frac{\pi(2l-1)}{6}\right)\right] \cos\left[y\sin\left(\frac{\pi(2l-1)}{6}\right)\right.$$
$$\left. - \frac{\pi(2l-1)k}{6}\right], \tag{5.102}$$

for $k = 0, ..., 5$. The graphs of the planar 6-dimensional cosexponential functions are shown in Fig. 5.4.

It can be checked that

$$\sum_{k=0}^{5} f_{6k}^{2}(y) = \frac{1}{3} + \frac{2}{3}\cosh\sqrt{3}y. \tag{5.103}$$

The addition theorems for the planar 6-dimensional cosexponential functions are

$$g_{60}(y+z) = g_{60}(y)g_{60}(z) - g_{61}(y)g_{65}(z) - g_{62}(y)g_{64}(z)$$
$$- g_{63}(y)g_{63}(z) - g_{64}(y)g_{62}(z) - g_{65}(y)g_{61}(z),$$

$$g_{61}(y+z) = g_{60}(y)g_{61}(z) + g_{61}(y)g_{60}(z) - g_{62}(y)g_{65}(z)$$
$$- g_{63}(y)g_{64}(z) - g_{64}(y)g_{63}(z) - g_{65}(y)g_{62}(z),$$

$$g_{62}(y+z) = g_{60}(y)g_{62}(z) + g_{61}(y)g_{61}(z) + g_{62}(y)g_{60}(z)$$
$$- g_{63}(y)g_{65}(z) - g_{64}(y)g_{64}(z) - g_{65}(y)g_{63}(z),$$

$$g_{63}(y+z) = g_{60}(y)g_{63}(z) + g_{61}(y)g_{62}(z) + g_{62}(y)g_{61}(z)$$
$$+ g_{63}(y)g_{60}(z) - g_{64}(y)g_{65}(z) - g_{65}(y)g_{64}(z),$$

$$g_{64}(y+z) = g_{60}(y)g_{64}(z) + g_{61}(y)g_{63}(z) + g_{62}(y)g_{62}(z)$$
$$+ g_{63}(y)g_{61}(z) + g_{64}(y)g_{60}(z) - g_{65}(y)g_{65}(z),$$

$$g_{65}(y+z) = g_{60}(y)g_{65}(z) + g_{61}(y)g_{64}(z) + g_{62}(y)g_{63}(z)$$
$$+ g_{63}(y)g_{62}(z) + g_{64}(y)g_{61}(z) + g_{65}(y)g_{60}(z). \tag{5.104}$$

It can be shown that

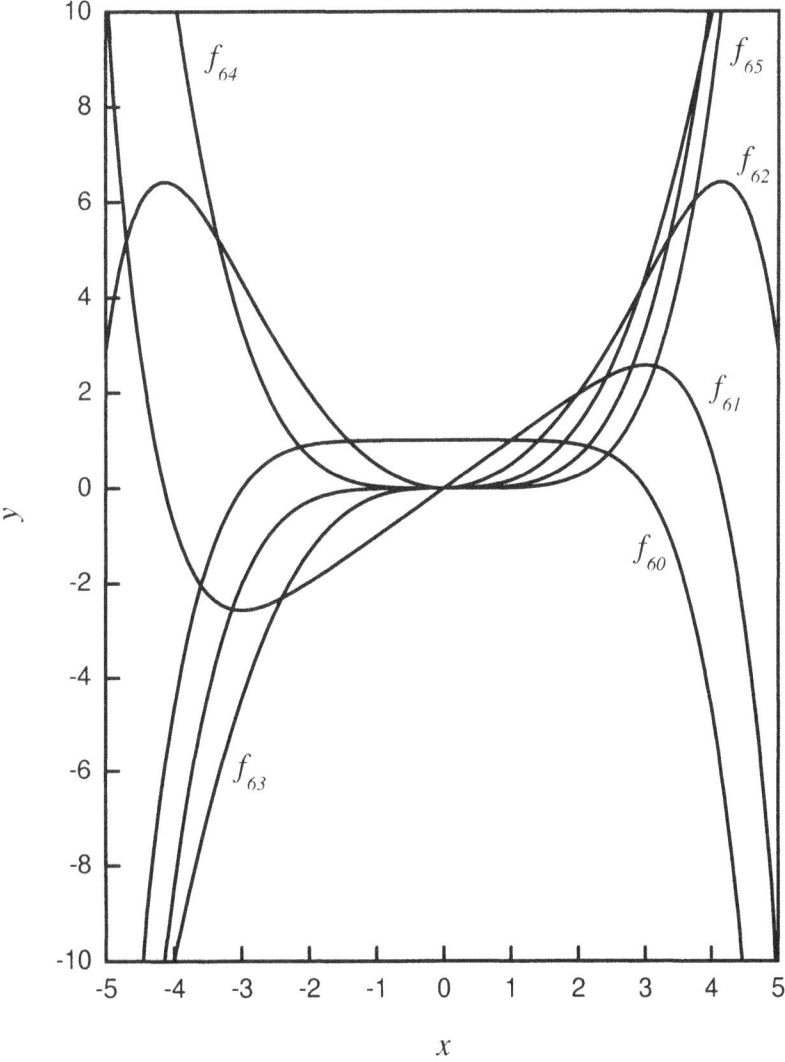

Figure 5.4: Planar cosexponential functions $f_{60}, f_{61}, f_{62}, f_{63}, f_{64}, f_{65}$.

$$\{f_{60}(y) + h_1 f_{61}(y) + h_2 f_{62}(y) + h_3 f_{63}(y) + h_4 f_{64}(y) + h_5 f_{65}(y)\}^l$$
$$= f_{60}(ly) + h_1 f_{61}(ly) + h_2 f_{62}(ly) + h_3 f_{63}(ly) + h_4 f_{64}(ly)$$
$$+ h_5 f_{65}(ly),$$
$$\{g_{60}(y) - g_{63}(y) + h_2\{g_{61}(y) - g_{64}(y)\} + h_4\{g_{62}(y) - g_{65}(y)\}\}^l$$
$$= g_{60}(ly) - g_{63}(ly) + h_2\{g_{61}(ly) - g_{64}(ly)\} + h_4\{g_{62}(ly)$$
$$- g_{65}(ly)\},$$
$$\{f_{60}(y) - f_{62}(y) + f_{64}(y) + h_3\{f_{61}(y) - f_{63}(y) + f_{65}(y)\}\}^l$$
$$= f_{60}(ly) - f_{62}(ly) + f_{64}(ly) + h_3\{f_{61}(ly) - f_{63}(ly) + f_{65}(ly)\},$$
$$\{g_{60}(y) + g_{63}(y) - h_2\{g_{62}(y) + g_{65}(y)\} + h_4\{g_{61}(y) + g_{64}(y)\}\}^l$$
$$= g_{60}(ly) + g_{63}(ly) - h_2\{g_{62}(ly) + g_{65}(ly)\} + h_4\{g_{61}(ly)$$
$$+ g_{64}(ly)\},$$
$$\{f_{60}(y) + h_1 f_{65}(y) - h_2 f_{64}(y) + h_3 f_{63}(y) - h_4 f_{62}(y) + h_5 f_{61}(y)\}^l$$
$$= f_{60}(ly) + h_1 f_{65}(ly) - h_2 f_{64}(ly) + h_3 f_{63}(ly) - h_4 f_{62}(ly)$$
$$+ h_5 f_{61}(ly).$$

$$(5.105)$$

The derivatives of the planar cosexponential functions are related by

$$\frac{df_{60}}{du} = -f_{65}, \quad \frac{df_{61}}{du} = f_{60}, \quad \frac{df_{62}}{du} = f_{61}, \quad \frac{df_{63}}{du} = f_{62}, \quad \frac{df_{64}}{du} = f_{63},$$
$$\frac{df_{65}}{du} = f_{64}.$$

$$(5.106)$$

5.2.4 Exponential and trigonometric forms of planar 6-complex numbers

The exponential and trigonometric forms of planar 6-complex numbers can be expressed with the aid of the hypercomplex bases

$$
\begin{pmatrix} e_1 \\ \tilde{e}_1 \\ e_2 \\ \tilde{e}_2 \\ e_3 \\ \tilde{e}_3 \end{pmatrix} =
\begin{pmatrix}
\frac{1}{3} & \frac{\sqrt{3}}{6} & \frac{1}{6} & 0 & -\frac{1}{6} & -\frac{\sqrt{3}}{6} \\
0 & \frac{1}{6} & \frac{\sqrt{3}}{6} & \frac{1}{3} & \frac{\sqrt{3}}{6} & \frac{1}{6} \\
\frac{1}{3} & 0 & -\frac{1}{3} & 0 & \frac{1}{3} & 0 \\
0 & \frac{1}{3} & 0 & -\frac{1}{3} & 0 & \frac{1}{3} \\
\frac{1}{3} & -\frac{\sqrt{3}}{6} & \frac{1}{6} & 0 & -\frac{1}{6} & \frac{\sqrt{3}}{6} \\
0 & \frac{1}{6} & -\frac{\sqrt{3}}{6} & \frac{1}{3} & -\frac{\sqrt{3}}{6} & \frac{1}{6}
\end{pmatrix}
\begin{pmatrix} 1 \\ h_1 \\ h_2 \\ h_3 \\ h_4 \\ h_5 \end{pmatrix}. \quad (5.107)
$$

The multiplication relations for the bases $e_k, \tilde{e}_k$ are

$$e_k^2 = e_k, \tilde{e}_k^2 = -e_k, e_k \tilde{e}_k = \tilde{e}_k, e_k e_l = 0, e_k \tilde{e}_l = 0, \tilde{e}_k \tilde{e}_l = 0,$$
$$k, l = 1, 2, 3, \ k \neq l. \quad (5.108)$$

The moduli of the bases $e_k, \tilde{e}_k$ are

$$|e_k| = \sqrt{\frac{1}{3}}, |\tilde{e}_k| = \sqrt{\frac{1}{3}}, \tag{5.109}$$

for $k = 1, 2, 3$. It can be shown that

$$x_0 + h_1 x_1 + h_2 x_2 + h_3 x_3 + h_4 x_4 + h_5 x_5 = \sum_{k=1}^{3} (e_k v_k + \tilde{e}_k \tilde{v}_k). \tag{5.110}$$

The ensemble $e_1, \tilde{e}_1, e_2, \tilde{e}_2, e_3, \tilde{e}_3$ will be called the canonical planar 6-complex base, and Eq. (5.110) gives the canonical form of the planar 6-complex number.

The exponential form of the 6-complex number u is

$$u = \rho \exp \left\{ \frac{1}{3}(h_2 - h_4) \ln \tan \psi_1 + \frac{1}{6}(\sqrt{3}h_1 - h_2 + h_4 - \sqrt{3}h_5) \ln \tan \psi_2 \right.$$
$$\left. + \tilde{e}_1 \phi_1 + \tilde{e}_2 \phi_2 + \tilde{e}_3 \phi_3 \right\}. \tag{5.111}$$

The trigonometric form of the 6-complex number u is

$$u = d\sqrt{3} \left(1 + \frac{1}{\tan^2 \psi_1} + \frac{1}{\tan^2 \psi_2} \right)^{-1/2}$$
$$\left(e_1 + \frac{e_2}{\tan \psi_1} + \frac{e_3}{\tan \psi_2} \right) \exp \left(\tilde{e}_1 \phi_1 + \tilde{e}_2 \phi_2 + \tilde{e}_3 \phi_3 \right). \tag{5.112}$$

The modulus d and the amplitude ρ are related by

$$d = \rho \frac{2^{1/3}}{\sqrt{6}} (\tan \psi_1 \tan \psi_2)^{1/3} \left(1 + \frac{1}{\tan^2 \psi_1} + \frac{1}{\tan^2 \psi_2} \right)^{1/2}. \tag{5.113}$$

5.2.5 Elementary functions of a planar 6-complex variable

The logarithm and power functions of the 6-complex number u exist for all $x_0, ..., x_5$ and are

$$\ln u = \ln \rho + \frac{1}{3}(h_2 - h_4) \ln \tan \psi_1$$
$$+ \frac{1}{6}(\sqrt{3}h_1 - h_2 + h_4 - \sqrt{3}h_5) \ln \tan \psi_2$$
$$+ \tilde{e}_1 \phi_1 + \tilde{e}_2 \phi_2 + \tilde{e}_3 \phi_3, \tag{5.114}$$

$$u^m = \sum_{k=1}^{3} \rho_k^m (e_k \cos m\phi_k + \tilde{e}_k \sin m\phi_k). \tag{5.115}$$

The exponential of the 6-complex variable u is

$$e^u = \sum_{k=1}^{3} e^{v_k} \left(e_k \cos \tilde{v}_k + \tilde{e}_k \sin \tilde{v}_k \right). \tag{5.116}$$

The trigonometric functions of the 6-complex variable u are

$$\cos u = \sum_{k=1}^{3} \left(e_k \cos v_k \cosh \tilde{v}_k - \tilde{e}_k \sin v_k \sinh \tilde{v}_k \right), \tag{5.117}$$

$$\sin u = \sum_{k=1}^{3} \left(e_k \sin v_k \cosh \tilde{v}_k + \tilde{e}_k \cos v_k \sinh \tilde{v}_k \right). \tag{5.118}$$

The hyperbolic functions of the 6-complex variable u are

$$\cosh u = \sum_{k=1}^{3} \left(e_k \cosh v_k \cos \tilde{v}_k + \tilde{e}_k \sinh v_k \sin \tilde{v}_k \right), \tag{5.119}$$

$$\sinh u = \sum_{k=1}^{3} \left(e_k \sinh v_k \cos \tilde{v}_k + \tilde{e}_k \cosh v_k \sin \tilde{v}_k \right). \tag{5.120}$$

5.2.6 Power series of planar 6-complex numbers

A power series of the 6-complex variable u is a series of the form

$$a_0 + a_1 u + a_2 u^2 + \cdots + a_l u^l + \cdots. \tag{5.121}$$

Since

$$|au^l| \leq 3^{l/2} |a| |u|^l, \tag{5.122}$$

the series is absolutely convergent for

$$|u| < c, \tag{5.123}$$

where

$$c = \lim_{l \to \infty} \frac{|a_l|}{\sqrt{3}|a_{l+1}|}. \tag{5.124}$$

If $a_l = \sum_{p=0}^{5} h_p a_{lp}$, and

$$A_{lk} = \sum_{p=0}^{5} a_{lp} \cos \frac{\pi(2k-1)p}{6}, \tag{5.125}$$

$$\tilde{A}_{lk} = \sum_{p=0}^{5} a_{lp} \sin \frac{\pi(2k-1)p}{6}, \tag{5.126}$$

where $k = 1, 2, 3$, the series (5.121) can be written as

$$\sum_{l=0}^{\infty} \left[\sum_{k=1}^{3} (e_k A_{lk} + \tilde{e}_k \tilde{A}_{lk})(e_k v_k + \tilde{e}_k \tilde{v}_k)^l \right]. \tag{5.127}$$

The series is absolutely convergent for

$$\rho_k < c_k, k = 1, 2, 3, \tag{5.128}$$

where

$$c_k = \lim_{l \to \infty} \frac{\left[A_{lk}^2 + \tilde{A}_{lk}^2 \right]^{1/2}}{\left[A_{l+1,k}^2 + \tilde{A}_{l+1,k}^2 \right]^{1/2}}. \tag{5.129}$$

5.2.7   Analytic functions of a planar 6-complex variable

If $f(u) = \sum_{k=0}^{5} h_k P_k(x_0, ..., x_5)$, then

$$\frac{\partial P_0}{\partial x_0} = \frac{\partial P_1}{\partial x_1} = \frac{\partial P_2}{\partial x_2} = \frac{\partial P_3}{\partial x_3} = \frac{\partial P_4}{\partial x_4} = \frac{\partial P_5}{\partial x_5}, \tag{5.130}$$

$$\frac{\partial P_1}{\partial x_0} = \frac{\partial P_2}{\partial x_1} = \frac{\partial P_3}{\partial x_2} = \frac{\partial P_4}{\partial x_3} = \frac{\partial P_5}{\partial x_4} = -\frac{\partial P_0}{\partial x_5}, \tag{5.131}$$

$$\frac{\partial P_2}{\partial x_0} = \frac{\partial P_3}{\partial x_1} = \frac{\partial P_4}{\partial x_2} = \frac{\partial P_5}{\partial x_3} = -\frac{\partial P_0}{\partial x_4} = -\frac{\partial P_1}{\partial x_5}, \tag{5.132}$$

$$\frac{\partial P_3}{\partial x_0} = \frac{\partial P_4}{\partial x_1} = \frac{\partial P_5}{\partial x_2} = -\frac{\partial P_0}{\partial x_3} = -\frac{\partial P_1}{\partial x_4} = -\frac{\partial P_2}{\partial x_5}, \tag{5.133}$$

$$\frac{\partial P_4}{\partial x_0} = \frac{\partial P_5}{\partial x_1} = -\frac{\partial P_0}{\partial x_2} = -\frac{\partial P_1}{\partial x_3} = -\frac{\partial P_2}{\partial x_4} = -\frac{\partial P_3}{\partial x_5}, \tag{5.134}$$

$$\frac{\partial P_5}{\partial x_0} = -\frac{\partial P_0}{\partial x_1} = -\frac{\partial P_1}{\partial x_2} = -\frac{\partial P_2}{\partial x_3} = -\frac{\partial P_3}{\partial x_4} = -\frac{\partial P_4}{\partial x_5}, \tag{5.135}$$

and

$$\frac{\partial^2 P_k}{\partial x_0 \partial x_l} = \frac{\partial^2 P_k}{\partial x_1 \partial x_{l-1}} = \cdots = \frac{\partial^2 P_k}{\partial x_{[l/2]} \partial x_{l-[l/2]}}$$

$$= -\frac{\partial^2 P_k}{\partial x_{l+1} \partial x_5} = -\frac{\partial^2 P_k}{\partial x_{l+2} \partial x_4} = \cdots = -\frac{\partial^2 P_k}{\partial x_{l+1+[(4-l)/2]} \partial x_{5-[(4-l)/2]}}. \tag{5.136}$$

5.2.8   Integrals of planar 6-complex functions

If $f(u)$ is an analytic 6-complex function, then

$$\oint_\Gamma \frac{f(u)du}{u-u_0} = 2\pi f(u_0)\left\{\tilde{e}_1 \text{ int}(u_{0\xi_1\eta_1},\Gamma_{\xi_1\eta_1}) + \tilde{e}_2 \text{ int}(u_{0\xi_2\eta_2},\Gamma_{\xi_2\eta_2})\right.$$
$$\left. + \tilde{e}_3 \text{ int}(u_{0\xi_3\eta_3},\Gamma_{\xi_3\eta_3})\right\}, \tag{5.137}$$

where $u_{0\xi_k\eta_k}$ and $\Gamma_{\xi_k\eta_k}$ are respectively the projections of the point u_0 and of the loop Γ on the plane defined by the axes ξ_k and η_k, $k = 1, 2, 3$.

5.2.9   Factorization of planar 6-complex polynomials

A polynomial of degree m of the planar 6-complex variable u has the form

$$P_m(u) = u^m + a_1 u^{m-1} + \cdots + a_{m-1}u + a_m, \tag{5.138}$$

where a_l, for $l = 1, ..., m$, are 6-complex constants. If $a_l = \sum_{p=0}^5 h_p a_{lp}$, and with the notations of Eqs. (5.125)-(5.126) applied for $l = 1, \cdots, m$, the polynomial $P_m(u)$ can be written as

$$P_m = \sum_{k=1}^3 \left[(e_k v_k + \tilde{e}_k \tilde{v}_k)^m + \sum_{l=1}^m (e_k A_{lk} + \tilde{e}_k \tilde{A}_{lk})(e_k v_k + \tilde{e}_k \tilde{v}_k)^{m-l}\right], \tag{5.139}$$

where the constants $A_{lk}, \tilde{A}_{lk}$ are real numbers.

The polynomial $P_m(u)$ can be written as a product of factors

$$P_m(u) = \prod_{p=1}^m (u - u_p), \tag{5.140}$$

where

$$u_p = \sum_{k=1}^3 (e_k v_{kp} + \tilde{e}_k \tilde{v}_{kp}), \tag{5.141}$$

for $p = 1, ..., m$. The quantities $e_k v_{kp} + \tilde{e}_k \tilde{v}_{kp}$, $p = 1, ..., m, k = 1, 2, 3$, are the roots of the corresponding polynomial in Eq. (5.139) and are real numbers. Since these roots may be ordered arbitrarily, the polynomial $P_m(u)$ can be written in many different ways as a product of linear factors.

If $P(u) = u^2 + 1$, the degree is $m = 2$, the coefficients of the polynomial are $a_1 = 0, a_2 = 1$, the coefficients defined in Eqs. (5.125)-(5.126) are $A_{21} = 1, \tilde{A}_{21} = 0, A_{22} = 1, \tilde{A}_{22} = 0, A_{23} = 1, \tilde{A}_{23} = 0$. The expression, Eq. (5.139), is $P(u)=(e_1 v_1 + \tilde{e}_1 \tilde{v}_1)^2 + e_1 + (e_2 v_2 + \tilde{e}_2 \tilde{v}_2)^2 + e_2 + (e_3 v_3 + \tilde{e}_3 \tilde{v}_3)^2 + e_3$.

The factorization of $P(u)$, Eq. (5.140), is $P(u) = (u - u_1)(u - u_2)$, where the roots are $u_1 = \pm\tilde{e}_1 \pm \tilde{e}_2 \pm \tilde{e}_3, u_2 = -u_1$. If $\tilde{e}_1, \tilde{e}_2, \tilde{e}_3$ are expressed with the aid of Eq. (5.107) in terms of h_1, h_2, h_3, h_4, h_5, the factorizations of $P(u)$ are obtained as

$$
\begin{aligned}
u^2 + 1 &= \left[u + \tfrac{1}{3}(2h_1 + h_3 + 2h_5)\right]\left[u - \tfrac{1}{3}(2h_1 + h_3 + 2h_5)\right], \\
u^2 + 1 &= \left[u + \tfrac{1}{3}(h_1 + \sqrt{3}h_2 - h_3 + \sqrt{3}h_4 + h_5)\right] \\
&\quad \left[u - \tfrac{1}{3}(h_1 + \sqrt{3}h_2 - h_3 + \sqrt{3}h_4 + h_5)\right], \\
u^2 + 1 &= (u + h_3)(u - h_3), \\
u^2 + 1 &= \left[u + \tfrac{1}{3}(-h_1 + \sqrt{3}h_2 + h_3 + \sqrt{3}h_4 - h_5)\right] \\
&\quad \left[u - \tfrac{1}{3}(-h_1 + \sqrt{3}h_2 + h_3 + \sqrt{3}h_4 - h_5)\right].
\end{aligned}
\tag{5.142}
$$

It can be checked that $(\pm\tilde{e}_1 \pm \tilde{e}_2 + \pm\tilde{e}_3)^2 = -e_1 - e_2 - e_3 = -1$.

5.2.10 Representation of planar 6-complex numbers by irreducible matrices

If the unitary matrix written in Eq. (5.85) is called T, the matric TUT^{-1} provides an irreducible representation [7] of the planar hypercomplex number u,

$$
TUT^{-1} = \begin{pmatrix} V_1 & 0 & 0 \\ 0 & V_2 & 0 \\ 0 & 0 & V_3 \end{pmatrix},
\tag{5.143}
$$

where U is the matrix in Eq. (5.96) used to represent the 6-complex number u, and the matrices V_k are

$$
V_k = \begin{pmatrix} v_k & \tilde{v}_k \\ -\tilde{v}_k & v_k \end{pmatrix},
\tag{5.144}
$$

for $k = 1, 2, 3$.

Chapter 6

Commutative Complex Numbers in n Dimensions

Two systems of complex numbers in n dimensions are described in this chapter, for which the multiplication is associative and commutative, which can be written in exponential and trigonometric forms, and for which the concepts of analytic n-complex function, contour integration and residue can be defined. The n-complex numbers introduced in this chapter have the form $u = x_0 + h_1 x_1 + h_2 x_2 + \cdots + h_{n-1} x_{n-1}$, the variables $x_0, ..., x_{n-1}$ being real numbers.

The multiplication rules for the complex units $h_1, ..., h_{n-1}$ discussed in Sec. 6.1 are $h_j h_k = h_{j+k}$ if $0 \leq j + k \leq n - 1$, and $h_j h_k = h_{j+k-n}$ if $n \leq j + k \leq 2n - 2$. The product of two n-complex numbers is equal to zero if both numbers are equal to zero, or if the numbers belong to certain n-dimensional hyperplanes described further in this chapter. If the n-complex number $u = x_0 + h_1 x_1 + h_2 x_2 + \cdots + h_{n-1} x_{n-1}$ is represented by the point A of coordinates $x_0, x_1, ..., x_{n-1}$, the position of the point A can be described, in an even number of dimensions, by the modulus $d = (x_0^2 + x_1^2 + \cdots + x_{n-1}^2)^{1/2}$, by $n/2 - 1$ azimuthal angles ϕ_k, by $n/2 - 2$ planar angles ψ_{k-1}, and by 2 polar angles θ_+, θ_-. In an odd number of dimensions, the position of the point A is described by d, by $(n-1)/2$ azimuthal angles ϕ_k, by $(n-3)/2$ planar angles ψ_{k-1}, and by 1 polar angle θ_+. An amplitude ρ can be defined for even n as $\rho^n = v_+ v_- \rho_1^2 \cdots \rho_{n/2-1}^2$, and for odd n as $\rho^n = v_+ \rho_1^2 \cdots \rho_{(n-1)/2}^2$, where $v_+ = x_0 + \cdots + x_{n-1}, v_- = x_0 - x_1 + \cdots + x_{n-2} - x_{n-1}$, and ρ_k are radii in orthogonal two-dimensional planes defined further in Sec. 6.1. The amplitude ρ, the variables v_+, v_-, the radii ρ_k, the variables $(1/\sqrt{2}) \tan \theta_+, (1/\sqrt{2}) \tan \theta_-, \tan \psi_{k-1}$ are multiplicative, and the azimuthal angles ϕ_k are additive upon the multiplication of n-complex numbers. Because of the role of the axis v_+ and, in an even number of dimensions, of the axis v_-, in the description of the position of the point A with the aid of the polar angle θ_+ and, in an even number of dimensions, of the polar angle θ_-, the hypercomplex numbers studied in Sec. 6.1 will be called polar n-complex number, to distinguish them from the planar n-complex numbers, which exist in an even number of dimensions.

The exponential function of a polar n-complex number can be expanded in terms of the polar n-dimensional cosexponential functions $g_{nk}(y) = \sum_{p=0}^{\infty} y^{k+pn}/(k+pn)!$, $k = 0, 1, ..., n-1$. It is shown that $g_{nk}(y) = \frac{1}{n} \sum_{l=0}^{n-1} \exp\{y\cos(2\pi l/n)\} \cos\{y\sin(2\pi l/n) - 2\pi kl/n\}$, $k = 0, 1, ..., n-1$. Addition theorems and other relations are obtained for the polar n-dimensional cosexponential functions.

The exponential form of a polar n-complex number, which in an even number of dimensions n can be defined for $x_0 + \cdots + x_{n-1} > 0, x_0 - x_1 + \cdots + x_{n-2} - x_{n-1} > 0$, is $u = \rho \exp\left\{\sum_{p=1}^{n-1} h_p \left[(1/n)\ln\sqrt{2}/\tan\theta_+ + ((-1)^p/n)\right.\right.$
$\left.\left.\ln\sqrt{2}/\tan\theta_- - (2/n)\sum_{k=2}^{n/2-1}\cos(2\pi kp/n)\ln\tan\psi_{k-1}\right]\right\}\exp\left(\sum_{k=1}^{n/2-1}\tilde{e}_k\phi_k\right)$,
where $\tilde{e}_k = (2/n)\sum_{p=1}^{n-1}h_p\sin(2\pi pk/n)$. In an odd number of dimensions n, the exponential form exists for $x_0 + \cdots + x_{n-1} > 0$, and is given by
$u = \rho\exp\left\{\sum_{p=1}^{n-1}h_p\left[(1/n)\ln\sqrt{2}/\tan\theta_+ - (2/n)\sum_{k=2}^{(n-1)/2}\cos(2\pi kp/n)\right.\right.$
$\left.\left.\ln\tan\psi_{k-1}\right]\right\}\exp\left(\sum_{k=1}^{(n-1)/2}\tilde{e}_k\phi_k\right)$. A trigonometric form also exists for an n-complex number u, when u is written as the product of the modulus d, of a factor depending on the polar and planar angles $\theta_+, \theta_-, \psi_{k-1}$ and of an exponential factor depending on the azimuthal angles ϕ_k.

Expressions are given for the elementary functions of polar n-complex variable. The functions $f(u)$ of n-complex variable which are defined by power series have derivatives independent of the direction of approach to the point under consideration. If the n-complex function $f(u)$ of the n-complex variable u is written in terms of the real functions $P_k(x_0, ..., x_{n-1})$, $k = 0, ..., n-1$, then relations of equality exist between partial derivatives of the functions P_k. The integral $\int_A^B f(u)du$ of an n-complex function between two points A, B is independent of the path connecting A, B, in regions where f is regular. If $f(u)$ is an analytic n-complex function, then $\oint_\Gamma f(u)du/(u-u_0)$
$= 2\pi f(u_0)\sum_{k=1}^{[(n-1)/2]}\tilde{e}_k$ int$(u_{0\xi_k\eta_k}, \Gamma_{\xi_k\eta_k})$, where the functional int takes the values 0 or 1 depending on the relation between $u_{0\xi_k\eta_k}$ and $\Gamma_{\xi_k\eta_k}$, which are respectively the projections of the point u_0 and of the loop Γ on the plane defined by the orthogonal axes ξ_k and η_k, as expained further in this work.

A polar n-complex polynomial can be written as a product of linear or quadratic factors, although the factorization may not be unique.

Particular cases for $n = 2, 3, 4, 5, 6$ of the polar n-complex numbers described in Sec. 6.1 have been studied in previous chapters.

The multiplication rules for the complex units $h_1, ..., h_{n-1}$ discussed in Sec. 6.2 are $h_j h_k = h_{j+k}$ if $0 \leq j + k \leq n - 1$, and $h_j h_k = -h_{j+k-n}$ if $n \leq j + k \leq 2n - 2$, where $h_0 = 1$. The product of two n-complex numbers is equal to zero if both numbers are equal to zero, or if the numbers belong to certain n-dimensional hyperplanes described further in Sec. 6.2. If the n-complex number $u = x_0 + h_1 x_1 + h_2 x_2 + \cdots + h_{n-1} x_{n-1}$ is represented by the point A of coordinates $x_0, x_1, ..., x_{n-1}$, the position of the point A can be described, in an even number of dimensions, by the modulus $d = (x_0^2 + x_1^2 + \cdots + x_{n-1}^2)^{1/2}$, by $n/2$ azimuthal angles ϕ_k and by $n/2 - 1$ planar angles ψ_{k-1}. An amplitude ρ can be defined as $\rho^n = \rho_1^2 \cdots \rho_{n/2}^2$, where ρ_k are radii in orthogonal two-dimensional planes defined further in Sec. 6.2. The amplitude ρ, the radii ρ_k and the variables $\tan \psi_{k-1}$ are multiplicative, and the azimuthal angles ϕ_k are additive upon the multiplication of n-complex numbers. Because the description of the position of the point A requires, in addition to the azimuthal angles ϕ_k, only the planar angles ψ_{k-1}, the hypercomplex numbers studied in Sec. 6.2 will be called planar n-complex number, to distinguish them from the polar n-complex numbers, which in an even number of dimensions required two polar angles, and in an odd number of dimensions required one polar angle.

The exponential function of an n-complex number can be expanded in terms of the planar n-dimensional cosexponential functions $f_{nk}(y) = \sum_{p=0}^{\infty} (-1)^p y^{k+pn}/(k + pn)!, k = 0, 1, ..., n - 1$. It is shown that $f_{nk}(y) = (1/n) \sum_{l=1}^{n} \exp\{y \cos(\pi(2l - 1)/n)\} \cos\{y \sin(\pi(2l - 1)/n) - \pi k(2l - 1)/n\}$, $k = 0, 1, ..., n - 1$. Addition theorems and other relations are obtained for the planar n-dimensional cosexponential functions.

The exponential form of a planar n-complex number, which can be defined for all values of $x_0, ..., x_{n-1}$, is given by $u = \rho \exp\left\{\sum_{p=1}^{n-1} h_p\left[-(2/n)\right.\right.$ $\left.\sum_{k=2}^{n/2} \cos(\pi(2k - 1)p/n) \ln \tan \psi_{k-1}\right]\right\} \exp\left(\sum_{k=1}^{n/2} \tilde{e}_k \phi_k\right)$, where $\tilde{e}_k = (2/n)$ $\sum_{p=1}^{n-1} h_p \sin(\pi(2k - 1)p/n)$. A trigonometric form also exists for an n-complex number, being given by $u = d(n/2)^{1/2} \left(1 + 1/\tan^2 \psi_1 + 1/\tan^2 \psi_2\right.$ $\left. + \cdots + 1/\tan^2 \psi_{n/2-1}\right)^{-1/2} \left(e_1 + \sum_{k=2}^{n/2} e_k/\tan \psi_{k-1}\right) \exp\left(\sum_{k=1}^{n/2} \tilde{e}_k \phi_k\right)$.

Expressions are given for the elementary functions of planar n-complex variable. The functions $f(u)$ of planar n-complex variable which are defined by power series have derivatives independent of the direction of approach to the point under consideration. If the n-complex function $f(u)$ of the n-complex variable u is written in terms of the real functions $P_k(x_0, ..., x_{n-1})$, $k = 0, ..., n - 1$, then relations of equality exist between partial derivatives of the functions P_k. The integral $\int_A^B f(u)du$ of an n-complex function between two points A, B is independent of the path connecting A, B, in regions where f is regular. If $f(u)$ is an analytic n-complex function, then $\oint_\Gamma f(u)du/(u - u_0) = 2\pi f(u_0) \sum_{k=1}^{n/2} \tilde{e}_k \; \text{int}(u_{0\xi_k\eta_k}, \Gamma_{\xi_k\eta_k})$, where the functional int takes the values 0 or 1 depending on the relation between $u_{0\xi_k\eta_k}$ and $\Gamma_{\xi_k\eta_k}$, which are respectively the projections of the point u_0 and of the loop Γ on the plane defined by the orthogonal axes ξ_k and η_k, as expained further in Sec. 6.2.

A planar n-complex polynomial can always be written as a product of linear factors, although the factorization may not be unique.

For $n = 2$, the n-complex numbers discussed in this paper become the usual 2-dimensional complex numbers $x + iy$. Particular cases for $n = 4$ and $n = 6$ of the planar n-complex numbers described in this paper have been studied in previous chapters.

6.1   Polar complex numbers in n dimensions

6.1.1   Operations with polar n-complex numbers

A complex number in n dimensions is determined by its n components $(x_0, x_1, ..., x_{n-1})$. The polar n-complex numbers and their operations discussed in this section can be represented by writing the n-complex number $(x_0, x_1, ..., x_{n-1})$ as $u = x_0 + h_1 x_1 + h_2 x_2 + \cdots + h_{n-1} x_{n-1}$, where $h_1, h_2, \cdots, h_{n-1}$ are bases for which the multiplication rules are

$$h_j h_k = h_l, \; l = j + k - n[(j + k)/n], \tag{6.1}$$

for $j, k, l = 0, 1, ..., n - 1$. In Eq. (6.1), $[(j + k)/n]$ denotes the integer part of $(j + k)/n$, the integer part being defined as $[a] \leq a < [a] + 1$, so that $0 \leq j + k - n[(j + k)/n] \leq n - 1$. In this chapter, brackets larger than the regular brackets [ ] do not have the meaning of integer part. The significance of the composition laws in Eq. (6.1) can be understood by representing the bases h_j, h_k by points on a circle at the angles $\alpha_j = 2\pi j/n, \alpha_k = 2\pi k/n$, as shown in Fig. 6.1, and the product $h_j h_k$ by the point of the circle at the angle $2\pi(j + k)/n$. If $2\pi \leq 2\pi(j + k)/n < 4\pi$, the point represents the basis h_l of angle $\alpha_l = 2\pi(j + k - n)/n$.

Two n-complex numbers $u = x_0 + h_1 x_1 + h_2 x_2 + \cdots + h_{n-1} x_{n-1}$, $u' = x'_0 + h_1 x'_1 + h_2 x'_2 + \cdots + h_{n-1} x'_{n-1}$ are equal if and only if $x_i = x'_i, i = 0, 1, ..., n-1$. The sum of the n-complex numbers u and u' is

$$u + u' = x_0 + x'_0 + h_1(x_1 + x'_1) + \cdots + h_{n-1}(x_{n-1} + x'_{n-1}). \tag{6.2}$$

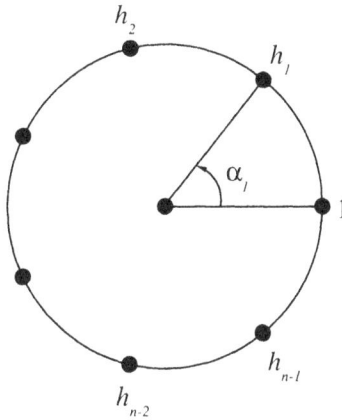

Figure 6.1: Representation of the hypercomplex bases $1, h_1, ..., h_{n-1}$ by points on a circle at the angles $\alpha_k = 2\pi k/n$. The product $h_j h_k$ will be represented by the point of the circle at the angle $2\pi(j+k)/n$, $i, k = 0, 1, ..., n-1$. If $2\pi \leq 2\pi(j+k)/n \leq 4\pi$, the point represents the basis h_l of angle $\alpha_l = 2\pi(j+k)/n - 2\pi$.

The product of the numbers u, u' is

$$
\begin{aligned}
uu' = & x_0 x_0' + x_1 x_{n-1}' + x_2 x_{n-2}' + x_3 x_{n-3}' + \cdots + x_{n-1} x_1' \\
& + h_1 (x_0 x_1' + x_1 x_0' + x_2 x_{n-1}' + x_3 x_{n-2}' + \cdots + x_{n-1} x_2') \\
& + h_2 (x_0 x_2' + x_1 x_1' + x_2 x_0' + x_3 x_{n-1}' + \cdots + x_{n-1} x_3') \\
& \vdots \\
& + h_{n-1} (x_0 x_{n-1}' + x_1 x_{n-2}' + x_2 x_{n-3}' + x_3 x_{n-4}' + \cdots + x_{n-1} x_0').
\end{aligned}
\tag{6.3}
$$

The product uu' can be written as

$$
uu' = \sum_{k=0}^{n-1} h_k \sum_{l=0}^{n-1} x_l x_{k-l+n[(n-k-1+l)/n]}'.
\tag{6.4}
$$

If u, u', u'' are n-complex numbers, the multiplication is associative

$$
(uu')u'' = u(u'u'')
\tag{6.5}
$$

and commutative

$$
uu' = u'u,
\tag{6.6}
$$

because the product of the bases, defined in Eq. (6.1), is associative and commutative. The fact that the multiplication is commutative can be seen also directly from Eq. (6.3). The n-complex zero is $0 + h_1 \cdot 0 + \cdots + h_{n-1} \cdot 0$, denoted simply 0, and the n-complex unity is $1 + h_1 \cdot 0 + \cdots + h_{n-1} \cdot 0$, denoted simply 1.

The inverse of the n-complex number $u = x_0 + h_1 x_1 + h_2 x_2 + \cdots + h_{n-1} x_{n-1}$ is the n-complex number $u' = x'_0 + h_1 x'_1 + h_2 x'_2 + \cdots + h_{n-1} x'_{n-1}$ having the property that

$$uu' = 1. \tag{6.7}$$

Written on components, the condition, Eq. (6.7), is

$$
\begin{aligned}
x_0 x'_0 + x_1 x'_{n-1} + x_2 x'_{n-2} + x_3 x'_{n-3} + \cdots + x_{n-1} x'_1 &= 1, \\
x_0 x'_1 + x_1 x'_0 + x_2 x'_{n-1} + x_3 x'_{n-2} + \cdots + x_{n-1} x'_2 &= 0, \\
x_0 x'_2 + x_1 x'_1 + x_2 x'_0 + x_3 x'_{n-1} + \cdots + x_{n-1} x'_3 &= 0, \\
&\vdots \\
x_0 x'_{n-1} + x_1 x'_{n-2} + x_2 x'_{n-3} + x_3 x'_{n-4} + \cdots + x_{n-1} x'_0 &= 0.
\end{aligned}
\tag{6.8}
$$

The system (6.8) has a solution provided that the determinant of the system,

$$\nu = \det(A), \tag{6.9}$$

is not equal to zero, $\nu \neq 0$, where

$$
A =
\begin{pmatrix}
x_0 & x_{n-1} & x_{n-2} & \cdots & x_1 \\
x_1 & x_0 & x_{n-1} & \cdots & x_2 \\
x_2 & x_1 & x_0 & \cdots & x_3 \\
\vdots & \vdots & \vdots & \cdots & \vdots \\
x_{n-1} & x_{n-2} & x_{n-3} & \cdots & x_0
\end{pmatrix}.
\tag{6.10}
$$

If $\nu > 0$, the quantity

$$\rho = \nu^{1/n} \tag{6.11}$$

will be called amplitude of the n-complex number $u = x_0 + h_1 x_1 + h_2 x_2 + \cdots + h_{n-1} x_{n-1}$. The quantity ν can be written as a product of linear factors

$$\nu = \prod_{k=0}^{n-1} \left(x_0 + \epsilon_k x_1 + \epsilon_k^2 x_2 + \cdots + \epsilon_k^{n-1} x_{n-1} \right), \tag{6.12}$$

where $\epsilon_k = e^{2\pi i k/n}$, i being the imaginary unit. The factors appearing in Eq. (6.12) are of the form

$$x_0 + \epsilon_k x_1 + \epsilon_k^2 x_2 + \cdots + \epsilon_k^{n-1} x_{n-1} = v_k + i\tilde{v}_k, \tag{6.13}$$

where

$$v_k = \sum_{p=0}^{n-1} x_p \cos \frac{2\pi kp}{n}, \tag{6.14}$$

$$\tilde{v}_k = \sum_{p=0}^{n-1} x_p \sin \frac{2\pi kp}{n}, \tag{6.15}$$

for $k = 1, 2, ..., n-1$ and, if n is even, $k \neq n/2$. For $k = 0$ the factor in Eq. (6.13) is

$$v_+ = x_0 + x_1 + \cdots + x_{n-1}, \tag{6.16}$$

and if n is even, for $k = n/2$ the factor in Eq. (6.13) is

$$v_- = x_0 - x_1 + \cdots + x_{n-2} - x_{n-1}. \tag{6.17}$$

It can be seen that $v_k = v_{n-k}, \tilde{v}_k = -\tilde{v}_{n-k}, k = 1, ..., [(n-1)/2]$. The variables $v_+, v_-, v_k, \tilde{v}_k, k = 1, ..., [(n-1)/2]$ will be called canonical polar n-complex variables. Therefore, the factors appear in Eq. (6.12) in complex-conjugate pairs of the form $v_k + i\tilde{v}_k$ and $v_{n-k} + i\tilde{v}_{n-k} = v_k - i\tilde{v}_k$, where $k = 1, ..., [(n-1)/2]$, so that the product ν is a real quantity. If n is an even number, the quantity ν is

$$\nu = v_+ v_- \prod_{k=1}^{n/2-1} (v_k^2 + \tilde{v}_k^2), \tag{6.18}$$

and if n is an odd number, ν is

$$\nu = v_+ \prod_{k=0}^{(n-1)/2} (v_k^2 + \tilde{v}_k^2). \tag{6.19}$$

Thus, in an even number of dimensions n, an n-complex number has an inverse unless it lies on one of the nodal hypersurfaces $x_0 + x_1 + \cdots + x_{n-1} = 0$, or $x_0 - x_1 + \cdots + x_{n-2} - x_{n-1} = 0$, or $v_1 = 0, \tilde{v}_1 = 0$, ..., or $v_{n/2-1} = 0, \tilde{v}_{n/2-1} = 0$. In an odd number of dimensions n, an n-complex number has an inverse unless it lies on one of the nodal hypersurfaces $x_0 + x_1 + \cdots + x_{n-1} = 0$, or $v_1 = 0, \tilde{v}_1 = 0$, ..., or $v_{(n-1)/2} = 0, \tilde{v}_{(n-1)/2} = 0$.

6.1.2 Geometric representation of polar n-complex numbers

The n-complex number $x_0 + h_1 x_1 + h_2 x_2 + \cdots + h_{n-1} x_{n-1}$ can be represented by the point A of coordinates $(x_0, x_1, ..., x_{n-1})$. If O is the origin of the n-dimensional space, the distance from the origin O to the point A of coordinates $(x_0, x_1, ..., x_{n-1})$ has the expression

$$d^2 = x_0^2 + x_1^2 + \cdots + x_{n-1}^2. \tag{6.20}$$

The quantity d will be called modulus of the n-complex number $u = x_0 + h_1 x_1 + h_2 x_2 + \cdots + h_{n-1} x_{n-1}$. The modulus of an n-complex number u will be designated by $d = |u|$.

The exponential and trigonometric forms of the n-complex number u can be obtained conveniently in a rotated system of axes defined by a transformation which, for even n, has the form

$$
\begin{pmatrix} \xi_+ \\ \xi_- \\ \vdots \\ \xi_k \\ \eta_k \\ \vdots \end{pmatrix}
=
\begin{pmatrix}
\frac{1}{\sqrt{n}} & \frac{1}{\sqrt{n}} & \cdots & \frac{1}{\sqrt{n}} & \frac{1}{\sqrt{n}} \\
\frac{1}{\sqrt{n}} & -\frac{1}{\sqrt{n}} & \cdots & \frac{1}{\sqrt{n}} & -\frac{1}{\sqrt{n}} \\
\vdots & \vdots & & \vdots & \vdots \\
\sqrt{\frac{2}{n}} & \sqrt{\frac{2}{n}}\cos\frac{2\pi k}{n} & \cdots & \sqrt{\frac{2}{n}}\cos\frac{2\pi(n-2)k}{n} & \sqrt{\frac{2}{n}}\cos\frac{2\pi(n-1)k}{n} \\
0 & \sqrt{\frac{2}{n}}\sin\frac{2\pi k}{n} & \cdots & \sqrt{\frac{2}{n}}\sin\frac{2\pi(n-2)k}{n} & \sqrt{\frac{2}{n}}\sin\frac{2\pi(n-1)k}{n} \\
\vdots & \vdots & & \vdots & \vdots
\end{pmatrix}
\begin{pmatrix} x_0 \\ x_1 \\ \vdots \\ \vdots \\ \vdots \\ x_{n-1} \end{pmatrix},
\tag{6.21}
$$

where $k = 1, 2, ..., n/2 - 1$. For odd n the rotation of the axes is described by the relations

$$
\begin{pmatrix} \xi_+ \\ \xi_1 \\ \eta_1 \\ \vdots \\ \xi_k \\ \eta_k \\ \vdots \end{pmatrix}
=
\begin{pmatrix}
\frac{1}{\sqrt{n}} & \frac{1}{\sqrt{n}} & \cdots & \frac{1}{\sqrt{n}} \\
\sqrt{\frac{2}{n}} & \sqrt{\frac{2}{n}}\cos\frac{2\pi}{n} & \cdots & \sqrt{\frac{2}{n}}\cos\frac{2\pi(n-1)}{n} \\
0 & \sqrt{\frac{2}{n}}\sin\frac{2\pi}{n} & \cdots & \sqrt{\frac{2}{n}}\sin\frac{2\pi(n-1)}{n} \\
\vdots & \vdots & & \vdots \\
\sqrt{\frac{2}{n}} & \sqrt{\frac{2}{n}}\cos\frac{2\pi k}{n} & \cdots & \sqrt{\frac{2}{n}}\cos\frac{2\pi(n-1)k}{n} \\
0 & \sqrt{\frac{2}{n}}\sin\frac{2\pi k}{n} & \cdots & \sqrt{\frac{2}{n}}\sin\frac{2\pi(n-1)k}{n} \\
\vdots & \vdots & & \vdots
\end{pmatrix}
\begin{pmatrix} x_0 \\ x_1 \\ x_2 \\ \vdots \\ \vdots \\ x_{n-1} \end{pmatrix},
$$

$$(6.22)$$

where $k = 0, 1, ..., (n-1)/2$. The lines of the matrices in Eqs. (6.21) or (6.22) give the components of the n basis vectors of the new system of axes. These vectors have unit length and are orthogonal to each other. By comparing Eqs. (6.14)-(6.17) and (6.21)-(6.22) it can be seen that

$$v_+ = \sqrt{n}\xi_+, \, v_- = \sqrt{n}\xi_-, v_k = \sqrt{\frac{n}{2}}\xi_k, \tilde{v}_k = \sqrt{\frac{n}{2}}\eta_k, \qquad (6.23)$$

i.e. the two sets of variables differ only by scale factors.

The sum of the squares of the variables $v_k, \tilde{v}_k$ is, for even n,

$$\sum_{k=1}^{n/2-1} (v_k^2 + \tilde{v}_k^2) = \frac{n-2}{2}(x_0^2 + \cdots + x_{n-1}^2) - 2(x_0x_2 + \cdots + x_{n-4}x_{n-2}$$
$$+x_1x_3 + \cdots + x_{n-3}x_{n-1}), \qquad (6.24)$$

and for odd n the sum is

$$\sum_{k=1}^{(n-1)/2} (v_k^2+\tilde{v}_k^2) = \frac{n-1}{2}(x_0^2+\cdots+x_{n-1}^2)-(x_0x_1+\cdots+x_{n-2}x_{n-1}).(6.25)$$

The relation (6.24) has been obtained with the aid of the identity, valid for even n,

$$\sum_{k=1}^{n/2-1} \cos\frac{2\pi pk}{n} = \begin{cases} -1, & \text{for even } p, \\ 0, & \text{for odd } p. \end{cases} \qquad (6.26)$$

The relation (6.25) has been obtained with the aid of the identity, valid for odd values of n,

$$\sum_{k=1}^{(n-1)/2} \cos\frac{2\pi pk}{n} = -\frac{1}{2}. \qquad (6.27)$$

From Eq. (6.24) it results that, for even n,

$$d^2 = \frac{1}{n}v_+^2 + \frac{1}{n}v_-^2 + \frac{2}{n}\sum_{k=1}^{n/2-1} \rho_k^2, \qquad (6.28)$$

and from Eq. (6.25) it results that, for odd n,

$$d^2 = \frac{1}{n}v_+^2 + \frac{2}{n}\sum_{k=1}^{(n-1)/2} \rho_k^2. \qquad (6.29)$$

The relations (6.28) and (6.29) show that the square of the distance d, Eq. (6.20), is the sum of the squares of the projections $v_+/\sqrt{n}, \rho_k\sqrt{2/n}$ and, for even n, of the square of $v_-/\sqrt{n}$. This is consistent with the fact that the transformation in Eqs. (6.21) or (6.22) is unitary.

The position of the point A of coordinates $(x_0, x_1, ..., x_{n-1})$ can be also described with the aid of the distance d, Eq. (6.20), and of $n-1$ angles defined further. Thus, in the plane of the axes $v_k, \tilde{v}_k$, the radius ρ_k and the azimuthal angle ϕ_k can be introduced by the relations

$$\rho_k^2 = v_k^2 + \tilde{v}_k^2, \ \cos\phi_k = v_k/\rho_k, \ \sin\phi_k = \tilde{v}_k/\rho_k, 0 \le \phi_k < 2\pi, \qquad (6.30)$$

so that there are $[(n-1)/2]$ azimuthal angles. If the projection of the point A on the plane of the axes $v_k, \tilde{v}_k$ is A_k, and the projection of the point A on the 4-dimensional space defined by the axes $v_1, \tilde{v}_1, v_k, \tilde{v}_k$ is A_{1k}, the angle ψ_{k-1} between the line OA_{1k} and the 2-dimensional plane defined by the axes $v_k, \tilde{v}_k$ is

$$\tan\psi_{k-1} = \rho_1/\rho_k, \qquad (6.31)$$

where $0 \le \psi_k \le \pi/2, k = 2, ..., [(n-1)/2]$, so that there are $[(n-3)/2]$ planar angles. Moreover, there is a polar angle θ_+, which can be defined as the angle between the line OA_{1+} and the axis v_+, where A_{1+} is the projection of the point A on the 3-dimensional space generated by the axes $v_1, \tilde{v}_1, v_+$,

$$\tan\theta_+ = \frac{\sqrt{2}\rho_1}{v_+}, \qquad (6.32)$$

where $0 \le \theta_+ \le \pi$, and in an even number of dimensions n there is also a polar angle θ_-, which can be defined as the angle between the line OA_{1-} and the axis v_-, where A_{1-} is the projection of the point A on the 3-dimensional space generated by the axes $v_1, \tilde{v}_1, v_-$,

$$\tan\theta_- = \frac{\sqrt{2}\rho_1}{v_-}, \qquad (6.33)$$

where $0 \le \theta_- \le \pi$. In Eqs. (6.32) and (6.33), the factor $\sqrt{2}$ appears from the ratio of the normalization factors in Eq. (6.23). Thus, the position of the point A is described, in an even number of dimensions, by the distance d, by $n/2 - 1$ azimuthal angles, by $n/2 - 2$ planar angles, and by 2 polar angles. In an odd number of dimensions, the position of the point A is described by $(n-1)/2$ azimuthal angles, by $(n-3)/2$ planar angles, and by 1 polar angle. These angles are shown in Fig. 6.2.

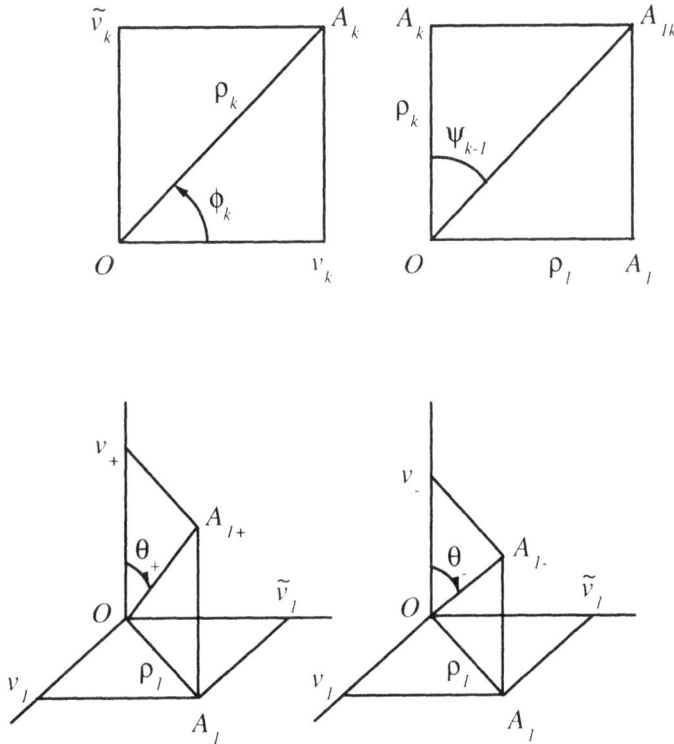

Figure 6.2: Radial distance ρ_k and azimuthal angle ϕ_k in the plane of the axes $v_k, \tilde{v}_k$, and planar angle ψ_{k-1} between the line OA_{1k} and the 2-dimensional plane defined by the axes $v_k, \tilde{v}_k$. A_k is the projection of the point A on the plane of the axes $v_k, \tilde{v}_k$, and A_{1k} is the projection of the point A on the 4-dimensional space defined by the axes $v_1, \tilde{v}_1, v_k, \tilde{v}_k$. The polar angle θ_+ is the angle between the line OA_{1+} and the axis v_+, where A_{1+} is the projection of the point A on the 3-dimensional space generated by the axes $v_1, \tilde{v}_1, v_+$. In an even number of dimensions n there is also a polar angle θ_-, which is the angle between the line OA_{1-} and the axis v_-, where A_{1-} is the projection of the point A on the 3-dimensional space generated by the axes $v_1, \tilde{v}_1, v_-$.

The variables ρ_k can be expressed in terms of d and the planar angles ψ_k as

$$\rho_k = \frac{\rho_1}{\tan \psi_{k-1}}, \tag{6.34}$$

for $k = 2, ..., [(n-1)/2]$, where, for even n,

$$\rho_1^2 = \frac{nd^2}{2} \left(\frac{1}{\tan^2 \theta_+} + \frac{1}{\tan^2 \theta_-} + 1 + \frac{1}{\tan^2 \psi_1} + \frac{1}{\tan^2 \psi_2} + \cdots \right.$$
$$\left. + \frac{1}{\tan^2 \psi_{n/2-2}} \right)^{-1}, \tag{6.35}$$

and for odd n

$$\rho_1^2 = \frac{nd^2}{2} \left(\frac{1}{\tan^2 \theta_+} + 1 + \frac{1}{\tan^2 \psi_1} + \frac{1}{\tan^2 \psi_2} + \cdots \right.$$
$$\left. + \frac{1}{\tan^2 \psi_{(n-3)/2}} \right)^{-1}. \tag{6.36}$$

If $u' = x_0' + h_1 x_1' + h_2 x_2' + \cdots + h_{n-1} x_{n-1}'$, $u'' = x_0'' + h_1 x_1'' + h_2 x_2'' + \cdots + h_{n-1} x_{n-1}''$ are n-complex numbers of parameters v_+', v_-', ρ_k', θ_+', θ_-', ψ_k', ϕ_k' and respectively v_+'', v_-'', ρ_k'', θ_+'', θ_-'', ψ_k'', ϕ_k'', then the parameters v_+, v_-, ρ_k, θ_+, θ_-, ψ_k, ϕ_k of the product n-complex number $u = u'u''$ are given by

$$v_+ = v_+' v_+'', \tag{6.37}$$

$$\rho_k = \rho_k' \rho_k'', \tag{6.38}$$

for $k = 1, ..., [(n-1)/2]$,

$$\tan \theta_+ = \frac{1}{\sqrt{2}} \tan \theta_+' \tan \theta_+'', \tag{6.39}$$

$$\tan \psi_k = \tan \psi_k' \tan \psi_k'', \tag{6.40}$$

for $k = 1, ..., [(n-3)/2]$,

$$\phi_k = \phi_k' + \phi_k'', \tag{6.41}$$

for $k = 1, ..., [(n-1)/2]$, and, if n is even,

$$v_- = v_-' v_-'', \tag{6.42}$$

$$\tan \theta_- = \frac{1}{\sqrt{2}} \tan \theta_-' \tan \theta_-''. \tag{6.43}$$

The Eqs. (6.37) and (6.42) can be checked directly, and Eqs. (6.38)-(6.41) and (6.43) are a consequence of the relations

$$v_k = v'_k v''_k - \tilde{v}'_k \tilde{v}''_k, \quad \tilde{v}_k = v'_k \tilde{v}''_k + \tilde{v}'_k v''_k, \tag{6.44}$$

and of the corresponding relations of definition. Then the product ν in Eqs. (6.18) and (6.19) has the property that

$$\nu = \nu' \nu'' \tag{6.45}$$

and, if $\nu' > 0, \nu'' > 0$, the amplitude ρ defined in Eq. (6.11) has the property that

$$\rho = \rho' \rho''. \tag{6.46}$$

The fact that the amplitude of the product is equal to the product of the amplitudes, as written in Eq. (6.46), can be demonstrated also by using a representation of the n-complex numbers by matrices, in which the n-complex number $u = x_0 + h_1 x_1 + h_2 x_2 + \cdots + h_{n-1} x_{n-1}$ is represented by the matrix

$$U = \begin{pmatrix} x_0 & x_1 & x_2 & \cdots & x_{n-1} \\ x_{n-1} & x_0 & x_1 & \cdots & x_{n-2} \\ x_{n-2} & x_{n-1} & x_0 & \cdots & x_{n-3} \\ \vdots & \vdots & \vdots & \cdots & \vdots \\ x_1 & x_2 & x_3 & \cdots & x_0 \end{pmatrix}. \tag{6.47}$$

The product $u = u'u''$ is represented by the matrix multiplication $U = U'U''$. The relation (6.45) is then a consequence of the fact the determinant of the product of matrices is equal to the product of the determinants of the factor matrices. The use of the representation of the n-complex numbers with matrices provides an alternative demonstration of the fact that the product of n-complex numbers is associative, as stated in Eq. (6.5).

According to Eqs. (6.37, (6.38), (6.42), (6.28) and (6.29), the modulus of the product uu' is, for even n,

$$|uu'|^2 = \frac{1}{n}(v_+ v'_+)^2 + \frac{1}{n}(v_- v'_-)^2 + \frac{2}{n} \sum_{k=1}^{n/2-1} (\rho_k \rho'_k)^2, \tag{6.48}$$

and for odd n

$$|uu'|^2 = \frac{1}{n}(v_+ v'_+)^2 + \frac{2}{n} \sum_{k=1}^{(n-1)/2} (\rho_k \rho'_k)^2. \tag{6.49}$$

Thus, if the product of two n-complex numbers is zero, $uu' = 0$, then $v_+ v'_+ = 0, \rho_k \rho'_k = 0, k = 1, ..., [(n-1)/2]$ and, if n is even, $v_- v'_- = 0$. This means that either $u = 0$, or $u' = 0$, or u, u' belong to orthogonal hypersurfaces in such a way that the afore-mentioned products of components should be equal to zero.

6.1.3   The polar n-dimensional cosexponential functions

The exponential function of a hypercomplex variable u and the addition theorem for the exponential function have been written in Eqs. (1.35)-(1.36). If $u = x_0 + h_1 x_1 + h_2 x_2 + \cdots + h_{n-1} x_{n-1}$, then $\exp u$ can be calculated as $\exp u = \exp x_0 \cdot \exp(h_1 x_1) \cdots \exp(h_{n-1} x_{n-1})$.

It can be seen with the aid of the representation in Fig. 6.1 that

$$h_k^{n+p} = h_k^p, \; p \text{ integer}, \tag{6.50}$$

for $k = 1, ..., n - 1$. Then $e^{h_k y}$ can be written as

$$e^{h_k y} = \sum_{p=0}^{n-1} h_{kp-n[kp/n]} g_{np}(y), \tag{6.51}$$

where the expression of the functions g_{nk}, which will be called polar cosexponential functions in n dimensions, is

$$g_{nk}(y) = \sum_{p=0}^{\infty} y^{k+pn}/(k+pn)!, \tag{6.52}$$

for $k = 0, 1, ..., n - 1$.

If n is even, the polar cosexponential functions of even index k are even functions, $g_{n,2p}(-y) = g_{n,2p}(y)$, $p = 0, 1, ..., n/2 - 1$, and the polar cosexponential functions of odd index are odd functions, $g_{n,2p+1}(-y) = -g_{n,2p+1}(y)$, $p = 0, 1, ..., n/2 - 1$. For odd values of n, the polar cosexponential functions do not have a definite parity. It can be checked that

$$\sum_{k=0}^{n-1} g_{nk}(y) = e^y \tag{6.53}$$

and, for even n,

$$\sum_{k=0}^{n-1} (-1)^k g_{nk}(y) = e^{-y}. \tag{6.54}$$

The expression of the polar n-dimensional cosexponential functions is

$$g_{nk}(y) = \frac{1}{n} \sum_{l=0}^{n-1} \exp\left[y \cos\left(\frac{2\pi l}{n}\right)\right] \cos\left[y \sin\left(\frac{2\pi l}{n}\right) - \frac{2\pi k l}{n}\right], \tag{6.55}$$

for $k = 0, 1, ..., n - 1$. In order to check that the function in Eq. (6.55) has the series expansion written in Eq. (6.52), the right-hand side of Eq. (6.55) will be written as

$$g_{nk}(y) = \frac{1}{n} \sum_{l=0}^{n-1} \text{Re} \left\{ \exp \left[\left(\cos \frac{2\pi l}{n} + i \sin \frac{2\pi l}{n} \right) y - i \frac{2\pi k l}{n} \right] \right\}, \quad (6.56)$$

for $k = 0, 1, ..., n - 1$, where $\text{Re}(a + ib) = a$, with a and b real numbers. The part of the exponential depending on y can be expanded in a series,

$$g_{nk}(y) = \frac{1}{n} \sum_{p=0}^{\infty} \sum_{l=0}^{n-1} \text{Re} \left\{ \frac{1}{p!} \exp \left[i \frac{2\pi l}{n} (p - k) \right] y^p \right\}, \quad (6.57)$$

for $k = 0, 1, ..., n - 1$. The expression of $g_{nk}(y)$ becomes

$$g_{nk}(y) = \frac{1}{n} \sum_{p=0}^{\infty} \sum_{l=0}^{n-1} \left\{ \frac{1}{p!} \cos \left[\frac{2\pi l}{n} (p - k) \right] y^p \right\}, \quad (6.58)$$

for $k = 0, 1, ..., n - 1$ and, since

$$\frac{1}{n} \sum_{l=0}^{n-1} \cos \frac{2\pi l}{n} (p - k) = \begin{cases} 1, & \text{if } p - k \text{ is a multiple of } n, \\ 0, & \text{otherwise,} \end{cases} \quad (6.59)$$

this yields indeed the expansion in Eq. (6.52).

It can be shown from Eq. (6.55) that

$$\sum_{k=0}^{n-1} g_{nk}^2(y) = \frac{1}{n} \sum_{l=0}^{n-1} \exp \left[2y \cos \left(\frac{2\pi l}{n} \right) \right]. \quad (6.60)$$

It can be seen that the right-hand side of Eq. (6.60) does not contain oscillatory terms. If n is a multiple of 4, it can be shown by replacing y by iy in Eq. (6.60) that

$$\sum_{k=0}^{n-1} (-1)^k g_{nk}^2(y) = \frac{2}{n} \left\{ 1 + \cos 2y + \sum_{l=1}^{n/4-1} \cos \left[2y \cos \left(\frac{2\pi l}{n} \right) \right] \right\}, \quad (6.61)$$

which does not contain exponential terms.

Addition theorems for the polar n-dimensional cosexponential functions can be obtained from the relation $\exp h_1(y + z) = \exp h_1 y \cdot \exp h_1 z$, by substituting the expression of the exponentials as given in Eq. (6.51) for $k = 1$, $e^{h_1 y} = g_{n0}(y) + h_1 g_{n1}(y) + \cdots + h_{n-1} g_{n,n-1}(y)$,

$$g_{nk}(y + z) = g_{n0}(y) g_{nk}(z) + g_{n1}(y) g_{n,k-1}(z) + \cdots + g_{nk}(y) g_{n0}(z)$$
$$+ g_{n,k+1}(y) g_{n,n-1}(z) + g_{n,k+2}(y) g_{n,n-2}(z) + \cdots$$
$$+ g_{n,n-1}(y) g_{n,k+1}(z), \quad (6.62)$$

where $k = 0, 1, ..., n - 1$. For $y = z$ the relations (6.62) take the form

$$g_{nk}(2y) = g_{n0}(y)g_{nk}(y) + g_{n1}(y)g_{n,k-1}(y) + \cdots + g_{nk}(y)g_{n0}(y)$$
$$+ g_{n,k+1}(y)g_{n,n-1}(y) + g_{n,k+2}(y)g_{n,n-2}(y) + \cdots$$
$$+ g_{n,n-1}(y)g_{n,k+1}(y), \tag{6.63}$$

where $k = 0, 1, ..., n - 1$. For $y = -z$ the relations (6.62) and (6.52) yield

$$g_{n0}(y)g_{n0}(-y) + g_{n1}(y)g_{n,n-1}(-y) + g_{n2}(y)g_{n,n-2}(-y) + \cdots$$
$$+ g_{n,n-1}(y)g_{n1}(-y) = 1, \tag{6.64}$$

$$g_{n0}(y)g_{nk}(-y) + g_{n1}(y)g_{n,k-1}(-y) + \cdots + g_{nk}(y)g_{n0}(-y)$$
$$+ g_{n,k+1}(y)g_{n,n-1}(-y) + g_{n,k+2}(y)g_{n,n-2}(-y) + \cdots$$
$$+ g_{n,n-1}(y)g_{n,k+1}(-y) = 0, \tag{6.65}$$

for $k = 1, ..., n - 1$.

From Eq. (6.51) it can be shown, for natural numbers l, that

$$\left(\sum_{p=0}^{n-1} h_{kp-n[kp/n]}g_{np}(y) \right)^l = \sum_{p=0}^{n-1} h_{kp-n[kp/n]}g_{np}(ly), \tag{6.66}$$

where $k = 0, 1, ..., n - 1$. For $k = 1$ the relation (6.66) is

$$\{g_{n0}(y) + h_1 g_{n1}(y) + \cdots + h_{n-1}g_{n,n-1}(y)\}^l = g_{n0}(ly) + h_1 g_{n1}(ly) + \cdots$$
$$+ h_{n,n-1}g_{n,n-1}(ly). \tag{6.67}$$

If

$$a_k = \sum_{p=0}^{n-1} g_{np}(y) \cos \left(\frac{2\pi kp}{n} \right), \tag{6.68}$$

for $k = 0, 1, ..., n - 1$, and

$$b_k = \sum_{p=0}^{n-1} g_{np}(y) \sin \left(\frac{2\pi kp}{n} \right), \tag{6.69}$$

for $k = 1, ..., n - 1$, where $g_{nk}(y)$ are the polar cosexponential functions in Eq. (6.55), it can be shown that

$$a_k = \exp \left[y \cos \left(\frac{2\pi k}{n} \right) \right] \cos \left[y \sin \left(\frac{2\pi k}{n} \right) \right], \tag{6.70}$$

where $k = 0, 1, ..., n - 1$,

$$b_k = \exp \left[y \cos \left(\frac{2\pi k}{n} \right) \right] \sin \left[y \sin \left(\frac{2\pi k}{n} \right) \right], \tag{6.71}$$

where $k = 1, ..., n - 1$. If

$$G_k^2 = a_k^2 + b_k^2, \tag{6.72}$$

for $k = 1, ..., n - 1$, then from Eqs. (6.70) and (6.71) it results that

$$G_k^2 = \exp\left[2y \cos\left(\frac{2\pi k}{n}\right)\right], \tag{6.73}$$

where $k = 1, ..., n - 1$. If

$$G_+ = g_{n0} + g_{n1} + \cdots + g_{n,n-1}, \tag{6.74}$$

from Eq. (6.68) it results that $G_+ = a_0$, so that $G_+ = e^y$, and, in an even number of dimensions n, if

$$G_- = g_{n0} - g_{n1} + \cdots + g_{n,n-2} - g_{n,n-1}, \tag{6.75}$$

from Eq. (6.68) it results that $G_- = a_{n/2}$, so that $G_{n/2} = e^{-y}$. Then with the aid of Eq. (6.26) applied for $p = 1$ it can be shown that the polar n-dimensional cosexponential functions have the property that, for even n,

$$G_+ G_- \prod_{k=1}^{n/2-1} G_k^2 = 1, \tag{6.76}$$

and in an odd number of dimensions, with the aid of Eq. (6.27) it can be shown that

$$G_+ \prod_{k=1}^{(n-1)/2} G_k^2 = 1. \tag{6.77}$$

The polar n-dimensional cosexponential functions are solutions of the nth-order differential equation

$$\frac{d^n \zeta}{du^n} = \zeta. \tag{6.78}$$

This equation has solutions of the form $\zeta(u) = A_0 g_{n0}(u) + A_1 g_{n1}(u) + \cdots + A_{n-1} g_{n,n-1}(u)$. It can be checked that the derivatives of the polar cosexponential functions are related by

$$\frac{dg_{n0}}{du} = g_{n,n-1}, \quad \frac{dg_{n1}}{du} = g_{n0}, \quad ...., \quad \frac{dg_{n,n-2}}{du} = g_{n,n-3},$$

$$\frac{dg_{n,n-1}}{du} = g_{n,n-2}. \tag{6.79}$$

6.1.4 Exponential and trigonometric forms of polar n-complex numbers

In order to obtain the exponential and trigonometric forms of n-complex numbers, a canonical base $e_+, e_-, e_1, \tilde{e}_1, ..., e_{n/2-1}, \tilde{e}_{n/2-1}$ for the polar n-complex numbers will be introduced for even n by the relations

$$
\begin{pmatrix} e_+ \\ e_- \\ \vdots \\ e_k \\ \tilde{e}_k \\ \vdots \end{pmatrix}
=
\begin{pmatrix}
\frac{1}{n} & \frac{1}{n} & \cdots & \frac{1}{n} & \frac{1}{n} \\
\frac{1}{n} & -\frac{1}{n} & \cdots & \frac{1}{n} & -\frac{1}{n} \\
\vdots & \vdots & & \vdots & \vdots \\
\frac{2}{n} & \frac{2}{n}\cos\frac{2\pi k}{n} & \cdots & \frac{2}{n}\cos\frac{2\pi(n-2)k}{n} & \frac{2}{n}\cos\frac{2\pi(n-1)k}{n} \\
0 & \frac{2}{n}\sin\frac{2\pi k}{n} & \cdots & \frac{2}{n}\sin\frac{2\pi(n-2)k}{n} & \frac{2}{n}\sin\frac{2\pi(n-1)k}{n} \\
\vdots & \vdots & \vdots & \vdots & \vdots
\end{pmatrix}
\begin{pmatrix} 1 \\ h_1 \\ \vdots \\ \vdots \\ \vdots \\ h_{n-1} \end{pmatrix},
\tag{6.80}
$$

where $k = 1, 2, ..., n/2 - 1$. For odd n, the canonical base $e_+, e_1, \tilde{e}_1, ...$ $e_{(n-1)/2}, \tilde{e}_{(n-1)/2}$ for the polar n-complex numbers will be introduced by the relations

$$
\begin{pmatrix} e_+ \\ e_1 \\ \tilde{e}_1 \\ \vdots \\ e_k \\ \tilde{e}_k \\ \vdots \end{pmatrix}
=
\begin{pmatrix}
\frac{1}{n} & \frac{1}{n} & \cdots & \frac{1}{n} \\
\frac{2}{n} & \frac{2}{n}\cos\frac{2\pi}{n} & \cdots & \frac{2}{n}\cos\frac{2\pi(n-1)}{n} \\
0 & \frac{2}{n}\sin\frac{2\pi}{n} & \cdots & \frac{2}{n}\sin\frac{2\pi(n-1)}{n} \\
\vdots & \vdots & & \vdots \\
\frac{2}{n} & \frac{2}{n}\cos\frac{2\pi k}{n} & \cdots & \frac{2}{n}\cos\frac{2\pi(n-1)k}{n} \\
0 & \frac{2}{n}\sin\frac{2\pi k}{n} & \cdots & \frac{2}{n}\sin\frac{2\pi(n-1)k}{n} \\
\vdots & \vdots & \vdots & \vdots
\end{pmatrix}
\begin{pmatrix} 1 \\ h_1 \\ h_2 \\ \vdots \\ \vdots \\ \vdots \\ h_{n-1} \end{pmatrix},
\tag{6.81}
$$

where $k = 0, 1, ..., (n-1)/2$.

The multiplication relations for the new bases are, for even n,

$$
e_+^2 = e_+, \; e_-^2 = e_-, \; e_+e_- = 0, \; e_+e_k = 0, \; e_+\tilde{e}_k = 0, \; e_-e_k = 0, \; e_-\tilde{e}_k = 0,
$$
$$
e_k^2 = e_k, \; \tilde{e}_k^2 = -e_k, \; e_k\tilde{e}_k = \tilde{e}_k, \; e_ke_l = 0, \; e_k\tilde{e}_l = 0, \; \tilde{e}_k\tilde{e}_l = 0, \; k \neq l,
$$
$$
\tag{6.82}
$$

where $k, l = 1, ..., n/2 - 1$. For odd n the multiplication relations are

$$e_+^2 = e_+, \; e_+ e_k = 0, \; e_+ \tilde{e}_k = 0,$$
$$e_k^2 = e_k, \; \tilde{e}_k^2 = -e_k, \; e_k \tilde{e}_k = \tilde{e}_k, \; e_k e_l = 0, \; e_k \tilde{e}_l = 0, \; \tilde{e}_k \tilde{e}_l = 0,$$
$$k \neq l, \tag{6.83}$$

where $k, l = 1, ..., (n-1)/2$. The moduli of the new bases are

$$|e_+| = \frac{1}{\sqrt{n}}, \; |e_-| = \frac{1}{\sqrt{n}}, \; |e_k| = \sqrt{\frac{2}{n}}, \; |\tilde{e}_k| = \sqrt{\frac{2}{n}}. \tag{6.84}$$

It can be shown that, for even n,

$$x_0 + h_1 x_1 + \cdots + h_{n-1} x_{n-1} = e_+ v_+ + e_- v_- + \sum_{k=1}^{n/2-1} (e_k v_k + \tilde{e}_k \tilde{v}_k), \tag{6.85}$$

and for odd n

$$x_0 + h_1 x_1 + \cdots + h_{n-1} x_{n-1} = e_+ v_+ + \sum_{k=1}^{(n-1)/2} (e_k v_k + \tilde{e}_k \tilde{v}_k). \tag{6.86}$$

The relations (6.85),(6.86) give the canonical form of a polar n-complex number.

Using the properties of the bases in Eqs. (6.82) and (6.83) it can be shown that

$$\exp(\tilde{e}_k \phi_k) = 1 - e_k + e_k \cos \phi_k + \tilde{e}_k \sin \phi_k, \tag{6.87}$$

$$\exp(e_k \ln \rho_k) = 1 - e_k + e_k \rho_k, \tag{6.88}$$

$$\exp(e_+ \ln v_+) = 1 - e_+ + e_+ v_+ \tag{6.89}$$

and, for even n,

$$\exp(e_- \ln v_-) = 1 - e_- + e_- v_-. \tag{6.90}$$

In Eq. (6.89), $\ln v_+$ exists as a real function provided that $v_+ = x_0 + x_1 + \cdots + x_{n-1} > 0$, which means that $0 < \theta_+ < \pi/2$, and for even n, $\ln v_-$ exists in Eq. (6.90) as a real function provided that $v_- = x_0 - x_1 + \cdots + x_{n-2} - x_{n-1} > 0$, which means that $0 < \theta_- < \pi/2$. By multiplying the relations (6.87)-(6.90) it results, for even n, that

$$\exp \left[e_+ \ln v_+ + e_- \ln v_- + \sum_{k=1}^{n/2-1} (e_k \ln \rho_k + \tilde{e}_k \phi_k) \right] = e_+ v_+ + e_- v_-$$
$$+ \sum_{k=1}^{n/2-1} (e_k v_k + \tilde{e}_k \tilde{v}_k), \tag{6.91}$$

where the fact has ben used that

$$e_+ + e_- + \sum_{k=1}^{n/2-1} e_k = 1, \tag{6.92}$$

the latter relation being a consequence of Eqs. (6.80) and (6.26). Similarly, by multiplying the relations (6.87)-(6.89) it results, for odd n, that

$$\exp\left[e_+ \ln v_+ + \sum_{k=1}^{(n-1)/2} (e_k \ln \rho_k + \tilde{e}_k \phi_k) \right] = e_+ v_+$$

$$+ \sum_{k=1}^{(n-1)/2} (e_k v_k + \tilde{e}_k \tilde{v}_k), \tag{6.93}$$

where the fact has ben used that

$$e_+ + \sum_{k=1}^{(n-1)/2} e_k = 1, \tag{6.94}$$

the latter relation being a consequence of Eqs. (6.81) and (6.27).

By comparing Eqs. (6.85) and (6.91), it can be seen that, for even n,

$$x_0 + h_1 x_1 + \cdots + h_{n-1} x_{n-1} = \exp\left[e_+ \ln v_+ + e_- \ln v_- \right.$$

$$\left. + \sum_{k=1}^{n/2-1} (e_k \ln \rho_k + \tilde{e}_k \phi_k) \right], \tag{6.95}$$

and by comparing Eqs. (6.86) and (6.93), it can be seen that, for odd n,

$$x_0 + h_1 x_1 + \cdots + h_{n-1} x_{n-1} = \exp\left[e_+ \ln v_+ \right.$$

$$\left. + \sum_{k=1}^{(n-1)/2} (e_k \ln \rho_k + \tilde{e}_k \phi_k) \right]. \tag{6.96}$$

Using the expression of the bases in Eqs. (6.80) and (6.81) yields, for even values of n, the exponential form of the n-complex number $u = x_0 + h_1 x_1 + \cdots + h_{n-1} x_{n-1}$ as

$$u = \rho \exp\left\{ \sum_{p=1}^{n-1} h_p \left[\frac{1}{n} \ln \frac{\sqrt{2}}{\tan \theta_+} + \frac{(-1)^p}{n} \ln \frac{\sqrt{2}}{\tan \theta_-} \right.\right.$$

$$\left.\left. - \frac{2}{n} \sum_{k=2}^{n/2-1} \cos\left(\frac{2\pi k p}{n} \right) \ln \tan \psi_{k-1} \right] + \sum_{k=1}^{n/2-1} \tilde{e}_k \phi_k \right\}, \tag{6.97}$$

where ρ is the amplitude defined in Eq. (6.11), which for even n has according to Eq. (6.18) the expression

$$\rho = \left(v_+ v_- \rho_1^2 \cdots \rho_{n/2-1}^2\right)^{1/n}. \tag{6.98}$$

For odd values of n, the exponential form of the n-complex number u is

$$u = \rho \exp\left\{\sum_{p=1}^{n-1} h_p\left[\frac{1}{n}\ln\frac{\sqrt{2}}{\tan\theta_+} - \frac{2}{n}\sum_{k=2}^{(n-1)/2}\cos\left(\frac{2\pi kp}{n}\right)\ln\tan\psi_{k-1}\right] \right.$$
$$\left. + \sum_{k=1}^{(n-1)/2}\tilde{e}_k\phi_k\right\}, \tag{6.99}$$

where for odd n, ρ has according to Eq. (6.19) the expression

$$\rho = \left(v_+ \rho_1^2 \cdots \rho_{(n-1)/2}^2\right)^{1/n}. \tag{6.100}$$

It can be checked with the aid of Eq. (6.87) that the n-complex number u can also be written, for even n, as

$$x_0 + h_1 x_1 + \cdots + h_{n-1}x_{n-1} = \left(e_+ v_+ + e_- v_- + \sum_{k=1}^{n/2-1} e_k\rho_k\right)$$
$$\exp\left(\sum_{k=1}^{n/2-1}\tilde{e}_k\phi_k\right), \tag{6.101}$$

and for odd n, as

$$x_0 + h_1 x_1 + \cdots + h_{n-1}x_{n-1} = \left(e_+ v_+ + \sum_{k=1}^{(n-1)/2} e_k\rho_k\right)$$
$$\exp\left(\sum_{k=1}^{(n-1)/2}\tilde{e}_k\phi_k\right). \tag{6.102}$$

Writing in Eqs. (6.101) and (6.102) the radius ρ_1, Eqs. (6.35) and (6.36), as a factor and expressing the variables in terms of the polar and planar angles with the aid of Eqs. (6.31)-(6.33) yields the trigonometric form of the n-complex number u, for even n, as

$$u = d\left(\frac{n}{2}\right)^{1/2}\left(\frac{1}{\tan^2\theta_+} + \frac{1}{\tan^2\theta_-} + 1 + \frac{1}{\tan^2\psi_1} + \frac{1}{\tan^2\psi_2} + \cdots\right.$$
$$\left. + \frac{1}{\tan^2\psi_{n/2-2}}\right)^{-1/2}\left(\frac{e_+\sqrt{2}}{\tan\theta_+} + \frac{e_-\sqrt{2}}{\tan\theta_-} + e_1 + \sum_{k=2}^{n/2-1}\frac{e_k}{\tan\psi_{k-1}}\right)$$

$$\exp\left(\sum_{k=1}^{n/2-1} \tilde{e}_k \phi_k\right),\qquad(6.103)$$

and for odd n as

$$u = d\left(\frac{n}{2}\right)^{1/2}\left(\frac{1}{\tan^2\theta_+} + 1 + \frac{1}{\tan^2\psi_1} + \frac{1}{\tan^2\psi_2} + \cdots\right.$$
$$\left. + \frac{1}{\tan^2\psi_{(n-3)/2}}\right)^{-1/2}$$
$$\left(\frac{e_+\sqrt{2}}{\tan\theta_+} + e_1 + \sum_{k=2}^{(n-1)/2}\frac{e_k}{\tan\psi_{k-1}}\right)\exp\left(\sum_{k=1}^{(n-1)/2}\tilde{e}_k\phi_k\right).\qquad(6.104)$$

In Eqs. (6.103) and 6.104), the n-complex number u, written in trigonometric form, is the product of the modulus d, of a part depending on the polar and planar angles $\theta_+, \theta_-, \psi_1, ..., \psi_{[(n-3)/2]}$, and of a factor depending on the azimuthal angles $\phi_1, ..., \phi_{[(n-1)/2]}$. Although the modulus of a product of n-complex numbers is not equal in general to the product of the moduli of the factors, it can be checked that the modulus of the factor in Eq. (6.103) is

$$\left|\frac{e_+\sqrt{2}}{\tan\theta_+} + \frac{e_-\sqrt{2}}{\tan\theta_-} + e_1 + \sum_{k=2}^{n/2-1}\frac{e_k}{\tan\psi_{k-1}}\right|$$
$$= \left(\frac{2}{n}\right)^{1/2}\left(\frac{1}{\tan^2\theta_+} + \frac{1}{\tan^2\theta_-} + 1 + \frac{1}{\tan^2\psi_1} + \frac{1}{\tan^2\psi_2} + \cdots\right.$$
$$\left. + \frac{1}{\tan^2\psi_{n/2-2}}\right)^{1/2},\qquad(6.105)$$

and the modulus of the factor in Eq. (6.104) is

$$\left|\frac{e_+\sqrt{2}}{\tan\theta_+} + e_1 + \sum_{k=2}^{(n-1)/2}\frac{e_k}{\tan\psi_{k-1}}\right|$$
$$= \left(\frac{2}{n}\right)^{1/2}\left(\frac{1}{\tan^2\theta_+} + 1 + \frac{1}{\tan^2\psi_1} + \frac{1}{\tan^2\psi_2} + \cdots\right.$$
$$\left. + \frac{1}{\tan^2\psi_{(n-3)/2}}\right)^{1/2}.\qquad(6.106)$$

Moreover, it can be checked that

$$\left|\exp\left[\sum_{k=1}^{[(n-1)/2]}\tilde{e}_k\phi_k\right]\right| = 1.\qquad(6.107)$$

The modulus d in Eqs. (6.103) and (6.104) can be expressed in terms of the amplitude ρ, for even n, as

$$d = \rho \frac{2^{(n-2)/2n}}{\sqrt{n}} \left(\tan \theta_+ \tan \theta_- \tan^2 \psi_1 \cdots \tan^2 \psi_{n/2-2} \right)^{1/n}$$

$$\left(\frac{1}{\tan^2 \theta_+} + \frac{1}{\tan^2 \theta_-} + 1 + \frac{1}{\tan^2 \psi_1} + \frac{1}{\tan^2 \psi_2} + \cdots \right.$$

$$\left. + \frac{1}{\tan^2 \psi_{n/2-2}} \right)^{1/2}, \tag{6.108}$$

and for odd n as

$$d = \rho \frac{2^{(n-1)/2n}}{\sqrt{n}} \left(\tan \theta_+ \tan^2 \psi_1 \cdots \tan^2 \psi_{(n-3)/2} \right)^{1/n}$$

$$\left(\frac{1}{\tan^2 \theta_+} + 1 + \frac{1}{\tan^2 \psi_1} + \frac{1}{\tan^2 \psi_2} + \cdots + \frac{1}{\tan^2 \psi_{(n-3)/2}} \right)^{1/2}. \tag{6.109}$$

6.1.5 Elementary functions of a polar n-complex variable

The logarithm u_1 of the n-complex number u, $u_1 = \ln u$, can be defined as the solution of the equation

$$u = e^{u_1}. \tag{6.110}$$

For even n the relation (6.91) shows that $\ln u$ exists as an n-complex function with real components if $v_+ = x_0 + x_1 + \cdots + x_{n-1} > 0$ and $v_- = x_0 - x_1 + \cdots + x_{n-2} - x_{n-1} > 0$, which means that $0 < \theta_+ < \pi/2, 0 < \theta_- < \pi/2$. For odd n the relation (6.93) shows that $\ln u$ exists as an n-complex function with real components if $v_+ = x_0 + x_1 + \cdots + x_{n-1} > 0$, which means that $0 < \theta_+ < \pi/2$. The expression of the logarithm, obtained from Eqs. (6.95) and (6.96), is, for even n,

$$\ln u = e_+ \ln v_+ + e_- \ln v_- + \sum_{k=1}^{n/2-1} (e_k \ln \rho_k + \tilde{e}_k \phi_k), \tag{6.111}$$

and for odd n the expression is

$$\ln u = e_+ \ln v_+ + \sum_{k=1}^{(n-1)/2} (e_k \ln \rho_k + \tilde{e}_k \phi_k). \tag{6.112}$$

An expression of the logarithm depending on the amplitude ρ can be obtained from the exponential forms in Eqs. (6.97) and (6.99), for even n, as

$$\ln u = \ln \rho + \sum_{p=1}^{n-1} h_p \left[\frac{1}{n} \ln \frac{\sqrt{2}}{\tan \theta_+} + \frac{(-1)^p}{n} \ln \frac{\sqrt{2}}{\tan \theta_-} \right.$$
$$\left. - \frac{2}{n} \sum_{k=2}^{n/2-1} \cos \left(\frac{2\pi kp}{n} \right) \ln \tan \psi_{k-1} \right] + \sum_{k=1}^{n/2-1} \tilde{e}_k \phi_k, \qquad (6.113)$$

and for odd n as

$$\ln u = \ln \rho + \sum_{p=1}^{n-1} h_p \left[\frac{1}{n} \ln \frac{\sqrt{2}}{\tan \theta_+} - \frac{2}{n} \sum_{k=2}^{(n-1)/2} \cos \left(\frac{2\pi kp}{n} \right) \ln \tan \psi_{k-1} \right]$$
$$+ \sum_{k=1}^{(n-1)/2} \tilde{e}_k \phi_k. \qquad (6.114)$$

The function $\ln u$ is multivalued because of the presence of the terms $\tilde{e}_k \phi_k$. It can be inferred from Eqs. (6.37)-(6.43) and (6.46) that

$$\ln(uu') = \ln u + \ln u', \qquad (6.115)$$

up to integer multiples of $2\pi \tilde{e}_k, k = 1, ..., [(n-1)/2]$.

The power function u^m can be defined for real values of m as

$$u^m = e^{m \ln u}. \qquad (6.116)$$

Using the expression of $\ln u$ in Eqs. (6.111) and (6.112) yields, for even values of n,

$$u^m = e_+ v_+^m + e_- v_-^m + \sum_{k=1}^{n/2-1} \rho_k^m (e_k \cos m\phi_k + \tilde{e}_k \sin m\phi_k), \qquad (6.117)$$

and for odd values of n

$$u^m = e_+ v_+^m + \sum_{k=1}^{(n-1)/2} \rho_k^m (e_k \cos m\phi_k + \tilde{e}_k \sin m\phi_k). \qquad (6.118)$$

For integer values of m, the relations (6.117) and (6.118) are valid for any $x_0, ..., x_{n-1}$. The power function is multivalued unless m is an integer. For integer m, it can be inferred from Eq. (6.115) that

$$(uu')^m = u^m u'^m. \qquad (6.119)$$

The trigonometric functions of the hypercomplex variable u and the addition theorems for these functions have been written in Eqs. (1.57)-(1.60). In order to obtain expressions for the trigonometric functions of n-complex variables, these will be expressed with the aid of the imaginary unit i as

$$\cos u = \frac{1}{2}(e^{iu} + e^{-iu}), \ \sin u = \frac{1}{2i}(e^{iu} - e^{-iu}). \tag{6.120}$$

The imaginary unit i is used for the convenience of notations, and it does not appear in the final results. The validity of Eq. (6.120) can be checked by comparing the series for the two sides of the relations. Since the expression of the exponential function $e^{h_k y}$ in terms of the units $1, h_1, ... h_{n-1}$ given in Eq. (6.51) depends on the polar cosexponential functions $g_{np}(y)$, the expression of the trigonometric functions will depend on the functions $g_{p+}^{(c)}(y) = (1/2)[g_{np}(iy) + g_{np}(-iy)]$ and $g_{p-}^{(c)}(y) = (1/2i)[g_{np}(iy) - g_{np}(-iy)]$,

$$\cos(h_k y) = \sum_{p=0}^{n-1} h_{kp-n[kp/n]} g_{p+}^{(c)}(y), \tag{6.121}$$

$$\sin(h_k y) = \sum_{p=0}^{n-1} h_{kp-n[kp/n]} g_{p-}^{(c)}(y), \tag{6.122}$$

where

$$g_{p+}^{(c)}(y) = \frac{1}{n} \sum_{l=0}^{n-1} \left\{ \cos\left[y\cos\left(\frac{2\pi l}{n}\right)\right] \cosh\left[y\sin\left(\frac{2\pi l}{n}\right)\right] \cos\left(\frac{2\pi lp}{n}\right) \right.$$
$$\left. - \sin\left[y\cos\left(\frac{2\pi l}{n}\right)\right] \sinh\left[y\sin\left(\frac{2\pi l}{n}\right)\right] \sin\left(\frac{2\pi lp}{n}\right) \right\}, \tag{6.123}$$

$$g_{p-}^{(c)}(y) = \frac{1}{n} \sum_{l=0}^{n-1} \left\{ \sin\left[y\cos\left(\frac{2\pi l}{n}\right)\right] \cosh\left[y\sin\left(\frac{2\pi l}{n}\right)\right] \cos\left(\frac{2\pi lp}{n}\right) \right.$$
$$\left. + \cos\left[y\cos\left(\frac{2\pi l}{n}\right)\right] \sinh\left[y\sin\left(\frac{2\pi l}{n}\right)\right] \sin\left(\frac{2\pi lp}{n}\right) \right\}. \tag{6.124}$$

The hyperbolic functions of the hypercomplex variable u and the addition theorems for these functions have been written in Eqs. (1.62)-(1.65). In order to obtain expressions for the hyperbolic functions of n-complex variables, these will be expressed as

$$\cosh u = \frac{1}{2}(e^u + e^{-u}), \ \sinh u = \frac{1}{2}(e^u - e^{-u}). \tag{6.125}$$

The validity of Eq. (6.125) can be checked by comparing the series for the two sides of the relations. Since the expression of the exponential function $e^{h_k y}$ in terms of the units $1, h_1, ... h_{n-1}$ given in Eq. (6.51) depends on the polar cosexponential functions $g_{np}(y)$, the expression of the hyperbolic functions will depend on the even part $g_{p+}(y) = (1/2)[g_{np}(y) + g_{np}(-y)]$ and on the odd part $g_{p-}(y) = (1/2)[g_{np}(y) - g_{np}(-y)]$ of g_{np},

$$\cosh(h_k y) = \sum_{p=0}^{n-1} h_{kp-n[kp/n]} g_{p+}(y), \tag{6.126}$$

$$\sinh(h_k y) = \sum_{p=0}^{n-1} h_{kp-n[kp/n]} g_{p-}(y), \tag{6.127}$$

where

$$g_{p+}(y) = \frac{1}{n} \sum_{l=0}^{n-1} \left\{ \cosh\left[y\cos\left(\frac{2\pi l}{n}\right)\right] \cos\left[y\sin\left(\frac{2\pi l}{n}\right)\right] \cos\left(\frac{2\pi lp}{n}\right) \right.$$
$$\left. + \sinh\left[y\cos\left(\frac{2\pi l}{n}\right)\right] \sin\left[y\sin\left(\frac{2\pi l}{n}\right)\right] \sin\left(\frac{2\pi lp}{n}\right) \right\}, \tag{6.128}$$

$$g_{p-}(y) = \frac{1}{n} \sum_{l=0}^{n-1} \left\{ \sinh\left[y\cos\left(\frac{2\pi l}{n}\right)\right] \cos\left[y\sin\left(\frac{2\pi l}{n}\right)\right] \cos\left(\frac{2\pi lp}{n}\right) \right.$$
$$\left. + \cosh\left[y\cos\left(\frac{2\pi l}{n}\right)\right] \sin\left[y\sin\left(\frac{2\pi l}{n}\right)\right] \sin\left(\frac{2\pi lp}{n}\right) \right\}. \tag{6.129}$$

The exponential, trigonometric and hyperbolic functions can also be expressed with the aid of the bases introduced in Eqs. (6.80) and (6.81). Using the expression of the n-complex number in Eq. (6.85), for even n, yields for the exponential of the n-complex variable u

$$e^u = e_+ e^{v+} + e_- e^{v-} + \sum_{k=1}^{n/2-1} e^{v_k} (e_k \cos \tilde{v}_k + \tilde{e}_k \sin \tilde{v}_k). \tag{6.130}$$

For odd n, the expression of the n-complex variable in Eq. (6.86) yileds for the exponential

$$e^u = e_+ e^{v+} + \sum_{k=1}^{(n-1)/2} e^{v_k} (e_k \cos \tilde{v}_k + \tilde{e}_k \sin \tilde{v}_k). \tag{6.131}$$

The trigonometric functions can be obtained from Eqs. (6.130) and (6.131 with the aid of Eqs. (6.120). The trigonometric functions of the n-complex variable u are, for even n,

$$\cos u = e_+ \cos v_+ + e_- \cos v_-$$
$$+ \sum_{k=1}^{n/2-1} \left(e_k \cos v_k \cosh \tilde{v}_k - \tilde{e}_k \sin v_k \sinh \tilde{v}_k \right), \qquad (6.132)$$

$$\sin u = e_+ \sin v_+ + e_- \sin v_-$$
$$+ \sum_{k=1}^{n/2-1} \left(e_k \sin v_k \cosh \tilde{v}_k + \tilde{e}_k \cos v_k \sinh \tilde{v}_k \right), \qquad (6.133)$$

and for odd n the trigonometric functions are

$$\cos u = e_+ \cos v_+ + \sum_{k=1}^{(n-1)/2} \left(e_k \cos v_k \cosh \tilde{v}_k - \tilde{e}_k \sin v_k \sinh \tilde{v}_k \right), \quad (6.134)$$

$$\sin u = e_+ \sin v_+ + \sum_{k=1}^{(n-1)/2} \left(e_k \sin v_k \cosh \tilde{v}_k + \tilde{e}_k \cos v_k \sinh \tilde{v}_k \right). \quad (6.135)$$

The hyperbolic functions can be obtained from Eqs. (6.130) and (6.131) with the aid of Eqs. (6.125). The hyperbolic functions of the n-complex variable u are, for even n,

$$\cosh u = e_+ \cosh v_+ + e_- \cosh v_-$$
$$+ \sum_{k=1}^{n/2-1} \left(e_k \cosh v_k \cos \tilde{v}_k + \tilde{e}_k \sinh v_k \sin \tilde{v}_k \right), \qquad (6.136)$$

$$\sinh u = e_+ \sinh v_+ + e_- \sinh v_-$$
$$+ \sum_{k=1}^{n/2-1} \left(e_k \sinh v_k \cos \tilde{v}_k + \tilde{e}_k \cosh v_k \sin \tilde{v}_k \right), \qquad (6.137)$$

and for odd n the hyperbolic functions are

$$\cosh u = e_+ \cosh v_+$$
$$+ \sum_{k=1}^{(n-1)/2} \left(e_k \cosh v_k \cos \tilde{v}_k + \tilde{e}_k \sinh v_k \sin \tilde{v}_k \right), \qquad (6.138)$$

$$\sinh u = e_+ \sinh v_+$$
$$+ \sum_{k=1}^{(n-1)/2} \left(e_k \sinh v_k \cos \tilde{v}_k + \tilde{e}_k \cosh v_k \sin \tilde{v}_k \right). \qquad (6.139)$$

6.1.6   Power series of polar n-complex numbers

An n-complex series is an infinite sum of the form

$$a_0 + a_1 + a_2 + \cdots + a_n + \cdots, \tag{6.140}$$

where the coefficients a_n are n-complex numbers. The convergence of the series (6.140) can be defined in terms of the convergence of its n real components. The convergence of a n-complex series can also be studied using n-complex variables. The main criterion for absolute convergence remains the comparison theorem, but this requires a number of inequalities which will be discussed further.

The modulus $d = |u|$ of an n-complex number u has been defined in Eq. (6.20). Since $|x_0| \leq |u|, |x_1| \leq |u|, ..., |x_{n-1}| \leq |u|$, a property of absolute convergence established via a comparison theorem based on the modulus of the series (6.140) will ensure the absolute convergence of each real component of that series.

The modulus of the sum $u_1 + u_2$ of the n-complex numbers u_1, u_2 fulfils the inequality

$$||u'| - |u''|| \leq |u' + u''| \leq |u'| + |u''|. \tag{6.141}$$

For the product, the relation is

$$|u'u''| \leq \sqrt{n}|u'||u''|, \tag{6.142}$$

as can be shown from Eqs. (6.28) and (6.29). The relation (6.142) replaces the relation of equality extant between 2-dimensional regular complex numbers. The equality in Eq. (6.142) takes place for $\rho_1 \rho_1' = 0, ..., \rho_{[(n-1)/2]} \rho_{[(n-1)/2]}' = 0$ and, for even n, for $v_+ v_-' = 0$, $v_- v_+' = 0$.

From Eq. (6.142) it results, for $u = u'$, that

$$|u^2| \leq \sqrt{n}|u|^2. \tag{6.143}$$

The relation in Eq. (6.143) becomes an equality for $\rho_1 = 0, ..., \rho_{[(n-1)/2]} = 0$ and, for even n, $v_+ = 0$ or $v_- = 0$. The inequality in Eq. (6.142) implies that

$$|u^l| \leq n^{(l-1)/2}|u|^l, \tag{6.144}$$

where l is a natural number. From Eqs. (6.142) and (6.144) it results that

$$|au^l| \leq n^{l/2}|a||u|^l. \tag{6.145}$$

A power series of the n-complex variable u is a series of the form

$$a_0 + a_1 u + a_2 u^2 + \cdots + a_l u^l + \cdots. \tag{6.146}$$

Since

$$\left| \sum_{l=0}^{\infty} a_l u^l \right| \leq \sum_{l=0}^{\infty} n^{l/2} |a_l| |u|^l, \tag{6.147}$$

a sufficient condition for the absolute convergence of this series is that

$$\lim_{l \to \infty} \frac{\sqrt{n} |a_{l+1}| |u|}{|a_l|} < 1. \tag{6.148}$$

Thus the series is absolutely convergent for

$$|u| < c, \tag{6.149}$$

where

$$c = \lim_{l \to \infty} \frac{|a_l|}{\sqrt{n} |a_{l+1}|}. \tag{6.150}$$

The convergence of the series (6.146) can be also studied with the aid of the formulas (6.117), (6.118) which for integer values of m are valid for any values of $x_0, ..., x_{n-1}$, as mentioned previously. If $a_l = \sum_{p=0}^{n-1} h_p a_{lp}$, and

$$A_{l+} = \sum_{p=0}^{n-1} a_{lp}, \tag{6.151}$$

$$A_{lk} = \sum_{p=0}^{n-1} a_{lp} \cos \frac{2\pi kp}{n}, \tag{6.152}$$

$$\tilde{A}_{lk} = \sum_{p=0}^{n-1} a_{lp} \sin \frac{2\pi kp}{n}, \tag{6.153}$$

for $k = 1, ..., [(n-1)/2]$, and for even n

$$A_{l-} = \sum_{p=0}^{n-1} (-1)^p a_{lp}, \tag{6.154}$$

the series (6.146) can be written, for even n, as

$$\sum_{l=0}^{\infty} \left[e_+ A_{l+} v_+^l + e_- A_{l-} v_-^l + \sum_{k=1}^{n/2-1} (e_k A_{lk} + \tilde{e}_k \tilde{A}_{lk})(e_k v_k + \tilde{e}_k \tilde{v}_k)^l \right], \tag{6.155}$$

and for odd n as

$$\sum_{l=0}^{\infty} \left[e_+ A_{l+} v_+^l + \sum_{k=1}^{(n-1)/2} (e_k A_{lk} + \tilde{e}_k \tilde{A}_{lk})(e_k v_k + \tilde{e}_k \tilde{v}_k)^l \right]. \tag{6.156}$$

The series in Eq. (6.146) is absolutely convergent for

$$|v_+| < c_+, \ |v_-| < c_-, \ \rho_k < c_k, \tag{6.157}$$

for $k = 1, ..., [(n-1)/2]$, where

$$c_+ = \lim_{l \to \infty} \frac{|A_{l+}|}{|A_{l+1,+}|}, \ c_- = \lim_{l \to \infty} \frac{|A_{l-}|}{|A_{l+1,-}|},$$

$$c_k = \lim_{l \to \infty} \frac{\left(A_{lk}^2 + \tilde{A}_{lk}^2 \right)^{1/2}}{\left(A_{l+1,k}^2 + \tilde{A}_{l+1,k}^2 \right)^{1/2}}. \tag{6.158}$$

The relations (6.157) show that the region of convergence of the series (6.146) is an n-dimensional cylinder.

It can be shown that, for even n, $c = (1/\sqrt{n}) \min(c_+, c_-, c_1, ..., c_{n/2-1})$, and for odd n $c = (1/\sqrt{n}) \min(c_+, c_1, ..., c_{(n-1)/2})$, where min designates the smallest of the numbers in the argument of this function. Using the expression of $|u|$ in Eqs. (6.28) or (6.29), it can be seen that the spherical region of convergence defined in Eqs. (6.149), (6.150) is a subset of the cylindrical region of convergence defined in Eqs. (6.157) and (6.158).

6.1.7   Analytic functions of polar n-complex variables

The analytic functions of the hypercomplex variable u and the series expansion of functions have been discussed in Eqs. (1.85)-(1.93). If the n-complex function $f(u)$ of the n-complex variable u is written in terms of the real functions $P_k(x_0, ..., x_{n-1}), k = 0, 1, ..., n-1$ of the real variables $x_0, x_1, ..., x_{n-1}$ as

$$f(u) = \sum_{k=0}^{n-1} h_k P_k(x_0, ..., x_{n-1}), \tag{6.159}$$

then relations of equality exist between the partial derivatives of the functions P_k. The derivative of the function f can be written as

$$\lim_{\Delta u \to 0} \frac{1}{\Delta u} \sum_{k=0}^{n-1} \left(h_k \sum_{l=0}^{n-1} \frac{\partial P_k}{\partial x_l} \Delta x_l \right), \tag{6.160}$$

where

$$\Delta u = \sum_{k=0}^{n-1} h_l \Delta x_l. \tag{6.161}$$

The relations between the partials derivatives of the functions P_k are obtained by setting successively in Eq. (6.160) $\Delta u = h_l \Delta x_l$, for $l = 0, 1, ..., n - 1$, and equating the resulting expressions. The relations are

$$\frac{\partial P_k}{\partial x_0} = \frac{\partial P_{k+1}}{\partial x_1} = \cdots = \frac{\partial P_{n-1}}{\partial x_{n-k-1}} = \frac{\partial P_0}{\partial x_{n-k}} = \cdots = \frac{\partial P_{k-1}}{\partial x_{n-1}}, \tag{6.162}$$

for $k = 0, 1, ..., n - 1$. The relations (6.162) are analogous to the Riemann relations for the real and imaginary components of a complex function. It can be shown from Eqs. (6.162) that the components P_k fulfil the second-order equations

$$\frac{\partial^2 P_k}{\partial x_0 \partial x_l} = \frac{\partial^2 P_k}{\partial x_1 \partial x_{l-1}} = \cdots = \frac{\partial^2 P_k}{\partial x_{[l/2]} \partial x_{l-[l/2]}}$$

$$= \frac{\partial^2 P_k}{\partial x_{l+1} \partial x_{n-1}} = \frac{\partial^2 P_k}{\partial x_{l+2} \partial x_{n-2}} = \cdots$$

$$= \frac{\partial^2 P_k}{\partial x_{l+1+[(n-l-2)/2]} \partial x_{n-1-[(n-l-2)/2]}}, \tag{6.163}$$

for $k, l = 0, 1, ..., n - 1$.

6.1.8 Integrals of polar n-complex functions

The singularities of n-complex functions arise from terms of the form $1/(u - u_0)^n$, with $n > 0$. Functions containing such terms are singular not only at $u = u_0$, but also at all points of the hypersurfaces passing through the pole u_0 and which are parallel to the nodal hypersurfaces.

The integral of an n-complex function between two points A, B along a path situated in a region free of singularities is independent of path, which means that the integral of an analytic function along a loop situated in a region free of singularities is zero,

$$\oint_\Gamma f(u) du = 0, \tag{6.164}$$

where it is supposed that a surface Σ spanning the closed loop Γ is not intersected by any of the hypersurfaces associated with the singularities of the function $f(u)$. Using the expression, Eq. (6.159), for $f(u)$ and the fact that

$$du = \sum_{k=0}^{n-1} h_k dx_k, \tag{6.165}$$

the explicit form of the integral in Eq. (6.164) is

$$\oint_\Gamma f(u)du = \oint_\Gamma \sum_{k=0}^{n-1} h_k \sum_{l=0}^{n-1} P_l dx_{k-l+n[(n-k-1+l)/n]}. \tag{6.166}$$

If the functions P_k are regular on a surface Σ spanning the loop Γ, the integral along the loop Γ can be transformed in an integral over the surface Σ of terms of the form $\partial P_l/\partial x_{k-m+n[(n-k+m-1)/n]} - \partial P_m/\partial x_{k-l+n[(n-k+l-1)/n]}$. These terms are equal to zero by Eqs. (6.162), and this proves Eq. (6.164).

The integral of the function $(u - u_0)^m$ on a closed loop Γ is equal to zero for m a positive or negative integer not equal to -1,

$$\oint_\Gamma (u - u_0)^m du = 0, \;\; m \text{ integer}, \; m \neq -1. \tag{6.167}$$

This is due to the fact that $\int (u - u_0)^m du = (u - u_0)^{m+1}/(m + 1)$, and to the fact that the function $(u - u_0)^{m+1}$ is singlevalued for m an integer.

The integral $\oint_\Gamma du/(u - u_0)$ can be calculated using the exponential form, Eqs. (6.97) and (6.99), for the difference $u - u_0$, which for even n is

$$u - u_0 = \rho \exp \left\{ \sum_{p=1}^{n-1} h_p \left[\frac{1}{n} \ln \frac{\sqrt{2}}{\tan \theta_+} + \frac{(-1)^p}{n} \ln \frac{\sqrt{2}}{\tan \theta_-} \right. \right.$$
$$\left. \left. -\frac{2}{n} \sum_{k=2}^{n/2-1} \cos \left(\frac{2\pi kp}{n} \right) \ln \tan \psi_{k-1} \right] + \sum_{k=1}^{n/2-1} \tilde{e}_k \phi_k \right\}, \tag{6.168}$$

and for odd n is

$$u - u_0 = \rho \exp \left\{ \sum_{p=1}^{n-1} h_p \left[\frac{1}{n} \ln \frac{\sqrt{2}}{\tan \theta_+} \right. \right.$$
$$\left. \left. -\frac{2}{n} \sum_{k=2}^{(n-1)/2} \cos \left(\frac{2\pi kp}{n} \right) \ln \tan \psi_{k-1} \right] + \sum_{k=1}^{(n-1)/2} \tilde{e}_k \phi_k \right\}. \tag{6.169}$$

Thus for even n the quantity $du/(u - u_0)$ is

$$\frac{du}{u - u_0} = \frac{d\rho}{\rho} + \sum_{p=1}^{n-1} h_p \left[\frac{1}{n} d \ln \frac{\sqrt{2}}{\tan \theta_+} + \frac{(-1)^p}{n} d \ln \frac{\sqrt{2}}{\tan \theta_-} \right.$$

$$-\frac{2}{n}\sum_{k=2}^{n/2-1}\cos\left(\frac{2\pi kp}{n}\right)d\ln\tan\psi_{k-1}\Bigg] + \sum_{k=1}^{n/2-1}\tilde{e}_k d\phi_k, \qquad (6.170)$$

and for odd n

$$\frac{du}{u-u_0} = \frac{d\rho}{\rho} + \sum_{p=1}^{n-1} h_p\left[\frac{1}{n}d\ln\frac{\sqrt{2}}{\tan\theta_+}\right.$$

$$\left.-\frac{2}{n}\sum_{k=2}^{(n-1)/2}\cos\left(\frac{2\pi kp}{n}\right)d\ln\tan\psi_{k-1}\right] + \sum_{k=1}^{(n-1)/2}\tilde{e}_k d\phi_k. \qquad (6.171)$$

Since $\rho, \ln(\sqrt{2}/\tan\theta_+), \ln(\sqrt{2}/\tan\theta_-), \ln(\tan\psi_{k-1})$ are singlevalued variables, it follows that $\oint_\Gamma d\rho/\rho = 0$, $\oint_\Gamma d(\ln\sqrt{2}/\tan\theta_+) = 0$, $\oint_\Gamma d(\ln\sqrt{2}/\tan\theta_-) = 0$, $\oint_\Gamma d(\ln\tan\psi_{k-1}) = 0$. On the other hand since, ϕ_k are cyclic variables, they may give contributions to the integral around the closed loop Γ.

The expression of $\oint_\Gamma du/(u-u_0)$ can be written with the aid of a functional which will be called $\mathrm{int}(M,C)$, defined for a point M and a closed curve C in a two-dimensional plane, such that

$$\mathrm{int}(M,C) = \begin{cases} 1 & \text{if } M \text{ is an interior point of } C, \\ 0 & \text{if } M \text{ is exterior to } C. \end{cases} \qquad (6.172)$$

With this notation the result of the integration on a closed path Γ can be written as

$$\oint_\Gamma \frac{du}{u-u_0} = \sum_{k=1}^{[(n-1)/2]} 2\pi\tilde{e}_k \, \mathrm{int}(u_{0\xi_k\eta_k}, \Gamma_{\xi_k\eta_k}), \qquad (6.173)$$

where $u_{0\xi_k\eta_k}$ and $\Gamma_{\xi_k\eta_k}$ are respectively the projections of the point u_0 and of the loop Γ on the plane defined by the axes ξ_k and η_k, as shown in Fig. 6.3.

If $f(u)$ is an analytic n-complex function which can be expanded in a series as written in Eq. (1.89), and the expansion holds on the curve Γ and on a surface spanning Γ, then from Eqs. (6.167) and (6.173) it follows that

$$\oint_\Gamma \frac{f(u)du}{u-u_0} = 2\pi f(u_0) \sum_{k=1}^{[(n-1)/2]} \tilde{e}_k \, \mathrm{int}(u_{0\xi_k\eta_k}, \Gamma_{\xi_k\eta_k}). \qquad (6.174)$$

Substituting in the right-hand side of Eq. (6.174) the expression of $f(u)$ in terms of the real components P_k, Eq. (6.159), yields

$$\oint_\Gamma \frac{f(u)du}{u-u_0} = \frac{2}{n} \sum_{k=1}^{[(n-1)/2]} \sum_{l,m=0}^{n-1} h_l \sin\left[\frac{2\pi(l-m)k}{n}\right] P_m(u_0)$$

$$\mathrm{int}(u_{0\xi_k\eta_k}, \Gamma_{\xi_k\eta_k}). \qquad (6.175)$$

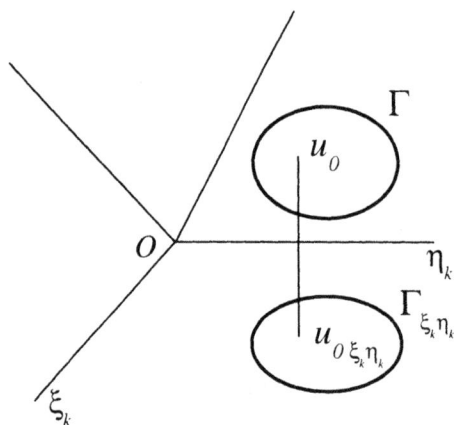

Figure 6.3: Integration path Γ and pole u_0, and their projections $\Gamma_{\xi_k \eta_k}$ and $u_{0\xi_k \eta_k}$ on the plane $\xi_k \eta_k$.

It the integral in Eq. (6.175) is written as

$$\oint_{\Gamma} \frac{f(u)du}{u - u_0} = \sum_{l=0}^{n-1} h_l I_l, \qquad (6.176)$$

it can be checked that

$$\sum_{l=0}^{n-1} I_l = 0. \qquad (6.177)$$

If $f(u)$ can be expanded as written in Eq. (1.89) on Γ and on a surface spanning Γ, then from Eqs. (6.167) and (6.173) it also results that

$$\oint_{\Gamma} \frac{f(u)du}{(u - u_0)^{n+1}} = \frac{2\pi}{n!} f^{(n)}(u_0) \sum_{k=1}^{[(n-1)/2]} \tilde{e}_k \, \text{int}(u_{0\xi_k \eta_k}, \Gamma_{\xi_k \eta_k}), \qquad (6.178)$$

where the fact has been used that the derivative $f^{(n)}(u_0)$ is related to the expansion coefficient in Eq. (1.89) according to Eq. (1.93).

If a function $f(u)$ is expanded in positive and negative powers of $u - u_l$, where u_l are n-complex constants, l being an index, the integral of f on a closed loop Γ is determined by the terms in the expansion of f which are of the form $r_l/(u - u_l)$,

$$f(u) = \cdots + \sum_l \frac{r_l}{u - u_l} + \cdots. \tag{6.179}$$

Then the integral of f on a closed loop Γ is

$$\oint_\Gamma f(u)du = 2\pi \sum_l \sum_{k=1}^{[(n-1)/2]} \tilde{e}_k \, \mathrm{int}(u_{l\xi_k\eta_k}, \Gamma_{\xi_k\eta_k})r_l. \tag{6.180}$$

6.1.9 Factorization of polar n-complex polynomials

A polynomial of degree m of the n-complex variable u has the form

$$P_m(u) = u^m + a_1 u^{m-1} + \cdots + a_{m-1}u + a_m, \tag{6.181}$$

where a_l, for $l = 1, ..., m$, are in general n-complex constants. If $a_l = \sum_{p=0}^{n-1} h_p a_{lp}$, and with the notations of Eqs. (6.151)-(6.154) applied for $l = 1, \cdots, m$, the polynomial $P_m(u)$ can be written, for even n, as

$$P_m = e_+ \left(v_+^m + \sum_{l=1}^m A_{l+} v_+^{m-l} \right) + e_- \left(v_-^m + \sum_{l=1}^m A_{l-} v_-^{m-l} \right)$$
$$+ \sum_{k=1}^{n/2-1} \left[(e_k v_k + \tilde{e}_k \tilde{v}_k)^m + \sum_{l=1}^m (e_k A_{lk} + \tilde{e}_k \tilde{A}_{lk})(e_k v_k + \tilde{e}_k \tilde{v}_k)^{m-l} \right], \tag{6.182}$$

where the constants $A_{l+}, A_{l-}, A_{lk}, \tilde{A}_{lk}$ are real numbers. For odd n the expression of the polynomial is

$$P_m = e_+ \left(v_+^m + \sum_{l=1}^m A_{l+} v_+^{m-l} \right)$$
$$+ \sum_{k=1}^{(n-1)/2} \left[(e_k v_k + \tilde{e}_k \tilde{v}_k)^m + \sum_{l=1}^m (e_k A_{lk} + \tilde{e}_k \tilde{A}_{lk})(e_k v_k + \tilde{e}_k \tilde{v}_k)^{m-l} \right]. \tag{6.183}$$

The polynomials of degree m in $e_k v_k + \tilde{e}_k \tilde{v}_k$ in Eqs. (6.182) and (6.183) can always be written as a product of linear factors of the form $e_k(v_k - v_{kp}) + \tilde{e}_k(\tilde{v}_k - \tilde{v}_{kp})$, where the constants $v_{kp}, \tilde{v}_{kp}$ are real. The polynomials of degree m with real coefficients in Eqs. (6.182) and (6.183) which are multiplied by e_+ and e_- can be written as a product of linear or quadratic factors with real coefficients, or as a product of linear factors which, if imaginary, appear always in complex conjugate pairs. Using the latter form for the simplicity of notations, the polynomial P_m can be written, for even n, as

$$P_m = e_+ \prod_{p=1}^{m}(v_+ - v_{p+}) + e_- \prod_{p=1}^{m}(v_- - v_{p-})$$
$$+ \sum_{k=1}^{n/2-1} \prod_{p=1}^{m}\{e_k(v_k - v_{kp}) + \tilde{e}_k(\tilde{v}_k - \tilde{v}_{kp})\}, \tag{6.184}$$

where the quantities v_{p+} appear always in complex conjugate pairs, and the quantities $\tilde{v}_{p-}$ appear always in complex conjugate pairs. For odd n the polynomial can be written as

$$P_m = e_+ \prod_{p=1}^{m}(v_+ - v_{p+}) + \sum_{k=1}^{(n-1)/2} \prod_{p=1}^{m}\{e_k(v_k - v_{kp}) + \tilde{e}_k(\tilde{v}_k - \tilde{v}_{kp})\}, \tag{6.185}$$

where the quantities v_{p+} appear always in complex conjugate pairs. Due to the relations (6.82),(6.83), the polynomial $P_m(u)$ can be written, for even n, as a product of factors of the form

$$P_m(u) = \prod_{p=1}^{m}\{e_+(v_+ - v_{p+}) + e_-(v_- - v_{p-})$$
$$+ \sum_{k=1}^{n/2-1}\{e_k(v_k - v_{kp}) + \tilde{e}_k(\tilde{v}_k - \tilde{v}_{kp})\}\}. \tag{6.186}$$

For odd n, the polynomial $P_m(u)$ can be written as the product

$$P_m(u) = \prod_{p=1}^{m}\{e_+(v_+ - v_{p+})$$
$$+ \sum_{k=1}^{(n-1)/2}\{e_k(v_k - v_{kp}) + \tilde{e}_k(\tilde{v}_k - \tilde{v}_{kp})\}\}. \tag{6.187}$$

These relations can be written with the aid of Eqs. (6.85) and (6.86) as

$$P_m(u) = \prod_{p=1}^{m}(u - u_p), \tag{6.188}$$

where, for even n,

$$u_p = e_+ v_{p+} + e_- v_{p-} + \sum_{k=1}^{n/2-1}(e_k v_{kp} + \tilde{e}_k \tilde{v}_{kp}), \tag{6.189}$$

and for odd n

$$u_p = e_+ v_{p+} + \sum_{k=1}^{(n-1)/2} \left(e_k v_{kp} + \tilde{e}_k \tilde{v}_{kp} \right), \qquad (6.190)$$

for $p = 1, ..., m$. The roots v_{p+}, the roots v_{p-} and, for a given k, the roots $e_k v_{k1} + \tilde{e}_k \tilde{v}_{k1}, ..., e_k v_{km} + \tilde{e}_k \tilde{v}_{km}$ defined in Eqs. (6.184) or (6.185) may be ordered arbitrarily. This means that Eqs. (6.189) or (6.190) give sets of m roots $u_1, ..., u_m$ of the polynomial $P_m(u)$, corresponding to the various ways in which the roots $v_{p+}, v_{p-}, e_k v_{kp} + \tilde{e}_k \tilde{v}_{kp}$ are ordered according to p in each group. Thus, while the n-complex components in Eq. (6.183) taken separately have unique factorizations, the polynomial $P_m(u)$ can be written in many different ways as a product of linear factors.

If $P(u) = u^2 - 1$, the degree is $m = 2$, the coefficients of the polynomial are $a_1 = 0, a_2 = -1$, the n-complex components of a_2 are $a_{20} = -1, a_{21} = 0, ..., a_{2,n-1} = 0$, the components $A_{2+}, A_{2-}, A_{2k}, \tilde{A}_{2k}$ calculated according to Eqs. (6.151)-(6.154) are $A_{2+} = -1, A_{2-} = -1, A_{2k} = -1, \tilde{A}_{2k} = 0, k = 1, ..., [(n-1)/2]$. The expression of $P(u)$ for even n, Eq. (6.182), is $e_+(v_+^2 - 1) + e_-(v_-^2 - 1) + \sum_{k=1}^{n/2-1} \{ (e_k v_k + \tilde{e}_k \tilde{v}_k)^2 - e_k \}$, and Eq. (6.184) has the form $u^2 - 1 = e_+(v_+ + 1)(v_+ - 1) + e_-(v_- + 1)(v_- - 1) + \sum_{k=1}^{n/2-1} \{ e_k(v_k + 1) + \tilde{e}_k \tilde{v}_k \} \{ e_k(v_k - 1) + \tilde{e}_k \tilde{v}_k \}$. For odd n, the expression of $P(u)$, Eq. (6.183), is $e_+(v_+^2 - 1) + \sum_{k=1}^{(n-1)/2} \{ (e_k v_k + \tilde{e}_k \tilde{v}_k)^2 - e_k \}$, and Eq. (6.185) has the form $u^2 - 1 = e_+(v_+ + 1)(v_+ - 1) + \sum_{k=1}^{(n-1)/2} \{ e_k(v_k + 1) + \tilde{e}_k \tilde{v}_k \} \{ e_k(v_k - 1) + \tilde{e}_k \tilde{v}_k \}$. The factorization in Eq. (6.188) is $u^2 - 1 = (u - u_1)(u - u_2)$, where for even n, $u_1 = \pm e_+ \pm e_- \pm e_1 \pm e_2 \pm \cdots \pm e_{n/2-1}, u_2 = -u_1$, so that there are $2^{n/2}$ independent sets of roots u_1, u_2 of $u^2 - 1$. It can be checked that $(\pm e_+ \pm e_- \pm e_1 \pm e_2 \pm \cdots \pm e_{n/2-1})^2 = e_+ + e_- + e_1 + e_2 + \cdots + e_{n/2-1} = 1$. For odd n, $u_1 = \pm e_+ \pm e_1 \pm e_2 \pm \cdots \pm e_{(n-1)/2}, u_2 = -u_1$, so that there are $2^{(n-1)/2}$ independent sets of roots u_1, u_2 of $u^2 - 1$. It can be checked that $(\pm e_+ \pm e_1 \pm e_2 \pm \cdots \pm e_{(n-1)/2})^2 = e_+ + e_1 + e_2 + \cdots + e_{(n-1)/2} = 1$.

6.1.10 Representation of polar n-complex numbers by irreducible matrices

If the unitary matrix written in Eq. (6.21), for even n, is called T_e, and the unitary matrix written in Eq. (6.22), for odd n, is called T_o, it can be shown that, for even n, the matrix $T_e U T_e^{-1}$ has the form

$$T_e U T_e^{-1} = \begin{pmatrix} v_+ & 0 & 0 & \cdots & 0 \\ 0 & v_- & 0 & \cdots & 0 \\ 0 & 0 & V_1 & \cdots & 0 \\ \vdots & \vdots & \vdots & \cdots & \vdots \\ 0 & 0 & 0 & \cdots & V_{n/2-1} \end{pmatrix} \tag{6.191}$$

and, for odd n, the matrix $T_o U T_o^{-1}$ has the form

$$T_o U T_o^{-1} = \begin{pmatrix} v_+ & 0 & 0 & \cdots & 0 \\ 0 & V_1 & 0 & \cdots & 0 \\ 0 & 0 & V_2 & \cdots & 0 \\ \vdots & \vdots & \vdots & \cdots & \vdots \\ 0 & 0 & 0 & \cdots & V_{(n-1)/2} \end{pmatrix}, \tag{6.192}$$

where U is the matrix in Eq. (6.47) used to represent the n-complex number u. In Eqs. (6.191) and (6.192), V_k are the matrices

$$V_k = \begin{pmatrix} v_k & \tilde{v}_k \\ -\tilde{v}_k & v_k \end{pmatrix}, \tag{6.193}$$

for $k = 1, ..., [(n-1)/2]$, where $v_k, \tilde{v}_k$ are the variables introduced in Eqs. (6.14) and (6.15), and the symbols 0 denote, according to the case, the real number zero, or one of the matrices

$$\begin{pmatrix} 0 \\ 0 \end{pmatrix} \quad \text{or} \quad \begin{pmatrix} 0 & 0 \\ 0 & 0 \end{pmatrix}. \tag{6.194}$$

The relations between the variables $v_k, \tilde{v}_k$ for the multiplication of n-complex numbers have been written in Eq. (6.44). The matrices $T_e U T_e^{-1}$ and $T_o U T_o^{-1}$ provide an irreducible representation [7] of the n-complex numbers u in terms of matrices with real coefficients.

6.2   Planar complex numbers in even n dimensions

6.2.1   Operations with planar n-complex numbers

A hypercomplex number in n dimensions is determined by its n components $(x_0, x_1, ..., x_{n-1})$. The planar n-complex numbers and their operations discussed in this section can be represented by writing the n-complex

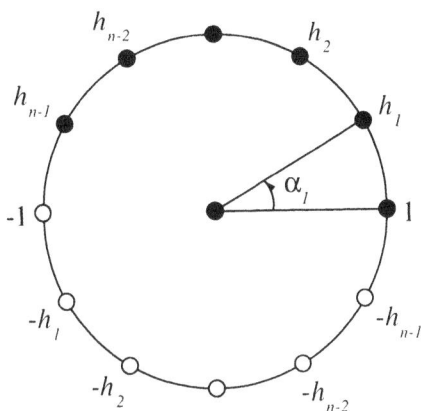

Figure 6.4: Representation of the hypercomplex bases $1, h_1, ..., h_{n-1}$ by points on a circle at the angles $\alpha_k = \pi k/n$. The product $h_j h_k$ will be represented by the point of the circle at the angle $\pi(j + k)/2n$, $j, k = 0, 1, ..., n - 1$. If $\pi \le \pi(j + k)/2n \le 2\pi$, the point is opposite to the basis h_l of angle $\alpha_l = \pi(j + k)/n - \pi$.

number $(x_0, x_1, ..., x_{n-1})$ as $u = x_0 + h_1 x_1 + h_2 x_2 + \cdots + h_{n-1} x_{n-1}$, where $h_1, h_2, \cdots, h_{n-1}$ are bases for which the multiplication rules are

$$h_j h_k = (-1)^{[(j+k)/n]} h_l, \quad l = j + k - n[(i + k)/n], \qquad (6.195)$$

for $j, k, l = 0, 1, ..., n - 1$, where $h_0 = 1$. In Eq. (6.195), $[(j + k)/n]$ denotes the integer part of $(j + k)/n$, the integer part being defined as $[a] \le a < [a] + 1$, so that $0 \le j + k - n[(j + k)/n] \le n - 1$. As already mentioned, brackets larger than the regular brackets [ ] do not have the meaning of integer part. The significance of the composition laws in Eq. (6.195) can be understood by representing the bases h_j, h_k by points on a circle at the angles $\alpha_j = \pi j/n, \alpha_k = \pi k/n$, as shown in Fig. 6.4, and the product $h_j h_k$ by the point of the circle at the angle $\pi(j + k)/n$. If $\pi \le \pi(j + k)/n < 2\pi$, the point is opposite to the basis h_l of angle $\alpha_l = \pi(j + k)/n - \pi$.

In an odd number of dimensions n, a transformation of coordinates according to

$$x_{2l} = x'_l, x_{2m-1} = -x'_{(n-1)/2+m}, \qquad (6.196)$$

and of the bases according to

$$h_{2l} = h'_l, h_{2m-1} = -h'_{(n-1)/2+m}, \tag{6.197}$$

where $l = 0, ..., (n-1)/2$, $m = 1, ..., (n-1)/2$, leaves the expression of an n-complex number unchanged,

$$\sum_{k=0}^{n-1} h_k x_k = \sum_{k=0}^{n-1} h'_k x'_k, \tag{6.198}$$

and the products of the bases h'_k are

$$h'_j h'_k = h'_l, \; l = j + k - n[(j+k)/n], \tag{6.199}$$

for $j, k, l = 0, 1, ..., n-1$. Thus, the n-complex numbers with the rules (6.195) are equivalent in an odd number of dimensions to the polar n-complex numbers described in the previous chapter. Therefore, in this section it will be supposed that n is an even number, unless otherwise stated.

Two n-complex numbers $u = x_0 + h_1 x_1 + h_2 x_2 + \cdots + h_{n-1} x_{n-1}$, $u' = x'_0 + h_1 x'_1 + h_2 x'_2 + \cdots + h_{n-1} x'_{n-1}$ are equal if and only if $x_j = x'_j, j = 0, 1, ..., n-1$. The sum of the n-complex numbers u and u' is

$$u + u' = x_0 + x'_0 + h_1(x_1 + x'_1) + \cdots + h_{n-1}(x_{n-1} + x'_{n-1}). \tag{6.200}$$

The product of the numbers u, u' is

$$\begin{aligned}
uu' = \; &x_0 x'_0 - x_1 x'_{n-1} - x_2 x'_{n-2} - x_3 x'_{n-3} - \cdots - x_{n-1} x'_1 \\
&+ h_1(x_0 x'_1 + x_1 x'_0 - x_2 x'_{n-1} - x_3 x'_{n-2} - \cdots - x_{n-1} x'_2) \\
&+ h_2(x_0 x'_2 + x_1 x'_1 + x_2 x'_0 - x_3 x'_{n-1} - \cdots - x_{n-1} x'_3) \\
&\vdots \\
&+ h_{n-1}(x_0 x'_{n-1} + x_1 x'_{n-2} + x_2 x'_{n-3} + x_3 x'_{n-4} + \cdots + x_{n-1} x'_0).
\end{aligned} \tag{6.201}$$

The product uu' can be written as

$$uu' = \sum_{k=0}^{n-1} h_k \sum_{l=0}^{n-1} (-1)^{[(n-k-1+l)/n]} x_l x'_{k-l+n[(n-k-1+l)/n]}. \tag{6.202}$$

If u, u', u'' are n-complex numbers, the multiplication is associative

$$(uu')u'' = u(u'u'') \tag{6.203}$$

and commutative

$$uu' = u'u, \tag{6.204}$$

because the product of the bases, defined in Eq. (6.195), is associative and commutative. The fact that the multiplication is commutative can be seen also directly from Eq. (6.201). The n-complex zero is $0 + h_1 \cdot 0 + \cdots + h_{n-1} \cdot 0$, denoted simply 0, and the n-complex unity is $1 + h_1 \cdot 0 + \cdots + h_{n-1} \cdot 0$, denoted simply 1.

The inverse of the n-complex number $u = x_0 + h_1 x_1 + h_2 x_2 + \cdots + h_{n-1} x_{n-1}$ is the n-complex number $u' = x'_0 + h_1 x'_1 + h_2 x'_2 + \cdots + h_{n-1} x'_{n-1}$ having the property that

$$uu' = 1. \tag{6.205}$$

Written on components, the condition, Eq. (6.205), is

$$
\begin{aligned}
&x_0 x'_0 - x_1 x'_{n-1} - x_2 x'_{n-2} - x_3 x'_{n-3} - \cdots - x_{n-1} x'_1 = 1, \\
&x_0 x'_1 + x_1 x'_0 - x_2 x'_{n-1} - x_3 x'_{n-2} - \cdots - x_{n-1} x'_2 = 0, \\
&x_0 x'_2 + x_1 x'_1 + x_2 x'_0 - x_3 x'_{n-1} - \cdots - x_{n-1} x'_3 = 0, \\
&\vdots \\
&x_0 x'_{n-1} + x_1 x'_{n-2} + x_2 x'_{n-3} + x_3 x'_{n-4} + \cdots + x_{n-1} x'_0 = 0.
\end{aligned}
\tag{6.206}
$$

The system (6.206) has a solution provided that the determinant of the system,

$$\nu = \det(A), \tag{6.207}$$

is not equal to zero, $\nu \neq 0$, where

$$
A = \begin{pmatrix}
x_0 & -x_{n-1} & -x_{n-2} & \cdots & -x_1 \\
x_1 & x_0 & -x_{n-1} & \cdots & -x_2 \\
x_2 & x_1 & x_0 & \cdots & -x_3 \\
\vdots & \vdots & \vdots & \cdots & \vdots \\
x_{n-1} & x_{n-2} & x_{n-3} & \cdots & x_0
\end{pmatrix}.
\tag{6.208}
$$

It will be shown that $\nu > 0$, and the quantity

$$\rho = \nu^{1/n} \tag{6.209}$$

will be called amplitude of the n-complex number $u = x_0 + h_1 x_1 + h_2 x_2 + \cdots + h_{n-1} x_{n-1}$. The quantity ν can be written as a product of linear factors

$$\nu = \prod_{k=1}^{n} \left(x_0 + \epsilon_k x_1 + \epsilon_k^2 x_2 + \cdots + \epsilon_k^{n-1} x_{n-1} \right), \tag{6.210}$$

where $\epsilon_k = e^{i\pi(2k-1)/n}$, $k = 1, ..., n$, and i being the imaginary unit. The factors appearing in Eq. (6.210) are of the form

$$x_0 + \epsilon_k x_1 + \epsilon_k^2 x_2 + \cdots + \epsilon_k^{n-1} x_{n-1} = v_k + i \tilde{v}_k, \tag{6.211}$$

where

$$v_k = \sum_{p=0}^{n-1} x_p \cos \frac{\pi(2k-1)p}{n}, \tag{6.212}$$

$$\tilde{v}_k = \sum_{p=0}^{n-1} x_p \sin \frac{\pi(2k-1)p}{n}, \tag{6.213}$$

for $k = 1, ..., n$. The variables $v_k, \tilde{v}_k, k = 1, ..., n/2$ will be called canonical polar n-complex variables. It can be seen that $v_k = v_{n-k+1}, \tilde{v}_k = -\tilde{v}_{n-k+1}$, for $k = 1, ..., n/2$. Therefore, the factors appear in Eq. (6.210) in complex-conjugate pairs of the form $v_k + i\tilde{v}_k$ and $v_{n-k+1} + i\tilde{v}_{n-k+1} = v_k - i\tilde{v}_k$, where $k = 1, ...n/2$, so that the determinant ν is a real and positive quantity, $\nu > 0$,

$$\nu = \prod_{k=1}^{n/2} \rho_k^2, \tag{6.214}$$

where

$$\rho_k^2 = v_k^2 + \tilde{v}_k^2. \tag{6.215}$$

Thus, an n-complex number has an inverse unless it lies on one of the nodal hypersurfaces $\rho_1 = 0$, or $\rho_2 = 0$, or ... or $\rho_{n/2} = 0$.

6.2.2 Geometric representation of planar n-complex numbers

The n-complex number $x_0 + h_1 x_1 + h_2 x_2 + \cdots + h_{n-1} x_{n-1}$ can be represented by the point A of coordinates $(x_0, x_1, ..., x_{n-1})$. If O is the origin of the n-dimensional space, the distance from the origin O to the point A of coordinates $(x_0, x_1, ..., x_{n-1})$ has the expression

$$d^2 = x_0^2 + x_1^2 + \cdots + x_{n-1}^2. \tag{6.216}$$

The quantity d will be called modulus of the n-complex number $u = x_0 + h_1 x_1 + h_2 x_2 + \cdots + h_{n-1} x_{n-1}$. The modulus of an n-complex number u will be designated by $d = |u|$.

The exponential and trigonometric forms of the n-complex number u can be obtained conveniently in a rotated system of axes defined by a transformation which has the form

$$
\begin{pmatrix} \vdots \\ \xi_k \\ \eta_k \\ \vdots \end{pmatrix} =
$$

$$
\begin{pmatrix} & \vdots & \vdots & & \vdots & \vdots \\ \cdots & \sqrt{\dfrac{2}{n}} & \sqrt{\dfrac{2}{n}}\cos\dfrac{\pi(2k-1)}{n} & \cdots & \sqrt{\dfrac{2}{n}}\cos\dfrac{\pi(2k-1)(n-2)}{n} & \sqrt{\dfrac{2}{n}}\cos\dfrac{\pi(2k-1)(n-1)}{n} \\ \cdots & 0 & \sqrt{\dfrac{2}{n}}\sin\dfrac{\pi(2k-1)}{n} & \cdots & \sqrt{\dfrac{2}{n}}\sin\dfrac{\pi(2k-1)(n-2)}{n} & \sqrt{\dfrac{2}{n}}\sin\dfrac{\pi(2k-1)(n-1)}{n} \\ & \vdots & \vdots & & \vdots & \vdots \end{pmatrix}
$$

$$
\begin{pmatrix} x_0 \\ \vdots \\ \vdots \\ x_{n-1} \end{pmatrix}, \tag{6.217}
$$

where $k = 1, 2, ..., n/2$. The lines of the matrices in Eq. (6.217) give the components of the n vectors of the new basis system of axes. These vectors have unit length and are orthogonal to each other. By comparing Eqs. (6.212)-(6.213) and (6.217) it can be seen that

$$
v_k = \sqrt{\frac{n}{2}}\xi_k, \ \tilde{v}_k = \sqrt{\frac{n}{2}}\eta_k, \tag{6.218}
$$

i.e. the two sets of variables differ only by a scale factor.

The sum of the squares of the variables $v_k, \tilde{v}_k$ is

$$
\sum_{k=1}^{n/2} (v_k^2 + \tilde{v}_k^2) = \frac{n}{2}d^2. \tag{6.219}
$$

The relation (6.219) has been obtained with the aid of the relation

$$
\sum_{k=1}^{n/2} \cos\frac{\pi(2k-1)p}{n} = 0, \tag{6.220}
$$

for $p = 1, ..., n-1$. From Eq. (6.219) it results that

$$
d^2 = \frac{2}{n}\sum_{k=1}^{n/2} \rho_k^2. \tag{6.221}
$$

The relation (6.221) shows that the square of the distance d, Eq. (6.216), is equal to the sum of the squares of the projections $\rho_k \sqrt{2/n}$. This is consistent with the fact that the transformation in Eq. (6.217) is unitary.

The position of the point A of coordinates $(x_0, x_1, ..., x_{n-1})$ can be also described with the aid of the distance d, Eq. (6.216), and of $n-1$ angles defined further. Thus, in the plane of the axes $v_k, \tilde{v}_k$, the azimuthal angle ϕ_k can be introduced by the relations

$$\cos \phi_k = v_k/\rho_k, \ \sin \phi_k = \tilde{v}_k/\rho_k, \tag{6.222}$$

where $0 \leq \phi_k < 2\pi$, $k = 1, ..., n/2$, so that there are $n/2$ azimuthal angles. The radial distance ρ_k in the plane of the axes $v_k, \tilde{v}_k$ has been defined in Eq. (6.215). If the projection of the point A on the plane of the axes $v_k, \tilde{v}_k$ is A_k, and the projection of the point A on the 4-dimensional space defined by the axes $v_1, \tilde{v}_1, v_k, \tilde{v}_k$ is A_{1k}, the angle ψ_{k-1} between the line OA_{1k} and the 2-dimensional plane defined by the axes $v_k, \tilde{v}_k$ is

$$\tan \psi_{k-1} = \rho_1/\rho_k, \tag{6.223}$$

where $0 \leq \psi_k \leq \pi/2, k = 2, ..., n/2$, so that there are $n/2 - 1$ planar angles. Thus, the position of the point A is described by the distance d, by $n/2$ azimuthal angles and by $n/2 - 1$ planar angles. These angles are shown in Fig. 6.5.

The variables ρ_k can be expressed in terms of d and the planar angles ψ_k as

$$\rho_k = \frac{\rho_1}{\tan \psi_{k-1}}, \tag{6.224}$$

for $k = 2, ..., n/2$, where

$$\rho_1^2 = \frac{nd^2}{2}\left(1 + \frac{1}{\tan^2 \psi_1} + \frac{1}{\tan^2 \psi_2} + \cdots + \frac{1}{\tan^2 \psi_{n/2-1}}\right)^{-1}. \tag{6.225}$$

If $u' = x_0' + h_1 x_1' + h_2 x_2' + \cdots + h_{n-1} x_{n-1}', u'' = x_0'' + h_1 x_1'' + h_2 x_2'' + \cdots + h_{n-1} x_{n-1}''$ are n-complex numbers of parameters $\rho_k', \psi_k', \phi_k'$ and respectively $\rho_k'', \psi_k'', \phi_k''$, then the parameters $v_+, \rho_k, \psi_k, \phi_k$ of the product n-complex number $u = u'u''$ are given by

$$\rho_k = \rho_k' \rho_k'', \tag{6.226}$$

for $k = 1, ..., n/2$,

$$\tan \psi_k = \tan \psi_k' \tan \psi_k'', \tag{6.227}$$

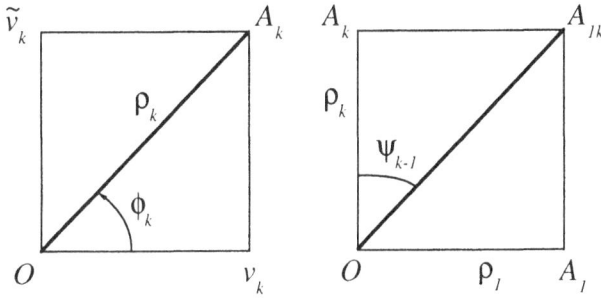

Figure 6.5: Radial distance ρ_k and azimuthal angle ϕ_k in the plane of the axes $v_k, \tilde{v}_k$, and planar angle ψ_{k-1} between the line OA_{1k} and the 2-dimensional plane defined by the axes $v_k, \tilde{v}_k$. A_k is the projection of the point A on the plane of the axes $v_k, \tilde{v}_k$, and A_{1k} is the projection of the point A on the 4-dimensional space defined by the axes $v_1, \tilde{v}_1, v_k, \tilde{v}_k$.

for $k = 1, ..., n/2 - 1$,

$$\phi_k = \phi'_k + \phi''_k, \tag{6.228}$$

for $k = 1, ..., n/2$. The Eqs. (6.226)-(6.228) are a consequence of the relations

$$v_k = v'_k v''_k - \tilde{v}'_k \tilde{v}''_k, \quad \tilde{v}_k = v'_k \tilde{v}''_k + \tilde{v}'_k v''_k, \tag{6.229}$$

and of the corresponding relations of definition. Then the product ν in Eq. (6.214) has the property that

$$\nu = \nu' \nu'', \tag{6.230}$$

and the amplitude ρ defined in Eq. (6.209) has the property that

$$\rho = \rho' \rho''. \tag{6.231}$$

The fact that the amplitude of the product is equal to the product of the amplitudes, as written in Eq. (6.231), can be demonstrated also by using a representation of the n-complex numbers by matrices, in which the n-complex number $u = x_0 + h_1 x_1 + h_2 x_2 + \cdots + h_{n-1} x_{n-1}$ is represented by the matrix

$$U = \begin{pmatrix} x_0 & x_1 & x_2 & \cdots & x_{n-1} \\ -x_{n-1} & x_0 & x_1 & \cdots & x_{n-2} \\ -x_{n-2} & -x_{n-1} & x_0 & \cdots & x_{n-3} \\ \vdots & \vdots & \vdots & \cdots & \vdots \\ -x_1 & -x_2 & -x_3 & \cdots & x_0 \end{pmatrix}. \tag{6.232}$$

The product $u = u'u''$ is be represented by the matrix multiplication $U = U'U''$. The relation (6.230) is then a consequence of the fact the determinant of the product of matrices is equal to the product of the determinants of the factor matrices. The use of the representation of the n-complex numbers with matrices provides an alternative demonstration of the fact that the product of n-complex numbers is associative, as stated in Eq. (6.203).

According to Eqs. (6.219 and (6.215), the modulus of the product uu' is given by

$$|uu'|^2 = \frac{2}{n} \sum_{k=1}^{n/2} (\rho_k \rho'_k)^2. \tag{6.233}$$

Thus, if the product of two n-complex numbers is zero, $uu' = 0$, then $\rho_k \rho'_k = 0, k = 1, ..., n/2$. This means that either $u = 0$, or $u' = 0$, or u, u' belong to orthogonal hypersurfaces in such a way that the afore-mentioned products of components should be equal to zero.

6.2.3　The planar n-dimensional cosexponential functions

The exponential function of a hypercomplex variable u and the addition theorem for the exponential function have been written in Eqs. (1.35)-(1.36). It can be seen with the aid of the representation in Fig. 6.4 that

$$h_k^{n+p} = (-1)^k h_k^p, \ p \text{ integer}, \tag{6.234}$$

where $k = 1, ..., n - 1$. For k even, $e^{h_k y}$ can be written as

$$e^{h_k y} = \sum_{p=0}^{n-1} (-1)^{[kp/n]} h_{kp-n[kp/n]} g_{np}(y), \tag{6.235}$$

where $h_0 = 1$, and where g_{np} are the polar n-dimensional cosexponential functions. For odd k, $e^{h_k y}$ is

$$e^{h_k y} = \sum_{p=0}^{n-1} (-1)^{[kp/n]} h_{kp-n[kp/n]} f_{np}(y), \tag{6.236}$$

where the functions f_{nk}, which will be called planar cosexponential functions in n dimensions, are

$$f_{nk}(y) = \sum_{p=0}^{\infty} (-1)^p \frac{y^{k+pn}}{(k+pn)!}, \tag{6.237}$$

for $k = 0, 1, ..., n-1$.

The planar cosexponential functions of even index k are even functions, $f_{n,2l}(-y) = f_{n,2l}(y)$, and the planar cosexponential functions of odd index are odd functions, $f_{n,2l+1}(-y) = -f_{n,2l+1}(y)$, $l = 0, ..., n/2 - 1$.

The planar n-dimensional cosexponential function $f_{nk}(y)$ is related to the polar n-dimensional cosexponential function $g_{nk}(y)$ discussed in the previous chapter by the relation

$$f_{nk}(y) = e^{-i\pi k/n} g_{nk}\left(e^{i\pi/n} y\right), \tag{6.238}$$

for $k = 0, ..., n-1$. The expression of the planar n-dimensional cosexponential functions is then

$$f_{nk}(y) = \frac{1}{n} \sum_{l=1}^{n} \exp\left[y \cos\left(\frac{\pi(2l-1)}{n}\right)\right] \cos\left[y \sin\left(\frac{\pi(2l-1)}{n}\right)\right.$$
$$\left. - \frac{\pi(2l-1)k}{n}\right], \tag{6.239}$$

for $k = 0, 1, ..., n-1$. The planar cosexponential function defined in Eq. (6.237) has the expression given in Eq. (6.239) for any natural value of n, this result not being restricted to even values of n. In order to check that the function in Eq. (6.239) has the series expansion written in Eq. (6.237), the right-hand side of Eq. (6.239) will be written as

$$f_{nk}(y) = \frac{1}{n} \sum_{l=1}^{n} \text{Re}\left\{\exp\left[\left(\cos\frac{\pi(2l-1)}{n} + i\sin\frac{\pi(2l-1)}{n}\right) y \right.\right.$$
$$\left.\left. - i\frac{\pi k(2l-1)}{n}\right]\right\}, \tag{6.240}$$

for $k = 0, 1, ..., n-1$, where $\text{Re}(a+ib) = a$, with a and b real numbers. The part of the exponential depending on y can be expanded in a series,

$$f_{nk}(y) = \frac{1}{n} \sum_{p=0}^{\infty} \sum_{l=1}^{n} \text{Re}\left\{\frac{1}{p!} \exp\left[i\frac{\pi(2l-1)}{n}(p-k)\right] y^p\right\}, \tag{6.241}$$

for $k = 0, 1, ..., n - 1$. The expression of $f_{nk}(y)$ becomes

$$f_{nk}(y) = \frac{1}{n} \sum_{p=0}^{\infty} \sum_{l=1}^{n} \left\{ \frac{1}{p!} \cos\left[\frac{\pi(2l-1)}{n}(p-k)\right] y^p \right\}, \tag{6.242}$$

where $k = 0, 1, ..., n - 1$ and, since

$$\frac{1}{n} \sum_{l=1}^{n} \cos\left[\frac{\pi(2l-1)}{n}(p-k)\right] = \begin{cases} 1, & \text{if } p - k \text{ is an even multiple of } n, \\ -1, & \text{if } p - k \text{ is an odd multiple of } n, \\ 0, & \text{otherwise,} \end{cases} \tag{6.243}$$

this yields indeed the expansion in Eq. (6.237).

It can be shown from Eq. (6.239) that

$$\sum_{k=0}^{n-1} f_{nk}^2(y) = \frac{1}{n} \sum_{l=1}^{n} \exp\left[2y \cos\left(\frac{\pi(2l-1)}{n}\right)\right]. \tag{6.244}$$

It can be seen that the right-hand side of Eq. (6.244) does not contain oscillatory terms. If n is a multiple of 4, it can be shown by replacing y by iy in Eq. (6.244) that

$$\sum_{k=0}^{n-1} (-1)^k f_{nk}^2(y) = \frac{4}{n} \sum_{l=1}^{n/4} \cos\left[2y \cos\left(\frac{\pi(2l-1)}{n}\right)\right], \tag{6.245}$$

which does not contain exponential terms.

For odd n, the planar n-dimensional cosexponential function $f_{nk}(y)$ is related to the n-dimensional cosexponential function $g_{nk}(y)$ discussed in the previous chapter also by the relation

$$f_{nk}(y) = (-1)^k g_{nk}(-y), \tag{6.246}$$

as can be seen by comparing the series for the two classes of functions. For values of the form $n = 4p + 2$, $p = 0, 1, 2, ...$, the planar n-dimensional cosexponential function $f_{nk}(y)$ is related to the n-dimensional cosexponential function $g_{nk}(y)$ by the relation

$$f_{nk}(y) = e^{-i\pi k/2} g_{nk}(iy). \tag{6.247}$$

Addition theorems for the planar n-dimensional cosexponential functions can be obtained from the relation $\exp h_1(y + z) = \exp h_1 y \cdot \exp h_1 z$, by substituting the expression of the exponentials as given in Eq. (6.236) for $k = 1$, $e^{h_1 y} = f_{n0}(y) + h_1 f_{n1}(y) + \cdots + h_{n-1} f_{n,n-1}(y)$,

$$f_{nk}(y + z) = f_{n0}(y) f_{nk}(z) + f_{n1}(y) f_{n,k-1}(z) + \cdots + f_{nk}(y) f_{n0}(z)$$
$$- f_{n,k+1}(y) f_{n,n-1}(z) - f_{n,k+2}(y) f_{n,n-2}(z) - \cdots$$
$$- f_{n,n-1}(y) f_{n,k+1}(z), \tag{6.248}$$

where $k = 0, 1, ..., n - 1$. For $y = z$ the relations (6.248) take the form

$$
\begin{aligned}
f_{nk}(2y) = &f_{n0}(y)f_{nk}(y) + f_{n1}(y)f_{n,k-1}(y) + \cdots + f_{nk}(y)f_{n0}(y) \\
&- f_{n,k+1}(y)f_{n,n-1}(y) - f_{n,k+2}(y)f_{n,n-2}(y) - \cdots \\
&- f_{n,n-1}(y)f_{n,k+1}(y),
\end{aligned} \tag{6.249}
$$

where $k = 0, 1, ..., n - 1$. For $y = -z$ the relations (6.248) and (6.237) yield

$$
\begin{aligned}
f_{n0}(y)f_{n0}(-y) - &f_{n1}(y)f_{n,n-1}(-y) - f_{n2}(y)f_{n,n-2}(-y) - \cdots \\
&- f_{n,n-1}(y)f_{n1}(-y) = 1,
\end{aligned} \tag{6.250}
$$

$$
\begin{aligned}
f_{n0}(y)f_{nk}(-y) + &f_{n1}(y)f_{n,k-1}(-y) + \cdots + f_{nk}(y)f_{n0}(-y) \\
&- f_{n,k+1}(y)f_{n,n-1}(-y) - f_{n,k+2}(y)f_{n,n-2}(-y) - \cdots \\
&- f_{n,n-1}(y)f_{n,k+1}(-y) = 0,
\end{aligned} \tag{6.251}
$$

where $k = 1, ..., n - 1$.

From Eq. (6.235) it can be shown, for even k and natural numbers l, that

$$
\left(\sum_{p=0}^{n-1} (-1)^{[kp/n]} h_{kp-n[kp/n]} g_{np}(y) \right)^l = \sum_{p=0}^{n-1} (-1)^{[kp/n]} h_{kp-n[kp/n]} g_{np}(ly),
$$
$$\tag{6.252}$$

where $k = 0, 1, ..., n - 1$. For odd k and natural numbers l, Eq. (6.236) implies

$$
\left(\sum_{p=0}^{n-1} (-1)^{[kp/n]} h_{kp-n[kp/n]} f_{np}(y) \right)^l = \sum_{p=0}^{n-1} (-1)^{[kp/n]} h_{kp-n[kp/n]} f_{np}(ly),
$$
$$\tag{6.253}$$

where $k = 0, 1, ..., n - 1$. For $k = 1$ the relation (6.253) is

$$
\begin{aligned}
\{ f_{n0}(y) + &h_1 f_{n1}(y) + \cdots + h_{n-1} f_{n,n-1}(y) \}^l \\
&= f_{n0}(ly) + h_1 f_{n1}(ly) + \cdots + h_{n-1} f_{n,n-1}(ly).
\end{aligned} \tag{6.254}
$$

If

$$
a_k = \sum_{p=0}^{n-1} f_{np}(y) \cos \left(\frac{\pi(2k-1)p}{n} \right), \tag{6.255}
$$

and

$$
b_k = \sum_{p=0}^{n-1} f_{np}(y) \sin \left(\frac{\pi(2k-1)p}{n} \right), \tag{6.256}
$$

for $k = 1, ..., n$, where $f_{np}(y)$ are the planar cosexponential functions in Eq. (6.239), it can be shown that

$$a_k = \exp\left[y\cos\left(\frac{\pi(2k-1)}{n}\right)\right]\cos\left[y\sin\left(\frac{\pi(2k-1)}{n}\right)\right], \qquad (6.257)$$

$$b_k = \exp\left[y\cos\left(\frac{\pi(2k-1)}{n}\right)\right]\sin\left[y\sin\left(\frac{\pi(2k-1)}{n}\right)\right], \qquad (6.258)$$

for $k = 1, ..., n$. If

$$G_k^2 = a_k^2 + b_k^2, \qquad (6.259)$$

from Eqs. (6.257) and (6.258) it results that

$$G_k^2 = \exp\left[2y\cos\left(\frac{\pi(2k-1)}{n}\right)\right], \qquad (6.260)$$

for $k = 1, ..., n$. Then the planar n-dimensional cosexponential functions have the property that

$$\prod_{p=1}^{n/2} G_p^2 = 1. \qquad (6.261)$$

The planar n-dimensional cosexponential functions are solutions of the nth-order differential equation

$$\frac{d^n\zeta}{du^n} = -\zeta. \qquad (6.262)$$

This equation has solutions of the form $\zeta(u) = A_0 f_{n0}(u) + A_1 f_{n1}(u) + \cdots + A_{n-1}f_{n,n-1}(u)$. It can be checked that the derivatives of the planar cosexponential functions are related by

$$\frac{df_{n0}}{du} = -f_{n,n-1}, \quad \frac{df_{n1}}{du} = f_{n0}, \quad ...., \quad \frac{df_{n,n-2}}{du} = f_{n,n-3},$$

$$\frac{df_{n,n-1}}{du} = f_{n,n-2}. \qquad (6.263)$$

6.2.4   Exponential and trigonometric forms of planar n-complex numbers

In order to obtain the exponential and trigonometric forms of n-complex numbers, a canonical base $e_1, \tilde{e}_1, ..., e_{n/2}, \tilde{e}_{n/2}$ for the planar n-complex numbers will be introduced by the relations

$$
\begin{pmatrix} \vdots \\ e_k \\ \tilde{e}_k \\ \vdots \end{pmatrix} =
$$

$$
\begin{pmatrix} \vdots & \vdots & & \vdots & \vdots \\ \frac{2}{n} & \frac{2}{n}\cos\frac{\pi(2k-1)}{n} & \cdots & \frac{2}{n}\cos\frac{\pi(2k-1)(n-2)}{n} & \frac{2}{n}\cos\frac{\pi(2k-1)(n-1)}{n} \\ 0 & \frac{2}{n}\sin\frac{\pi(2k-1)}{n} & \cdots & \frac{2}{n}\sin\frac{\pi(2k-1)(n-2)}{n} & \frac{2}{n}\sin\frac{\pi(2k-1)(n-1)}{n} \\ \vdots & \vdots & & \vdots & \vdots \end{pmatrix}
$$

$$
\begin{pmatrix} 1 \\ h_1 \\ \vdots \\ h_{n-1} \end{pmatrix}, \tag{6.264}
$$

where $k = 1, 2, ..., n/2$.

The multiplication relations for the bases $e_k, \tilde{e}_k$ are

$$
e_k^2 = e_k, \tilde{e}_k^2 = -e_k, e_k\tilde{e}_k = \tilde{e}_k, e_k e_l = 0, e_k\tilde{e}_l = 0, \tilde{e}_k\tilde{e}_l = 0, k \neq l, \tag{6.265}
$$

where $k, l = 1, ..., n/2$. The moduli of the bases $e_k, \tilde{e}_k$ are

$$
|e_k| = \sqrt{\frac{2}{n}}, |\tilde{e}_k| = \sqrt{\frac{2}{n}}. \tag{6.266}
$$

It can be shown that

$$
x_0 + h_1 x_1 + \cdots + h_{n-1}x_{n-1} = \sum_{k=1}^{n/2}(e_k v_k + \tilde{e}_k\tilde{v}_k). \tag{6.267}
$$

The relation (6.267 gives the canonical form of a planar n-complex number. Using the properties of the bases in Eqs. (6.264) it can be shown that

$$
\exp(\tilde{e}_k\phi_k) = 1 - e_k + e_k\cos\phi_k + \tilde{e}_k\sin\phi_k, \tag{6.268}
$$

$$
\exp(e_k\ln\rho_k) = 1 - e_k + e_k\rho_k. \tag{6.269}
$$

By multiplying the relations (6.268), (6.269) it results that

$$
\exp\left[\sum_{k=1}^{n/2}(e_k\ln\rho_k + \tilde{e}_k\phi_k)\right] = \sum_{k=1}^{n/2}(e_k v_k + \tilde{e}_k\tilde{v}_k), \tag{6.270}
$$

where the fact has ben used that

$$\sum_{k=1}^{n/2} e_k = 1, \tag{6.271}$$

the latter relation being a consequence of Eqs. (6.264) and (6.220).

By comparing Eqs. (6.267) and (6.270), it can be seen that

$$x_0 + h_1 x_1 + \cdots + h_{n-1} x_{n-1} = \exp\left[\sum_{k=1}^{n/2}(e_k \ln \rho_k + \tilde{e}_k \phi_k)\right]. \tag{6.272}$$

Using the expression of the bases in Eq. (6.264) yields the exponential form of the n-complex number $u = x_0 + h_1 x_1 + \cdots + h_{n-1} x_{n-1}$ as

$$u = \rho \exp\left\{\sum_{p=1}^{n-1} h_p\left[-\frac{2}{n}\sum_{k=2}^{n/2}\cos\left(\frac{\pi(2k-1)p}{n}\right)\ln\tan\psi_{k-1}\right]\right.$$
$$\left. + \sum_{k=1}^{n/2} \tilde{e}_k \phi_k \right\}, \tag{6.273}$$

where ρ is the amplitude defined in Eq. (6.209), and has according to Eq. (6.214) the expression

$$\rho = \left(\rho_1^2 \cdots \rho_{n/2}^2\right)^{1/n}. \tag{6.274}$$

It can be checked with the aid of Eq. (6.268) that the n-complex number u can also be written as

$$x_0 + h_1 x_1 + \cdots + h_{n-1} x_{n-1} = \left(\sum_{k=1}^{n/2} e_k \rho_k\right)\exp\left(\sum_{k=1}^{n/2} \tilde{e}_k \phi_k\right). \tag{6.275}$$

Writing in Eq. (6.275) the radius ρ_1, Eq. (6.225), as a factor and expressing the variables in terms of the planar angles with the aid of Eq. (6.223) yields the trigonometric form of the n-complex number u as

$$u = d\left(\frac{n}{2}\right)^{1/2}\left(1 + \frac{1}{\tan^2\psi_1} + \frac{1}{\tan^2\psi_2} + \cdots + \frac{1}{\tan^2\psi_{n/2-1}}\right)^{-1/2}$$
$$\left(e_1 + \sum_{k=2}^{n/2} \frac{e_k}{\tan\psi_{k-1}}\right)\exp\left(\sum_{k=1}^{n/2} \tilde{e}_k \phi_k\right). \tag{6.276}$$

In Eq. (6.276), the n-complex number u, written in trigonometric form, is the product of the modulus d, of a part depending on the planar angles

$\psi_1, ..., \psi_{n/2-1}$, and of a factor depending on the azimuthal angles $\phi_1, ..., \phi_{n/2}$. Although the modulus of a product of n-complex numbers is not equal in general to the product of the moduli of the factors, it can be checked that the modulus of the factors in Eq. (6.276) are

$$\left| e_1 + \sum_{k=2}^{n/2} \frac{e_k}{\tan \psi_{k-1}} \right|$$

$$= \left(\frac{2}{n} \right)^{1/2} \left(1 + \frac{1}{\tan^2 \psi_1} + \frac{1}{\tan^2 \psi_2} + \cdots + \frac{1}{\tan^2 \psi_{n/2-1}} \right)^{1/2},$$

$$\tag{6.277}$$

and

$$\left| \exp \left(\sum_{k=1}^{n/2} \tilde{e}_k \phi_k \right) \right| = 1. \tag{6.278}$$

The modulus d in Eqs. (6.276) can be expressed in terms of the amplitude ρ as

$$d = \rho \frac{2^{(n-2)/2n}}{\sqrt{n}} \left(\tan \psi_1 \cdots \tan \psi_{n/2-1} \right)^{2/n}$$

$$\left(1 + \frac{1}{\tan^2 \psi_1} + \frac{1}{\tan^2 \psi_2} + \cdots + \frac{1}{\tan^2 \psi_{n/2-1}} \right)^{1/2}. \tag{6.279}$$

6.2.5 Elementary functions of a planar n-complex variable

The logarithm u_1 of the n-complex number u, $u_1 = \ln u$, can be defined as the solution of the equation

$$u = e^{u_1}. \tag{6.280}$$

The relation (6.270) shows that $\ln u$ exists as an n-complex function with real components for all values of $x_0, ..., x_{n-1}$ for which $\rho \neq 0$. The expression of the logarithm, obtained from Eq. (6.272) is

$$\ln u = \sum_{k=1}^{n/2} (e_k \ln \rho_k + \tilde{e}_k \phi_k). \tag{6.281}$$

An expression of the logarithm depending on the amplitude ρ can be obtained from the exponential forms in Eq. (6.273) as

$$\ln u = \ln \rho + \sum_{p=1}^{n-1} h_p \left[-\frac{2}{n} \sum_{k=2}^{n/2} \cos \left(\frac{\pi(2k-1)p}{n} \right) \ln \tan \psi_{k-1} \right]$$

$$+ \sum_{k=1}^{n/2} \tilde{e}_k \phi_k. \tag{6.282}$$

The function $\ln u$ is multivalued because of the presence of the terms $\tilde{e}_k \phi_k$. It can be inferred from Eqs. (6.226)-(6.228) and (6.231) that

$$\ln(uu') = \ln u + \ln u', \tag{6.283}$$

up to integer multiples of $2\pi \tilde{e}_k, k = 1, ..., n/2$.

The power function u^m can be defined for real values of m as

$$u^m = e^{m \ln u}. \tag{6.284}$$

Using the expression of $\ln u$ in Eq. (6.281) yields

$$u^m = \sum_{k=1}^{n/2} \rho_k^m (e_k \cos m\phi_k + \tilde{e}_k \sin m\phi_k). \tag{6.285}$$

The power function is multivalued unless m is an integer. For integer m, it can be inferred from Eq. (6.283) that

$$(uu')^m = u^m u'^m. \tag{6.286}$$

The trigonometric functions of the hypercomplex variable u and the addition theorems for these functions have been written in Eqs. (1.57)-(1.60). In order to obtain expressions for the trigonometric functions of n-complex variables, these will be expressed with the aid of the imaginary unit i as

$$\cos u = \frac{1}{2}(e^{iu} + e^{-iu}), \ \sin u = \frac{1}{2i}(e^{iu} - e^{-iu}). \tag{6.287}$$

The imaginary unit i is used for the convenience of notations, and it does not appear in the final results. The validity of Eq. (6.287) can be checked by comparing the series for the two sides of the relations. Since the expression of the exponential function $e^{h_k y}$ in terms of the units $1, h_1, ... h_{n-1}$ given in Eq. (6.236) depends on the planar cosexponential functions $f_{np}(y)$,

the expression of the trigonometric functions will depend on the functions
$f_{p+}^{(c)}(y) = (1/2)[f_{np}(iy) + f_{np}(-iy)]$ and $f_{p-}^{(c)}(y) = (1/2i)[f_{np}(iy) - f_{np}(-iy)]$,

$$\cos(h_k y) = \sum_{p=0}^{n-1} (-1)^{[kp/n]} h_{kp-n[kp/n]} f_{p+}^{(c)}(y), \tag{6.288}$$

$$\sin(h_k y) = \sum_{p=0}^{n-1} (-1)^{[kp/n]} h_{kp-n[kp/n]} f_{p-}^{(c)}(y), \tag{6.289}$$

where

$$f_{p+}^{(c)}(y) = \frac{1}{n} \sum_{l=1}^{n} \left\{ \cos\left[y\cos\left(\frac{\pi(2l-1)}{n} \right) \right] \right.$$
$$\cosh\left[y\sin\left(\frac{\pi(2l-1)}{n} \right) \right] \cos\left(\frac{\pi(2l-1)p}{n} \right)$$
$$- \sin\left[y\cos\left(\frac{\pi(2l-1)}{n} \right) \right]$$
$$\left. \sinh\left[y\sin\left(\frac{\pi(2l-1)}{n} \right) \right] \sin\left(\frac{\pi(2l-1)p}{n} \right) \right\}, \tag{6.290}$$

$$f_{p-}^{(c)}(y) = \frac{1}{n} \sum_{l=1}^{n} \left\{ \sin\left[y\cos\left(\frac{\pi(2l-1)}{n} \right) \right] \right.$$
$$\cosh\left[y\sin\left(\frac{\pi(2l-1)}{n} \right) \right] \cos\left(\frac{\pi(2l-1)p}{n} \right)$$
$$+ \cos\left[y\cos\left(\frac{\pi(2l-1)}{n} \right) \right]$$
$$\left. \sinh\left[y\sin\left(\frac{\pi(2l-1)}{n} \right) \right] \sin\left(\frac{\pi(2l-1)p}{n} \right) \right\}. \tag{6.291}$$

The hyperbolic functions of the hypercomplex variable u and the addition theorems for these functions have been written in Eqs. (1.62)-(1.65). In order to obtain expressions for the hyperbolic functions of n-complex variables, these will be expressed as

$$\cosh u = \frac{1}{2}(e^u + e^{-u}), \quad \sinh u = \frac{1}{2}(e^u - e^{-u}). \tag{6.292}$$

The validity of Eq. (6.292) can be checked by comparing the series for the two sides of the relations. Since the expression of the exponential function $e^{h_k y}$ in terms of the units $1, h_1, \ldots h_{n-1}$ given in Eq. (6.236) depends on the planar cosexponential functions $f_{np}(y)$, the expression of the hyperbolic functions will depend on the even part $f_{p+}(y) = (1/2)[f_{np}(y) + f_{np}(-y)]$ and on the odd part $f_{p-}(y) = (1/2)[f_{np}(y) - f_{np}(-y)]$ of f_{np},

$$\cosh(h_k y) = \sum_{p=0}^{n-1} (-1)^{[kp/n]} h_{kp-n[kp/n]} f_{p+}(y), \tag{6.293}$$

$$\sinh(h_k y) = \sum_{p=0}^{n-1} (-1)^{[kp/n]} h_{kp-n[kp/n]} f_{p-}(y), \tag{6.294}$$

where

$$f_{p+}(y) = \frac{1}{n} \sum_{(2l-1)=1}^{n} \left\{ \cosh\left[y \cos\left(\frac{\pi(2l-1)}{n}\right)\right] \right.$$
$$\cos\left[y \sin\left(\frac{\pi(2l-1)}{n}\right)\right] \cos\left(\frac{\pi(2l-1)p}{n}\right)$$
$$+ \sinh\left[y \cos\left(\frac{\pi(2l-1)}{n}\right)\right]$$
$$\left. \sin\left[y \sin\left(\frac{\pi(2l-1)}{n}\right)\right] \sin\left(\frac{\pi(2l-1)p}{n}\right) \right\}, \tag{6.295}$$

$$f_{p-}(y) = \frac{1}{n} \sum_{l=1}^{n} \left\{ \sinh\left[y \cos\left(\frac{\pi(2l-1)}{n}\right)\right] \right.$$
$$\cos\left[y \sin\left(\frac{\pi(2l-1)}{n}\right)\right] \cos\left(\frac{\pi(2l-1)p}{n}\right)$$
$$+ \cosh\left[y \cos\left(\frac{\pi(2l-1)}{n}\right)\right]$$
$$\left. \sin\left[y \sin\left(\frac{\pi(2l-1)}{n}\right)\right] \sin\left(\frac{\pi(2l-1)p}{n}\right) \right\}. \tag{6.296}$$

The exponential, trigonometric and hyperbolic functions can also be expressed with the aid of the bases introduced in Eq. (6.264). Using the expression of the n-complex number in Eq. (6.267) yields for the exponential of the n-complex variable u

$$e^u = \sum_{k=1}^{n/2} e^{v_k} (e_k \cos \tilde{v}_k + \tilde{e}_k \sin \tilde{v}_k). \tag{6.297}$$

The trigonometric functions can be obtained from Eq. (6.297) with the aid of Eqs. (6.287). The trigonometric functions of the n-complex variable u are

$$\cos u = \sum_{k=1}^{n/2} (e_k \cos v_k \cosh \tilde{v}_k - \tilde{e}_k \sin v_k \sinh \tilde{v}_k), \tag{6.298}$$

$$\sin u = \sum_{k=1}^{n/2} \left(e_k \sin v_k \cosh \tilde{v}_k + \tilde{e}_k \cos v_k \sinh \tilde{v}_k \right). \qquad (6.299)$$

The hyperbolic functions can be obtained from Eq. (6.297) with the aid of Eqs. (6.292). The hyperbolic functions of the n-complex variable u are

$$\cosh u = \sum_{k=1}^{n/2} \left(e_k \cosh v_k \cos \tilde{v}_k + \tilde{e}_k \sinh v_k \sin \tilde{v}_k \right), \qquad (6.300)$$

$$\sinh u = \sum_{k=1}^{n/2} \left(e_k \sinh v_k \cos \tilde{v}_k + \tilde{e}_k \cosh v_k \sin \tilde{v}_k \right). \qquad (6.301)$$

6.2.6 Power series of planar n-complex numbers

An n-complex series is an infinite sum of the form

$$a_0 + a_1 + a_2 + \cdots + a_n + \cdots, \qquad (6.302)$$

where the coefficients a_n are n-complex numbers. The convergence of the series (6.302) can be defined in terms of the convergence of its n real components. The convergence of a n-complex series can also be studied using n-complex variables. The main criterion for absolute convergence remains the comparison theorem, but this requires a number of inequalities which will be discussed further.

The modulus $d = |u|$ of an n-complex number u has been defined in Eq. (6.216). Since $|x_0| \le |u|, |x_1| \le |u|, ..., |x_{n-1}| \le |u|$, a property of absolute convergence established via a comparison theorem based on the modulus of the series (6.302) will ensure the absolute convergence of each real component of that series.

The modulus of the sum $u_1 + u_2$ of the n-complex numbers u_1, u_2 fulfils the inequality

$$||u'| - |u''|| \le |u' + u''| \le |u'| + |u''|. \qquad (6.303)$$

For the product, the relation is

$$|u'u''| \le \sqrt{\frac{n}{2}} |u'||u''|, \qquad (6.304)$$

as can be shown from Eq. (6.221). The relation (6.304) replaces the relation of equality extant between 2-dimensional regular complex numbers.

For $u = u'$ Eq. (6.304) becomes

$$|u^2| \leq \sqrt{\frac{n}{2}}|u|^2, \tag{6.305}$$

and in general

$$|u^l| \leq \left(\frac{n}{2}\right)^{(l-1)/2}|u|^l, \tag{6.306}$$

where l is a natural number. From Eqs. (6.304) and (6.306) it results that

$$|au^l| \leq \left(\frac{n}{2}\right)^{l/2}|a||u|^l. \tag{6.307}$$

A power series of the n-complex variable u is a series of the form

$$a_0 + a_1 u + a_2 u^2 + \cdots + a_l u^l + \cdots. \tag{6.308}$$

Since

$$\left|\sum_{l=0}^{\infty} a_l u^l\right| \leq \sum_{l=0}^{\infty} (n/2)^{l/2}|a_l||u|^l, \tag{6.309}$$

a sufficient condition for the absolute convergence of this series is that

$$\lim_{l \to \infty} \frac{\sqrt{n/2}|a_{l+1}||u|}{|a_l|} < 1. \tag{6.310}$$

Thus the series is absolutely convergent for

$$|u| < c, \tag{6.311}$$

where

$$c = \lim_{l \to \infty} \frac{|a_l|}{\sqrt{n/2}|a_{l+1}|}. \tag{6.312}$$

The convergence of the series (6.308) can be also studied with the aid of the formula (6.285) which is valid for any values of $x_0, ..., x_{n-1}$, as mentioned previously. If $a_l = \sum_{p=0}^{n-1} h_p a_{lp}$, and

$$A_{lk} = \sum_{p=0}^{n-1} a_{lp} \cos \frac{\pi(2k-1)p}{n}, \tag{6.313}$$

$$\tilde{A}_{lk} = \sum_{p=0}^{n-1} a_{lp} \sin \frac{\pi(2k-1)p}{n}, \tag{6.314}$$

where $k = 1, ..., n/2$, the series (6.308) can be written as

$$\sum_{l=0}^{\infty} \left[\sum_{k=1}^{n/2} (e_k A_{lk} + \tilde{e}_k \tilde{A}_{lk})(e_k v_k + \tilde{e}_k \tilde{v}_k)^l \right].$$ (6.315)

The series in Eq. (6.315) can be regarded as the sum of the $n/2$ series obtained from each value of k, so that the series in Eq. (6.308) is absolutely convergent for

$$\rho_k < c_k,$$ (6.316)

for $k = 1, ..., n/2$, where

$$c_k = \lim_{l \to \infty} \frac{\left[A_{lk}^2 + \tilde{A}_{lk}^2 \right]^{1/2}}{\left[A_{l+1,k}^2 + \tilde{A}_{l+1,k}^2 \right]^{1/2}}.$$ (6.317)

The relations (6.316) show that the region of convergence of the series (6.308) is an n-dimensional cylinder.

It can be shown that $c = \sqrt{2/n} \, \min(c_+, c_-, c_1, ..., c_{n/2-1})$, where min designates the smallest of the numbers in the argument of this function. Using the expression of $|u|$ in Eq. (6.221), it can be seen that the spherical region of convergence defined in Eqs. (6.311), (6.312) is a subset of the cylindrical region of convergence defined in Eqs. (6.316) and (6.317).

6.2.7 Analytic functions of planar n-complex variables

The analytic functions of the hypercomplex variable u and the series expansion of functions have been discussed in Eqs. (1.85)-(1.93). If the n-complex function $f(u)$ of the n-complex variable u is written in terms of the real functions $P_k(x_0, ..., x_{n-1}), k = 0, 1, ..., n - 1$ of the real variables $x_0, x_1, ..., x_{n-1}$ as

$$f(u) = \sum_{k=0}^{n-1} h_k P_k(x_0, ..., x_{n-1}),$$ (6.318)

where $h_0 = 1$, then relations of equality exist between the partial derivatives of the functions P_k. The derivative of the function f can be written as

$$\lim_{\Delta u \to 0} \frac{1}{\Delta u} \sum_{k=0}^{n-1} \left(h_k \sum_{l=0}^{n-1} \frac{\partial P_k}{\partial x_l} \Delta x_l \right),$$ (6.319)

where

$$\Delta u = \sum_{k=0}^{n-1} h_l \Delta x_l. \tag{6.320}$$

The relations between the partials derivatives of the functions P_k are obtained by setting successively in Eq. (6.319) $\Delta u = h_l \Delta x_l$, for $l = 0, 1, ..., n - 1$, and equating the resulting expressions. The relations are

$$\frac{\partial P_k}{\partial x_0} = \frac{\partial P_{k+1}}{\partial x_1} = \cdots = \frac{\partial P_{n-1}}{\partial x_{n-k-1}} = -\frac{\partial P_0}{\partial x_{n-k}} = \cdots = -\frac{\partial P_{k-1}}{\partial x_{n-1}}, \tag{6.321}$$

for $k = 0, 1, ..., n - 1$. The relations (6.321) are analogous to the Riemann relations for the real and imaginary components of a complex function. It can be shown from Eqs. (6.321) that the components P_k fulfil the second-order equations

$$\frac{\partial^2 P_k}{\partial x_0 \partial x_l} = \frac{\partial^2 P_k}{\partial x_1 \partial x_{l-1}} = \cdots = \frac{\partial^2 P_k}{\partial x_{[l/2]} \partial x_{l-[l/2]}}$$

$$= -\frac{\partial^2 P_k}{\partial x_{l+1} \partial x_{n-1}} = -\frac{\partial^2 P_k}{\partial x_{l+2} \partial x_{n-2}} = \cdots$$

$$= -\frac{\partial^2 P_k}{\partial x_{l+1+[(n-l-2)/2]} \partial x_{n-1-[(n-l-2)/2]}}, \tag{6.322}$$

for $k, l = 0, 1, ..., n - 1$.

6.2.8   Integrals of planar n-complex functions

The singularities of n-complex functions arise from terms of the form $1/(u - u_0)^m$, with $m > 0$. Functions containing such terms are singular not only at $u = u_0$, but also at all points of the hypersurfaces passing through the pole u_0 and which are parallel to the nodal hypersurfaces.

The integral of an n-complex function between two points A, B along a path situated in a region free of singularities is independent of path, which means that the integral of an analytic function along a loop situated in a region free of singularities is zero,

$$\oint_\Gamma f(u)du = 0, \tag{6.323}$$

where it is supposed that a surface Σ spanning the closed loop Γ is not intersected by any of the hypersurfaces associated with the singularities of the function $f(u)$. Using the expression, Eq. (6.318), for $f(u)$ and the fact that

$$du = \sum_{k=0}^{n-1} h_k dx_k,$$
(6.324)

the explicit form of the integral in Eq. (6.323) is

$$\oint_\Gamma f(u)du = \oint_\Gamma \sum_{k=0}^{n-1} h_k \sum_{l=0}^{n-1} (-1)^{[(n-k-1+l)/n]} P_l dx_{k-l+n[(n-k-1+l)/n]}.$$
(6.325)

If the functions P_k are regular on a surface Σ spanning the loop Γ, the integral along Γ can be transformed in an integral over the surface Σ of terms of the form $\partial P_l/\partial x_{k-m+n[(n-k+m-1)/n]} - (-1)^s \partial P_m/\partial x_{k-l+n[(n-k+l-1)/n]}$, where $s = [(n-k+m-1)/n] - [(n-k+l-1)/n]$. These terms are equal to zero by Eqs. (6.321), and this proves Eq. (6.323).

The integral of the function $(u-u_0)^m$ on a closed loop Γ is equal to zero for m a positive or negative integer not equal to -1,

$$\oint_\Gamma (u-u_0)^m du = 0, \ m \text{ integer}, \ m \neq -1.$$
(6.326)

This is due to the fact that $\int (u-u_0)^m du = (u-u_0)^{m+1}/(m+1)$, and to the fact that the function $(u-u_0)^{m+1}$ is singlevalued for n an integer.

The integral $\oint_\Gamma du/(u-u_0)$ can be calculated using the exponential form, Eq. (6.273), for the difference $u-u_0$,

$$u-u_0 = \rho \exp \left\{ \sum_{p=1}^{n-1} h_p \left[-\frac{2}{n} \sum_{k=2}^{n/2} \cos\left(\frac{2\pi kp}{n}\right) \ln \tan \psi_{k-1} \right] + \sum_{k=1}^{n/2} \tilde{e}_k \phi_k \right\}.$$
(6.327)

Thus the quantity $du/(u-u_0)$ is

$$\frac{du}{u-u_0} = \frac{d\rho}{\rho} + \sum_{p=1}^{n-1} h_p \left[-\frac{2}{n} \sum_{k=2}^{n/2} \cos\left(\frac{2\pi kp}{n}\right) d\ln \tan \psi_{k-1} \right] + \sum_{k=1}^{n/2} \tilde{e}_k d\phi_k.$$
(6.328)

Since ρ and $\ln(\tan \psi_{k-1})$ are singlevalued variables, it follows that $\oint_\Gamma d\rho/\rho = 0$, and $\oint_\Gamma d(\ln \tan \psi_{k-1}) = 0$. On the other hand, since ϕ_k are cyclic variables, they may give contributions to the integral around the closed loop Γ.

The expression of $\oint_\Gamma du/(u - u_0)$ can be written with the aid of a functional which will be called $\mathrm{int}(M, C)$, defined for a point M and a closed curve C in a two-dimensional plane, such that

$$\mathrm{int}(M, C) = \begin{cases} 1 \text{ if } M \text{ is an interior point of } C, \\ 0 \text{ if } M \text{ is exterior to } C. \end{cases} \tag{6.329}$$

With this notation the result of the integration on a closed path Γ can be written as

$$\oint_\Gamma \frac{du}{u - u_0} = \sum_{k=1}^{n/2} 2\pi \tilde{e}_k \, \mathrm{int}(u_{0\xi_k\eta_k}, \Gamma_{\xi_k\eta_k}), \tag{6.330}$$

where $u_{0\xi_k\eta_k}$ and $\Gamma_{\xi_k\eta_k}$ are respectively the projections of the point u_0 and of the loop Γ on the plane defined by the axes ξ_k and η_k, as shown in Fig. 6.6.

If $f(u)$ is an analytic n-complex function which can be expanded in a series as written in Eq. (1.89), and the expansion holds on the curve Γ and on a surface spanning Γ, then from Eqs. (6.326) and (6.330) it follows that

$$\oint_\Gamma \frac{f(u)du}{u - u_0} = 2\pi f(u_0) \sum_{k=1}^{n/2} \tilde{e}_k \, \mathrm{int}(u_{0\xi_k\eta_k}, \Gamma_{\xi_k\eta_k}). \tag{6.331}$$

Substituting in the right-hand side of Eq. (6.331) the expression of $f(u)$ in terms of the real components P_k, Eq. (6.318), yields

$$\oint_\Gamma \frac{f(u)du}{u - u_0} = \frac{2}{n} \sum_{k=1}^{n/2} \sum_{l=0}^{n-1} h_l$$

$$\sum_{p=1}^{n-1} (-1)^{[(l-p)/n]} \sin\left[\frac{\pi(2k - 1)p}{n}\right] P_{n-p+l-n[(n-p+l)/n]}(u_0)$$

$$\mathrm{int}(u_{0\xi_k\eta_k}, \Gamma_{\xi_k\eta_k}). \tag{6.332}$$

It the integral in Eq. (6.332) is written as

$$\oint_\Gamma \frac{f(u)du}{u - u_0} = \sum_{l=0}^{n-1} h_l I_l, \tag{6.333}$$

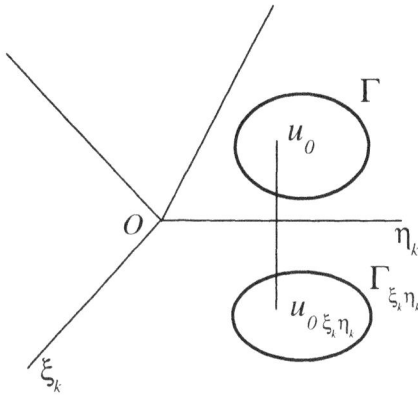

Figure 6.6: Integration path Γ and pole u_0, and their projections $\Gamma_{\xi_k \eta_k}$ and $u_{0\xi_k \eta_k}$ on the plane $\xi_k \eta_k$.

it can be checked that

$$\sum_{l=0}^{n-1} I_l = 0. \tag{6.334}$$

If $f(u)$ can be expanded as written in Eq. (1.89) on Γ and on a surface spanning Γ, then from Eqs. (6.326) and (6.330) it also results that

$$\oint_\Gamma \frac{f(u)du}{(u-u_0)^{n+1}} = \frac{2\pi}{n!} f^{(n)}(u_0) \sum_{k=1}^{[(n-1)/2]} \tilde{e}_k \, \mathrm{int}(u_{0\xi_k \eta_k}, \Gamma_{\xi_k \eta_k}), \tag{6.335}$$

where the fact has been used that the derivative $f^{(n)}(u_0)$ is related to the expansion coefficient in Eq. (1.89) according to Eq. (1.93).

If a function $f(u)$ is expanded in positive and negative powers of $u - u_l$, where u_l are n-complex constants, l being an index, the integral of f on a closed loop Γ is determined by the terms in the expansion of f which are of the form $r_l/(u-u_l)$,

$$f(u) = \cdots + \sum_l \frac{r_l}{u-u_l} + \cdots. \tag{6.336}$$

Then the integral of f on a closed loop Γ is

$$\oint_\Gamma f(u)du = 2\pi \sum_l \sum_{k=1}^{n/2} \tilde{e}_k \, \mathrm{int}(u_{l\xi_k \eta_k}, \Gamma_{\xi_k \eta_k})r_l. \tag{6.337}$$

6.2.9   Factorization of planar n-complex polynomials

A polynomial of degree m of the n-complex variable u has the form

$$P_m(u) = u^m + a_1 u^{m-1} + \cdots + a_{m-1} u + a_m, \tag{6.338}$$

where a_l, for $l = 1, ..., m$, are in general n-complex constants. If $a_l = \sum_{p=0}^{n-1} h_p a_{lp}$, and with the notations of Eqs. (6.313), (6.314) applied for $l = 1, \cdots, m$, the polynomial $P_m(u)$ can be written as

$$P_m = \sum_{k=1}^{n/2} \left\{ (e_k v_k + \tilde{e}_k \tilde{v}_k)^m + \sum_{l=1}^{m} (e_k A_{lk} + \tilde{e}_k \tilde{A}_{lk})(e_k v_k + \tilde{e}_k \tilde{v}_k)^{m-l} \right\}, \tag{6.339}$$

where the constants $A_{lk}, \tilde{A}_{lk}$ are real numbers.

The polynomials of degree m in $e_k v_k + \tilde{e}_k \tilde{v}_k$ in Eq. (6.339) can always be written as a product of linear factors of the form $e_k(v_k - v_{kp}) + \tilde{e}_k(\tilde{v}_k - \tilde{v}_{kp})$, where the constants $v_{kp}, \tilde{v}_{kp}$ are real,

$$(e_k v_k + \tilde{e}_k \tilde{v}_k)^m + \sum_{l=1}^{m} (e_k A_{lk} + \tilde{e}_k \tilde{A}_{lk})(e_k v_k + \tilde{e}_k \tilde{v}_k)^{m-l}$$

$$= \prod_{p=1}^{m} \left\{ e_k(v_k - v_{kp}) + \tilde{e}_k(\tilde{v}_k - \tilde{v}_{kp}) \right\}. \tag{6.340}$$

Then the polynomial P_m can be written as

$$P_m = \sum_{k=1}^{n/2} \prod_{p=1}^{m} \left\{ e_k(v_k - v_{kp}) + \tilde{e}_k(\tilde{v}_k - \tilde{v}_{kp}) \right\}. \tag{6.341}$$

Due to the relations (6.265), the polynomial $P_m(u)$ can be written as a product of factors of the form

$$P_m(u) = \prod_{p=1}^{m} \left\{ \sum_{k=1}^{n/2} \left\{ e_k(v_k - v_{kp}) + \tilde{e}_k(\tilde{v}_k - \tilde{v}_{kp}) \right\} \right\}. \tag{6.342}$$

This relation can be written with the aid of Eq. (6.267) as

$$P_m(u) = \prod_{p=1}^{m} (u - u_p), \tag{6.343}$$

where

$$u_p = \sum_{k=1}^{n/2} \left(e_k v_{kp} + \tilde{e}_k \tilde{v}_{kp} \right), \tag{6.344}$$

for $p = 1, ..., m$. For a given k, the roots $e_k v_{k1} + \tilde{e}_k \tilde{v}_{k1}, ..., e_k v_{km} + \tilde{e}_k \tilde{v}_{km}$ defined in Eq. (6.340) may be ordered arbitrarily. This means that Eq. (6.344) gives sets of m roots $u_1, ..., u_m$ of the polynomial $P_m(u)$, corresponding to the various ways in which the roots $e_k v_{kp} + \tilde{e}_k \tilde{v}_{kp}$ are ordered according to p for each value of k. Thus, while the n-complex components in Eq. (6.340) taken separately have unique factorizations, the polynomial $P_m(u)$ can be written in many different ways as a product of linear factors.

If $P(u) = u^2 + 1$, the degree is $m = 2$, the coefficients of the polynomial are $a_1 = 0, a_2 = 1$, the n-complex components of a_2 are $a_{20} = 1, a_{21} = 0, ..., a_{2,n-1} = 0$, the components $A_{2k}, \tilde{A}_{2k}$ calculated according to Eqs. (6.313), (6.314) are $A_{2k} = 1, \tilde{A}_{2k} = 0, k = 1, ..., n/2$. The left-hand side of Eq. (6.340) has the form $(e_k v_k + \tilde{e}_k \tilde{v}_k)^2 + e_k$, and since $e_k = -\tilde{e}_k^2$, the right-hand side of Eq. (6.340) is $\{e_k v_k + \tilde{e}_k(\tilde{v}_k + 1)\} \{e_k v_k + \tilde{e}_k(\tilde{v}_k - 1)\}$, so that $v_{kp} = 0, \tilde{v}_{kp} = \pm 1, k = 1, ..., n/2, p = 1, 2$. Then Eq. (6.341) has the form $u^2 + 1 = \sum_{k=1}^{n/2} \{e_k v_k + \tilde{e}_k(\tilde{v}_k + 1)\} \{e_k v_k + \tilde{e}_k(\tilde{v}_k - 1)\}$. The factorization in Eq. (6.343) is $u^2 + 1 = (u - u_1)(u - u_2)$, where $u_1 = \pm \tilde{e}_1 \pm \tilde{e}_2 \pm \cdots \pm \tilde{e}_{n/2}, u_2 = -u_1$, so that there are $2^{n/2-1}$ independent sets of roots u_1, u_2 of $u^2 + 1$. It can be checked that $(\pm \tilde{e}_1 \pm \tilde{e}_2 \pm \cdots \pm \tilde{e}_{n/2})^2 = -e_1 - e_2 - \cdots - e_{n/2} = -1$.

6.2.10 Representation of planar n-complex numbers by irreducible matrices

If the unitary matrix written in Eq. (6.217) is called T, it can be shown that the matrix TUT^{-1} has the form

$$TUT^{-1} = \begin{pmatrix} V_1 & 0 & \cdots & 0 \\ 0 & V_2 & \cdots & 0 \\ \vdots & \vdots & \cdots & \vdots \\ 0 & 0 & \cdots & V_{n/2} \end{pmatrix}, \tag{6.345}$$

where U is the matrix in Eq. (6.232) used to represent the n-complex number u. In Eq. (6.345), the matrices V_k are the matrices

$$V_k = \begin{pmatrix} v_k & \tilde{v}_k \\ -\tilde{v}_k & v_k \end{pmatrix}, \tag{6.346}$$

for $k = 1, ..., n/2$, where $v_k, \tilde{v}_k$ are the variables introduced in Eqs. (6.212) and (6.213), and the symbols 0 denote the matrix

$$\begin{pmatrix} 0 & 0 \\ 0 & 0 \end{pmatrix}. \tag{6.347}$$

The relations between the variables $v_k, \tilde{v}_k$ for the multiplication of n-complex numbers have been written in Eq. (6.229). The matrix TUT^{-1} provides an irreducible representation [7] of the n-complex number u in terms of matrices with real coefficients. For $n = 2$, Eqs. (6.212) and (6.213) give $v_1 = x_0, \tilde{v}_1 = x_1$, and Eq. (6.264) gives $e_1 = 1, \tilde{e}_1 = h_1$, where according to Eq. (6.195) $h_1^2 = -1$, so that the matrix V_1, Eq. (6.346), is

$$v_1 = \begin{pmatrix} x_0 & x_1 \\ -x_1 & x_0 \end{pmatrix}, \tag{6.348}$$

which shows that, for $n = 2$, the hypercomplex numbers $x_0 + h_1 x_1$ are identical to the usual 2-dimensional complex numbers $x + iy$.

Bibliography

[1] G. Birkhoff and S. MacLane, *Modern Algebra* (Macmillan, New York, Third Edition 1965), p. 222.

[2] B. L. van der Waerden, *Modern Algebra* (F. Ungar, New York, Third Edition 1950), vol. II, p. 133.

[3] O. Taussky, Algebra, in *Handbook of Physics*, edited by E. U. Condon and H. Odishaw (McGraw-Hill, New York, Second Edition 1958), p. I-22.

[4] D. Kaledin, arXiv:alg-geom/9612016; K. Scheicher, R. F. Tichy, and K. W. Tomantschger, Anzeiger Abt. II 134, 3 (1997); S. De Leo and P. Rotelli, arXiv:funct-an/9701004, 9703002; M. Verbitsky, arXiv:alg-geom/9703016; S. De Leo, arXiv:physics/9703033; J. D. E. Grant and I. A. B. Strachan, arXiv:solv-int/9808019; D. M. J. Calderbank and P. Tod, arXiv:math.DG/9911121; L. Ornea and P. Piccinni, arXiv:math.DG/0001066.

[5] S. Olariu, arXiv:math.OA/0007180, math.CV/0008119-0008125; Int. J. Math. Math. Sci. vol. 25, p. 429 (2001).

[6] E. T. Whittaker and G. N. Watson *A Course of Modern Analysis*, (Cambridge University Press, Fourth Edition 1958), p. 83.

[7] E. Wigner, *Group Theory* (Academic Press, New York, 1959), p. 73.

Index